ROUTLEDGE LIBRARY EDITIONS:
ECONOMIC GEOGRAPHY

Volume 11

THE WESTERN EUROPEAN
ECONOMY

THE WESTERN EUROPEAN ECONOMY
A geography of post-war development

ALLAN M. WILLIAMS

Routledge
Taylor & Francis Group

LONDON AND NEW YORK

First published in 1987

This edition first published in 2015
by Routledge
2 Park Square, Milton Park, Abingdon, Oxon, OX14 4RN

and by Routledge
711 Third Avenue, New York, NY 10017

Routledge is an imprint of the Taylor & Francis Group, an informa business

© 1987 Allan Williams

British Library Cataloguing in Publication Data
A catalogue record for this book is available from the British Library

ISBN: 978-1-138-85764-3 (Set)
eISBN: 978-1-315-71580-3 (Set)
ISBN: 978-1-138-85959-3 (Volume 11)
eISBN: 978-1-315-71709-8 (Volume 11)
Pb ISBN: 978-1-138-85960-9 (Volume 11)

Publisher's Note
The publisher has gone to great lengths to ensure the quality of this reprint but
points out that some imperfections in the original copies may be apparent.

Disclaimer
The publisher has made every effort to trace copyright holders and would
welcome correspondence from those they have been unable to trace.

The Western European Economy

A geography of post-war development

Allan M. Williams

HUTCHINSON

London   Melbourne   Auckland   Johannesburg

BARNES & NOBLE BOOKS

Totowa, New Jersey

Hutchinson Education

An imprint of Century Hutchinson Ltd

62–65 Chandos Place, London WC2N 4NW

Century Hutchinson Australia Pty Ltd
PO Box 496, 16–22 Church Street, Hawthorn,
Victoria 3122, Australia

Century Hutchinson New Zealand Ltd
PO Box 40–086, Glenfield, Auckland 10,
New Zealand

Century Hutchinson South Africa (Pty) Ltd
PO Box 337, Berglvei 2012, South Africa

First published 1987

First published in the USA 1987 by
BARNES & NOBLE BOOKS
81 ADAMS DRIVE
TOTOWA, NEW JERSEY, 07512

Set in Linotron Times
by Hope Services, Abingdon
Printed and bound in Great Britain by
Mackays of Chatham

British Library Cataloguing in Publication Data
Williams, Allan M.
 The Western European economy: a geography
 of post-war development.
 1. Europe—Economic conditions
 2. 1945–
 I. Title
 330.94′055   HC240
ISBN: (UK) 0 09 173161 5
ISBN: (US) 0 389 20772 1

Contents

Figures

Tables

Preface

The need to study and understand changes in the western European economy has never been greater than is the case in the late 1980s. More than forty years after the end of the Second World War, Europe has experienced a remarkable economic expansion followed, since the mid-1970s, by sharp recession. In this time the role of the region has been transformed in its relationship with the USA, Japan, the Eastern 'block' and the 'newly industrializing countries' (NICs) of Africa, Latin America and, especially, Asia. There have also been changes within Europe both in terms of the balance of economic strength between the North and the Mediterranean, and amongst individual states. Not least, it seems that the pace and magnitude of events have forced European affairs upon our attention in new and more urgent ways.

The European Community (EC), in the course of little more than a decade, has expanded from the original 'club' of six to the present twelve members, including more than two-thirds of the major states in western Europe. This has increased the role of international economic policy making, relative to the power of individual states. However, the EC has proven incapable of internal reorganization to meet the growing need to reform its economic policies; at present these dominate agriculture, overshadow manufacturing and ignore service activities. Enlargement of the EC has, however, confirmed the changing structure of the European economy. With the accession of Greece in 1981, and of Spain and Portugal in 1986, the centre of economic power can hardly be said to have shifted to the Mediterranean, but this region has experienced some of the more profound economic changes of the last four decades. Yet, only recently have we begun to appreciate, say, the extent of commercialization of Spanish agriculture or the strength of Italian manufactured exports. Meanwhile, beyond the EC, the Scandinavian and Alpine states continue to provide examples of alternative models of economic and social development.

In the shorter run, however – certainly since the early 1970s – the economic outlook for Mediterranean and Scandinavian Europe, in common with most other capitalist regions, has dimmed considerably with the onset of global economic crisis. The symptoms of the crisis can be seen in many different forms, ranging from Britain's inner city riots, to the inability of some Portuguese firms to pay any wages for more than twelve months, to the bankruptcy of Sweden's Kockum, Europe's largest shipyard. It is not surprising, therefore, that the search for 'solutions' to high unemployment and stagnant growth has become increasingly international as the laggards, such as the UK or Belgium, have sought to learn from the experiences of more successful economies, such as the Federal Republic of Germany or the Netherlands. For these reasons, it has become more important to re-examine the sources, forms and consequences of economic development in western Europe at large, in individual states and in particular regions.

Reassessment of the European experience is also pressing because of the need for empirical analyses to catch up with new theoretical perspectives. Economic geography increasingly has become politico-economic geography, incorporating analyses of state policies, social classes and internationalization tendencies alongside more traditional concerns such as firm formation, company organization, and regional policies. There has been some lowering of traditional disciplinary boundaries and a shift to a more holistic viewpoint, whereby the theories of a more integrated (and largely critical) social science are brought to bear on particular features of development. A number of examples of such approaches to economic geography exist, and they are well-illustrated by the writings of Massey (1978 and 1984 provide useful benchmarks in the development of her work).* In short, there is a need to review the changing economic geography of post-war Europe in a new light, whereby study of patterns (of firms, coal mines, etc.) is replaced by analysis of the processes which produce these – that is, capital accumulation and the tensions between capital, labour and the state. These categories have to be viewed as anything but homogeneous, for they are organized differently at the international and the interregional levels and the relationships between them change over time. Capital assumes many forms – ownership may be indigenous or multinational, private or public – and each has its own imperatives, potential mobility and forms of organization. Labour, too, is variable and can be disaggregated by age, sex, geographical origin, level of collective organization, and militancy. Similarly, the state represents a multitude of interests and has to fulfil a number of functions. This may give rise to conflicts both between states, and between central and local government, not to mention international bodies, such as the EC or the General Agreement on Trade and Tariffs.

From what has already been stated, it is clear that the aims of this book could not have been a comprehensive review of the economic geography of western Europe, whereby each region in each country and each sector was treated similarly and at equal length. This is neither possible – except in the most cursory fashion – nor necessary. Instead, the objective is to outline the major features of the economic development of western Europe since 1945, illustrated by reference to particular countries or regions which had a key role in events, or which typify the changes that have occurred. This immediately begs two questions: why 1945, and what is understood by

---

* Full references quoted in the text can be found in the Bibliography beginning on p. 317.

western Europe? The first of these is relatively easy to respond to, for 1945 did mark a real turning point in the European economy. It saw the end of massive war damage and disruption, and the start of reconstruction and, eventually, phenomenal economic growth within the framework of a new international economic order. This was, of course, conditioned by the (existing) accumulated pattern of development, especially that resulting from different imperial histories and phases of industrialization; but 1945 still marks one of the key turning points in recent economic history. In contrast, defining western Europe is a more difficult task. Excluding the non-capitalist countries of the Council for Mutual Economic Assistance (COMECON) and Albania is not too difficult. Of course, their very existence has conditioned the form of development in western Europe (especially with intensified competition in recent years), but their relationships with Western economies are relatively weak. After this, Iceland (on account of location), Turkey (as lying mainly within Asia) and Yugoslavia (for its hybrid East–West economic system) were also excluded, although poor data availability was also a consideration. This leaves as the focus of this volume the seventeen countries shown on the accompanying map: the current twelve members of the EC, the Alpine states of Austria and Switzerland, and the non-EC states of Scandinavia (Finland, Norway and Sweden).

The basic approach of the volume is to present a broad review of the major features of economic development. In particular, it focuses on the relationship between processes of development and spatial structure. Each of these depends on, and is conditioned by, the other in a myriad of ways. For example, previous phases of industrialization may have produced patterns of factories and labour markets which no longer accord with the needs of modern production; yet, the investment tied up in these inherited structures – and the possibility of regional political pressures – conditions the responses of capital, labour and the state to the need to produce new structures. Some of these ever-changing relationships will be unfolded in this volume. At the same time the analytical framework will have to extend beyond the continent, for Europe is no economic island. It has long been part of an Atlantic economy, so that 'when America sneezes Europe catches a cold'. Furthermore, the emergence of Japan and of the NICs has made it necessary to think in terms of the global economy.

The structure of the book flows from this framework. It is divided into three parts. The first sets the broader context, offering a review of the international economy: Chapter 1 outlines the evolution of the European economy since 1945 and Chapter 2 considers some of the salient features of capital, labour and the state. The second part considers changes in the major sectors of the economy: energy, primary, secondary and tertiary activities are discussed in Chapters 3–6. The level of analysis here is primarily the state, although, of course, states are aggregates of regional

Western Europe

economies. Nevertheless, for pragmatic reasons (there are seventeen countries opposed to over 500 Level II regions for the EC alone) as well as for analytical reasons – most economic policies are formulated at the state level – this seems appropriate. The final part examines the differential role of regions in these aggregate changes – emphasizing both the impact on them of, and their contributions to, the processes of change. Chapter 7 offers a working typology of the regions and of regional policies, while Chapters 8 and 9 examine selected regional case studies. It must be stressed, however, that the structure of the book does not imply that a unidirectional causal analysis from international to national to regional scales is possible. These scales are, in a real sense, indivisible and, for example, regional features determine the shape of both national and international scale processes. For example, the fact that there were labour reserves available in north-eastern Italy in the 1970s influenced the development of that region, of the Italian economy (via decentralization of production from the north-west) and of the European economy as a whole (in the way Italian firms maintained their export-competitiveness). These constantly changing relationships between different scales are a recurring theme in this volume.

Finally, I am pleased to acknowledge the help of a number of people who have greatly eased the task of writing this book. They include Ray Hudson and Mark Blacksell who commented on the initial outline, Jim Lewis who was an endless source of new ideas and of novel ways of obtaining information, Eleonore Kofman who read the final manuscript and Mark Cohen who has been a patient and encouraging editor. At a more practical but an invaluable level Alan Townsend provided data for employment change in the UK, Terry Bacon has imaginatively transformed my rough drafts of diagrams, while Jean Baker, Joan Fry, Heather Hughes and Tracy Reeves have produced typescripts from my even rougher draft manuscripts. I am also grateful to the University of Exeter for two terms of study leave in 1984–5: this book was written after that time, but those months provided breathing space to broaden my reading and clear some of the other commitments piled on my desk. Despite this, however, the usual disclaimer must be stated; the final product has benefited from many people's attention but the responsibility for its errors and inadequacies are mine alone.

Allan M. Williams
Exeter
November 1986

Part One

The International Context

1 The post-war international economy

Introduction

Europe's economy in 1945 appeared weak and fragmented following the recession of the 1930s and the destruction of the Second World War. Yet, 15 years later, almost all of western Europe's national economies had been carried along by a sustained wave of economic recovery. This was a period of remarkable expansion and transformation of the European economy but it proved to be relatively short-lived. By the late 1960s the economies of western Europe were already faltering, and the 1970s, punctuated by two oil price crises, were uncomfortable years of adjustment prior to descent into the second great economic recession of the twentieth century. Only since the mid-1980s has there been evidence of economic recovery.

This chapter seeks to unravel the economic and political changes which governed the course of post-war European development; but, first, it notes how, in the longer course of history, the European economy evolved up to the Second World War. In the fourteenth, fifteenth and sixteenth centuries the centres of economic power in Europe lay around the Mediterranean, especially in Italy (Venice), Spain and Portugal, which established the bases of overseas empires in Africa, Asia and Latin America. Their position was usurped by the rise of Dutch, and later British, commercial trade. Then, during 'the long sixteenth century' from about 1450 to 1640 (Wallerstein 1979), the UK came to dominate trade in the European and world economies. Its position was reinforced by the emergence of industrial capitalism in the eighteenth century as Britain became 'the first industrial nation' (Mathias 1969).

The UK maintained its leading role through both the water-powered and coalfield-based phases of industrialization. Then, after about 1850, new centres of industrialization emerged in Europe; to begin with in the Sambre–Meuse and Scheldt regions of Belgium and in France and, thereafter, in the Ruhr, Alsace and Lower Rhone. By the close of the nineteenth century, industrialization was established in many areas of the UK, France, Belgium, western Germany and the Netherlands. More selectively, it was established in southern Scandinavia, northern Italy, eastern Austria and Catalonia (Pollard 1981). Many northern European countries also expanded or consolidated their colonial empires in these years and, indeed, access to protected colonial markets was an important ingredient in their economic development. The UK remained the dominant economy but it was about to be surpassed by the United States of America and challenged by Germany. Within western Europe only Iberia (excepting Catalonia and, to a lesser extent, the Basque country and north-west Portugal), northern Scandinavia, southern Italy, Greece and Ireland remained largely untouched by industrialization. Europe's premier financial centre continued to be London.

The early twentieth century was dominated by two major developments. Continued rapid expansion in the USA meant that it replaced the UK as the leading power in the world economy, although Japan also grew rapidly in this period. However, the growth potential of North America could not sustain a stagnant European economy which had been disrupted by the First World War; laden with debts, and in the face of over-production crises, it retreated into protectionism. The 1920s were years of irregular growth and recovery which ended, calamitously, in world recession. In the UK, for example, agriculture and manufacturing were both depressed so that unemployment peaked in 1933 at over 3 million people, while in Germany hyper-inflation and unemployment nurtured the rise of fascism. European recovery from this nadir was slow and was far from complete when, in 1939, the Second World War began.

Europe in 1945

The Second World War was one of the major turning points in world economic history, marking a transition between very distinctive eras. In some ways it devastated the economies of Europe, but it also provided conditions for US dominance and the creation of a new international economic order. Within Europe it led to a major reshaping of the international roles of individual states. This culminated, at the national level, in the relative decline of the UK and the growth of the Federal Republic of Germany and, at the international level, in the formation of such supranational bodies as the European Economic Community (EEC) and the European Free Trade Association (EFTA).

In 1945, however, devastation rather than the potential for recovery was most obvious. Yet the potential was real, for the economies of the UK, Japan and the USA had been restructured to meet wartime needs. There had been shifts from consumption to investment, while output had increased significantly – for example, by some 20 per cent in the UK (Aldcroft 1980). Among the defeated powers, however, production had been shattered by the close of the war, having been reduced in Italy to the levels of 1900. Germany had ruthlessly exploited the European continent, having diverted some 42 billion dollars in levies and credits to its war machine. As a result the standard of living in France, for example, had been reduced by 50 per cent in just six years (Pollard 1981).

War destruction was widespread in 1945. Much capital stock (factories, schools, roads, etc.) had been destroyed – for example, 20 per cent of all houses in West Germany, and some 6–9 per cent in Austria, France and the Netherlands. However, industrial capital fared better than housing; losses were cancelled out by wartime additions, so that net industrial capital changed little in this period (Dyas and Thanheiser 1976).

This aggregate picture is deceptive since there had been serious losses in some sectors in some regions among the major combatants. Chief among these was transport, for the railway network had been severely disrupted and the merchant fleet had been reduced to 40 per cent of its 1939 level (Aldcroft 1980). Even in those countries, such as the UK, where capital stock had increased, there were serious arrears on maintenance and renewal of old capital stock; long-term economic requirements had been sacrificed to short-term strategic needs. One serious and lasting consequence of this was that it allowed the USA to extend its technological lead over Europe, where research and development had been virtually abandoned save in the realm of armaments. In addition, many European colonial powers, such as the UK, France and the Netherlands, lost some of their overseas assets and shipping earnings during this period. In the UK these assets had been sold to pay for the war, notably the lifeline of imports from the USA.

It was not simply capital stock which had been destroyed between 1939 and 1946; the stock of labour had also been reduced and disrupted. In total, some 40 million people are estimated to have died in the Second World War. Russia, with 20 million, was worst affected, but Germany lost 6 million. As with all wars the losses were selective, being greatest amongst young men. Beyond these savage losses, Europe's labour supply had been further disrupted with some 30 million workers having been forcibly dispersed, mainly to serve Germany's war needs.

Disruption of the economy did not end with the hostilities. Instead the 'Cold War' division of Europe into East and West was to disrupt many traditional trading links. Germany was affected most: until 1948 it was divided into four occupation zones which operated as separate economies (Milward 1984), and that occupied by the USSR was subject to heavy requisitioning of equipment (Owen-Smith 1983). With the Cold War the divide between East and West Germany became permanent. Furthermore, Europe's southern flank continued to be troubled, and a civil war proved as damaging to Greece as had the German occupation.

There were other shocks in store for the European powers, for the Second World War further loosened the already weakening hold of colonialism in Africa and Asia (see Figure 1). The USA, in particular, was determined to break down the old systems of imperial preferences and protected colonial markets after 1945. By about 1950 the Dutch West Indies, China, the Indian sub-continent and South-East Asia had all gained some measure of independence and, by the late 1960s, decolonization had been extended to much of Africa – bar the Portuguese and Spanish possessions. In all, some 57 new independent states were established between 1945 and 1965 although, in many cases, traditional economic ties

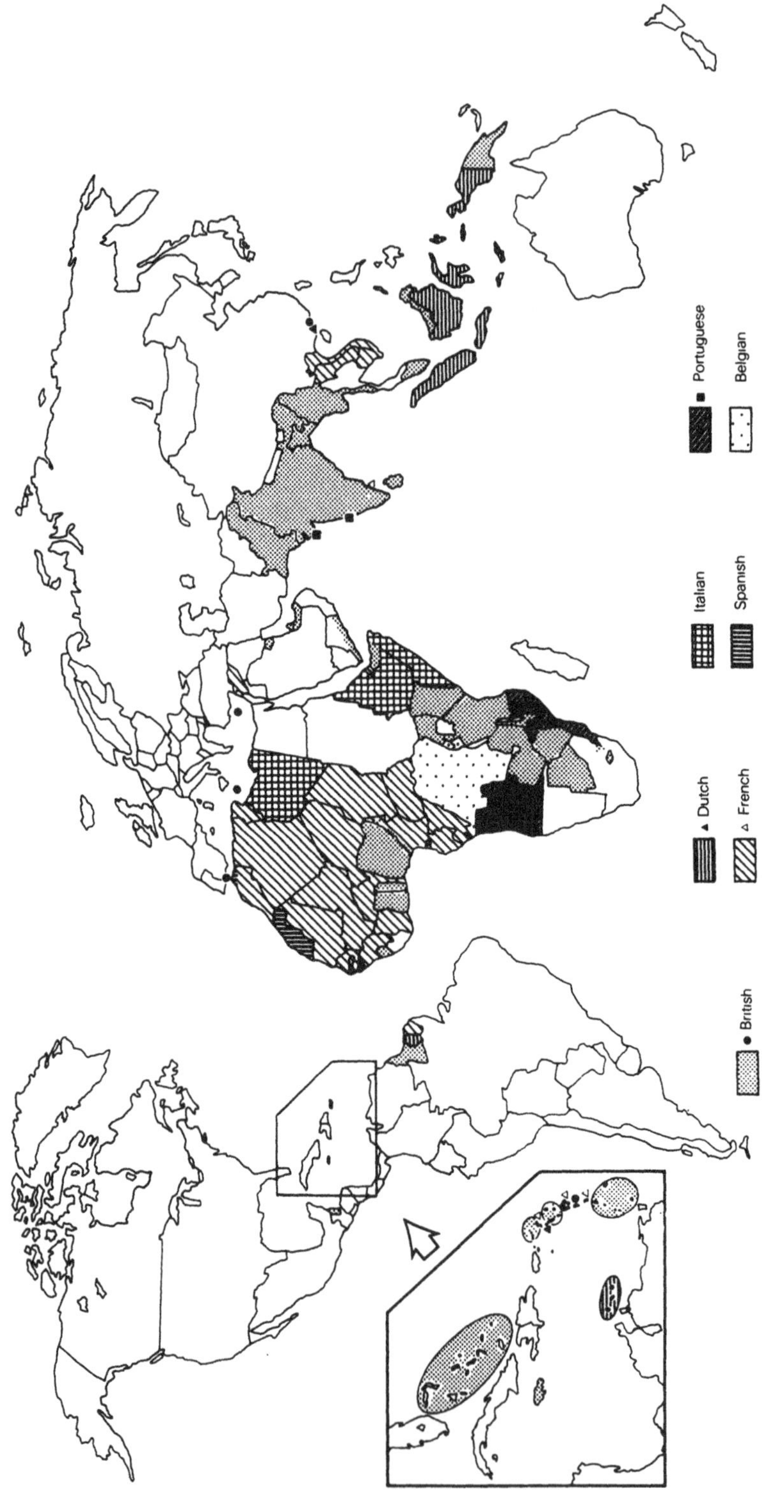
British
Dutch
French
Italian
Spanish
Portuguese
Belgian

Figure 1 Major European colonies in 1939

with and dependency on the old imperial powers remained dominant. Nevertheless, these political changes did prelude the emergence of new international economic relationships.

In summary, then, most of Europe suffered widespread disruption in the years before 1945, and even a 'victor', such as the UK, was severely affected. Only four major western European countries remained outside the hostilities. Neutral Portugal benefited, to some extent, from supplying both sets of combatants, but Spain was still in process of rebuilding after its civil war and was, anyway, already retreating into autarky and isolationism (Williams 1984). Sweden and Switzerland had also remained neutral and, in 1945, given the weakness of the combatants, held some comparative economic advantage in European trade.

Europe had been devastated but it is important not to exaggerate this; otherwise, it would be impossible to comprehend the recovery that was to occur in the following decade. Although capital stock had been damaged, in both the UK and Germany, there had been expansion of some high-technology branches of engineering and chemicals. These would provide a basis for growth in the 1950s. Meanwhile, although railways, bridges and ships had been lost, much of the network of roads, rail, canals and harbours remained intact, so that international trade could be quickly rebuilt. Wartime wastage of the labour force was also ameliorated in some ways; millions had died, but millions – especially women – had been brought into the labour force for the first time. This would be important in the growth and organization of many new manufacturing industries, especially given changes in production methods. Finally, Europe was debt-laden in 1945, but the effects of this were soon to be lessened. Much USA aid to the UK had been in the form of loans rather than grants, while Marshall Aid (see p. 26) would provide a boost to European recovery. Within Europe, Germany had been saddled with huge reparations to the Allies but, excepting the Russian sector, only a small part of these was ever extracted; this assisted the recovery of West Germany, a key economy in the continent. The Second World War and its aftermath dramatically reshaped the economy of western Europe and, without doubt, greatly weakened it; nevertheless, as events would reveal, there was still scope for rapid recovery and growth.

The new international economic order and European recovery

In a sense, the major economic consequences of the Second World War were evident outside, rather than within, Europe; for example, in decolonization and the rise of 'newly industrialized countries' (NICs), in the economic revival of Japan and, above all, in the dominance of the United States. The USA emerged from inter-war isolationism and accepted the mantle of the world's leading economic power. In addition to the weakness of West Germany and the UK, the USA's relative strength was based on a number of wartime events: the accumulation of gold reserves (following a transatlantic gold drain from Europe), the takeover of European companies' markets in the less-developed countries (such as that for pharmaceuticals in Latin America, previously dominated by Germany), the establishment of a research and development lead over the rest of the world, and the strengthening of a 'planned war economy which had seen 17 million new jobs created' (Arnell and Nygren 1980). Together, these gave the USA the ability to dominate and shape the world economy; this was to be the single most important feature of the next 25 years, until its dominance was broken.

Bretton Woods, the IMF and GATT

Even before the war was concluded, the USA had taken the first crucial steps in creating an institutional framework for a new international economic order. The Bretton Woods conference in 1944 had established the dollar and sterling as the major reserve currencies for international trade. The dollar had the key role and it was to be convertible to gold at a fixed rate of 35 dollars per ounce; all other currencies were to be made fully convertible with the dollar, and with each other, as soon as possible after the war. In order to regulate these new monetary arrangements, two new institutions were created. The International Monetary Fund (IMF) had responsibility for overseeing the fixing of exchange rates and could make loans to individual countries to finance temporary balance of payments deficits, although it could also impose conditions on borrower governments. As voting power within the IMF depended on each country's contribution to its finances, it was – with growing controversy – dominated by the USA. At the same time, the International Bank for Reconstruction and Development (IBRD) was created, to channel investment funds, initially to western Europe, and later to the less-developed countries. This was seen as an instrument of longer-term economic planning, complementary to the IMF's short-term managerial role.

Revival of trade was also to be pursued through liberalization measures and the dismantling of traditional protectionist measures; these had been considerably strengthened in the inter-war years. In 1947, therefore, the General Agreement on Tariffs and Trade (GATT) set in motion a series of conferences which attempted to cut tariffs; 23 countries were initially involved in these talks but their number had reached over 90 by the 1980s. The negotiations, to date, have proceeded through three main stages: the Dillon, Kennedy and Tokyo 'rounds'. Although agriculture and services were removed from the remit of these negotiations at an early stage, substantial progress has been made with respect to manufactured goods. For example, the 1962–7 'Kennedy round' ended with an average 35 per cent cut in tariffs on some 60,000 trade products. The subsequent 'Tokyo round' saw tariffs on most industrial goods (bar some particularly sensitive products such as textiles and iron and steel) reduced to very low levels, while some inroads were made into the much thornier problem of non-tariff trade barriers. While most western European states became involved in these negotiations at an early stage, Spain, Portugal and Greece preferred more autarkic development strategies and, initially, remained outside them (Tsoukalis 1981). A further round of GATT negotiations began in 1986 amid strong pressures from the developed countries to include services on the agenda.

Survival without disaster

The new institutional framework would, in time, facilitate substantial growth; but, in 1945, 'survival without disaster' was required (Scammell 1980). Given gross shortages of raw materials, most economies were tightly controlled and trade operated mostly on bilateral lines. Not surprisingly, given its strength (it accounted for one-third of world exports in 1947), the USA dominated European trade. This, however, caused its own problems, for the USA's trade surplus led to acute dollar shortages in Europe (Tumlir 1983) which hampered trade and recovery. Matters were compounded by the difficulties experienced in establishing the new international monetary system, particularly in respect of sterling's role as a key currency. In 1945 the UK had received 3.75 billion dollars for reconstruction purposes under the Anglo-American Loan Agreement, in return for agreeing to sterling becoming fully convertible with other currencies within a year. This would permit a multilateral (as opposed to bilateral) payments system to be built around it. The British Government would have preferred a longer transition period for sterling, but economic weakness made the UK susceptible to US pressure and its own trade requirements (Scammell 1980). Immediately, the difficulties of sterling became apparent: most of the UK's imports were from the US but most exports were to countries which did not have convertible currencies. The result was a rapid and heavy net drain on the UK's dollar and gold reserves. This was compounded by the fact that countries with accumulated sterling surpluses from the war years also chose to convert these to dollars. As a result, the Anglo-American loan was nearly exhausted by 1947 and, in August of that year, the British Government suspended sterling's convertibility.

In the event, 1947 was to prove the low point for Europe. In addition to the sterling crisis, the German economy was still severely hampered by restitution payments; and, to crown all these difficulties, the winter of 1947 was very severe, leading to widespread energy shortfalls. However, after 1948 Europe began to recover and, again, the prime mover in this was the USA. Faced with slow recovery in the UK, with Germany's ruinous state, and with the presence of strong communist parties in Italy and France at a time of growing tensions with the USSR, the USA launched the European Recovery Programme, otherwise known as Marshall Aid. Altogether the USA provided 13 billion dollars to western European states, mostly as grants, to facilitate trade and recovery. Some three-quarters of the funds went to the countries most severely affected by the war: France, Italy, Netherlands, the UK and the Federal Republic of Germany.

To some extent, recovery was under way before the initiation of Marshall Aid (Aldcroft 1980) – for example, in 1948 output in France was almost equal to that of 1938, while in the UK it had advanced 12 per cent over the same period (Scammell 1980). Nevertheless, Marshall Aid was of critical importance. In the short run it provided a breathing space for Europe and paid for essential raw-material and technology imports. It only amounted to about 5 per cent of the GNP of the receiving countries, but this financed about two-thirds of their merchandise imports from the Dollar Area between 1947 and 1950. In the longer term, Marshall Aid was important in encouraging trade and acceptance of GATT liberalization proposals. Without the aid package, it is also unlikely that full convertibility of western European currencies could have been achieved as early as 1959.

By the early 1950s, therefore, European recovery was under way, operating within an increasingly effective international monetary system, and partly funded by American aid. At this stage, recovery was still very uneven with, for example, the UK being several steps ahead of Italy and West Germany. However, all the economies were now showing positive signs (even Italy after 1949) and both the aggregate shape of the western European economy and the relative strengths of its constituent states would change fundamentally during the next two decades.

An age of expansion, 1948–70

There were early signs that European recovery was under way for, by the early 1950s, both the Federal Republic of Germany and the UK had recorded trade surpluses. On a broader front, the trading position of the USA was already deteriorating: by 1953, its trade with the other developed economies had come into rough balance. Even the onset of the East–West Cold War in the mid-1950s failed to halt recovery. By this stage, trading links within Europe were firmly re-established and there was less dependency on North America. Instead, Europe in the 1950s was able to move steadily to a restoration of the international economic system, and the European Monetary Agreement in 1958 finally confirmed a return to full currency convertibility and abandonment of bilateral payments agreements.

The context for growth was established and, between the late 1940s and early 1970s, there was an unprecedented boom in the European economy. Between 1950 and 1970 the annual growth in Gross Domestic Product (GDP) per capita was 4.4 per cent, and this outstripped world economic growth (Aldcroft 1980). Over the longer term, Bairoch (1976) suggests, per capita incomes grew by 4.5 per cent a year after 1950 compared with only 1 per cent a year between 1800 and 1950. Europe's position in the world economy strengthened and it even recovered some of the relative losses sustained during two world wars and the intervening depression.

Reasons for expansion

No single factor can explain this economic transformation; rather, there was a unique conjuncture of events which provided an unrivalled context for growth. The principal elements were the new monetary system and a revival of trade, high domestic and international demand, partial closing of the USA's technology lead, high investment ratios, an abundant supply of labour, and individual state policies. Together these shaped both aggregate European growth and the fortunes of individual countries (see Table 1).

Table 1 *Percentage annual growth in GNP per capita, 1960–83*

	1960–70	1973–83
Austria	4.5	2.8
Belgium	4.8	1.8
Denmark	4.7	1.8
Finland	4.6	2.7
France	5.7	2.5
F.R. Germany	4.4	2.1
Greece	6.9	3.0
Ireland	4.2	3.2
Italy	5.3	2.2
Netherlands	5.5	1.5
Norway	4.9	3.7
Portugal	6.2	—
Spain	7.1	1.8
Sweden	4.4	1.3
Switzerland	4.3	0.7
United Kingdom	2.9	1.1

SOURCE: World Development Report (1985).

The importance of the new international monetary system has already been stressed (see pp. 24–5), not least in permitting the re-establishment of trade. In the 1950s and 60s exports grew by around 8–9 per cent a year (Aldcroft 1980), and in most countries expansion was export-led. A growing US trade deficit made an important contribution to this as it provided expanding opportunities for European exporters (Boltho 1982).

Relationships with the less-developed countries (LDCs) also favoured exports from Europe. During these decades there was a process of unequal exchange between the advanced capitalist states and the LDCs (Amin 1976). Prices for raw materials were relatively unfavourable compared with those for manufactured goods, and the terms of trade led to progressive deterioration of the position of the LDCs. This benefited Europe which secured both relatively cheap industrial inputs and food, and high prices for its manufacturing exports. These trade conditions were particularly important for some of the smaller economies, and in the Scandinavian states, for example, between 25 and 40 per cent of output was exported (Andersen and Akerholm 1982). Similarly, in Italy it was the growing importance of exports over domestic demand which was crucial in that country's economic 'miracle' between 1958 and 1963 (Rey 1982). As a result, by the 1960s western Europe – despite decolonization – had re-established its role in world trade, and accounted for nearly 40 per cent of world exports. This was crucial given that technology-based increases in economies of scale, in such industries as car and aircraft assembly (see pp. 152–3), made it necessary to enter international markets.

This is not to deny the importance of domestic demand which was at a high level throughout this period. In part, demand was fuelled by the need to restore war damage, but rationing during and immediately after the war meant there was a high level of pent-up, frustrated demand. This was released in the early 1950s and provided a long, sustained boom, especially in the fields of consumer durables, car ownership and tourism. Steady income growth was, of course, important in sustaining demand. Wages in many countries became subject to a 'ratchet' effect: trade union pressure ensured that wages rose and that they never fell back at times of crisis. Together, the growth of domestic and international markets created a climate of confidence for investors and opportunities to innovate and launch new products.

The scope for innovation in western Europe was considerable, for the war years had opened up a large technology, and research and development, gap with the USA in both consumer goods and producer goods (such as machine tools). Adoption of US production methods – often linked with US managerial methods – therefore permitted not just widening but also deepening of the capital base in western Europe. This was the period when revolutionary advances were made in such fields as jet engines and synthetic textiles, and there seemed to be a whole set of interrelated and mutually reinforcing innovations in many industries (Denison 1967).

Technological advances would not have led to such a sustained growth of output had not the conditions of production been favourable, in terms of capital and labour availability and organization. These are discussed more fully in the following chapter but some salient features can be noted here. There were very high levels of fixed investment in this period, reflecting the, as yet, high rates of profits to be made in most branches of the

European economy. Investment, as a percentage of GNP in western Europe, rose from an average of only 9.6 per cent in 1928–38 to some 16.8 per cent by 1950–70 (Boltho 1982). This was facilitated by development of new multinational forms of capital as well as by expansion of direct state investment in most countries. The supply of labour was also important, whether drawn from national reserves or involving international migration. Kindleberger (1967) argues that cheap labour was the key to maintaining low costs and high profits in western Europe. Paine (1982) concurs in this view, believing that the absence of labour market constraints on the supply side, combined with expansionary demand, made for a 'virtuous' circle of growth; that is, low cost supply encouraged demand and vice versa.

Finally, it is pertinent to note the role of the state in the economic expansion of some countries, although, again, this is a theme elaborated more fully in the next chapter. Different governments adopted different economic strategies and, in part, this explains some of the variations within Europe in these decades. For example, the West German strategy was pursuit of a balance of payments surplus and price stability within a 'social market' framework (market operations bounded by social constraints), and this contributed to high growth rates and modest inflation. France was committed to a strategy which prioritized growth, and this was achieved, although at the expense of high inflation rates. In contrast, the priority in the UK was low unemployment, to be achieved through Keynesian demand-management; in practice this led to stop–go policies, low growth and recurring balance of payments crises.

The 1950s and 60s were years of strong economic growth in western Europe, but already, as Table 1 shows, there were divergent performances among the constituent states. It was not only the neutrals such as Sweden and Switzerland which experienced rapid growth; so too did some of the ex-combatants such as West Germany and Italy. On the other hand some neutrals such as Ireland, and ex-combatants such as Belgium, experienced sluggish economic growth. It is useful to consider briefly some of the major features of the individual economies.

The laggards
The classic laggard is, of course, the UK. While the economy still expanded during these decades, Milward (1984) considers that a watershed had been passed as early as 1949. In that year many European countries had balance of payments difficulties, but the UK rode the storm least well and sterling had to be substantially devalued. The reasons for the UK's poor performance are complex, but they include adherence to a global rather than a European strategy, large-scale exports of capital, debilitating stop–go policies, and sluggish productivity (Hudson and Williams 1986). The weakness of the UK also retarded the Irish economy which was heavily dependent on it for labour migration, trade and investment. It was

only after entry to the European Community that this dependency was loosened.

The other laggard was Belgium, especially in the 1950s. During the 1949 crisis it too had devalued its currency, but the Belgian franc was only reduced by 12 per cent which, with hindsight, was too little to restore competitivity. It had other traits in common with the UK, including large-scale export of capital overseas (especially to the Congo), and the channelling of domestic investment in 'defensive' strategies for traditional industries, rather than into modern industrial sectors (Van Rijckeghem 1982). It was only in the late 1950s and the 1960s that Belgium experienced more rapid growth, based on foreign investment in and around Antwerp (see p. 301), the availability of labour reserves in Flanders, and the trading advantages of EC membership.

Northern Europe's growth economies

The 'success' story of post-war economic recovery is, of course, West Germany. The government's immediate policy objective was stability (an obsession stemming from the experience of hyper-inflation in the 1930s). Towards this end, a 'social market' strategy was adopted, combining central planning with market principles (Owen-Smith 1983); given the experiences of the 1930s and 40s, labour readily acquiesced to this strategy. By 1953 German trade was in surplus and, thereafter, exports led a long and sustained economic expansion; exceptional competitivity was based on a productivity-conscious, non-militant labour force which was boosted by large-scale immigration, at first from East Germany and later from southern Europe and Turkey. Also important was the flexibility of medium-scale plants. While West Germany has its own multinational companies, it has not experienced the same level of capital exports as the UK, and there has been a commitment to domestic investment and research and development programmes. This country has become the driving-force of western European development. It is the largest economy, has a high potential to import/export, and is an important purchaser of services (Emminger 1981).

France has also experienced high growth rates, and there were '30 glorious years' after 1945 when it was out-performed only by Japan. High levels of investment, linked to adoption of new technology, has been a key element (Sautter 1982). Constraints on production have also been eased by the availability of labour released from agriculture, or imported from southern Europe or ex-colonies such as Algeria and Tunisia. The centralized state has played an important role in the national economy, wholeheartedly pursuing growth objectives, exercising a coordinating role through a system of National Plans, and providing direct investment. Domestic demand, too, has been important, for France has become a mass consumer society in a very short time; for example, between 1960 and 1977 the proportion of households without a car, refrigerator or TV fell from 48 to 4 per cent (Holmes and Fawcett 1983). Only in the 1980s has manufacturing competitivity begun to weaken.

Economic expansion in the Netherlands was based initially on 'catching-up' in the manufacturing field, for the Netherlands had a relatively low industrial base in 1945. Formation of the Benelux Customs Union (see p. 34) was important in this – not least because wages were relatively low, giving it a competitive edge over its co-member, Belgium (Van Rijckeghem 1982). In the 1960s, growth was further stimulated by the Netherlands' locational advantages within the EEC (see p. 160), enabling it to attract foreign investment, and by wages remaining relatively low; the latter was encouraged by large-scale imports of labour from ex-colonies and the Mediterranean. Discovery of natural gas also boosted economic growth.

There has been strong growth, too, in the Alpine states. Initially Austria was an 'occupied' state, and it achieved full independence only at the cost of stressing its neutrality. Consequently it was not able to join the European Community and, instead, became a member of the European Free Trade Association (EFTA). Its rapid growth has been based on industrialization from a low base, allied to strong exports. This was facilitated by very high investment levels, second only to those of Japan and Norway in the 1960s (Knapp 1981). In turn, high investment ratios (implying reduced short-term consumption) were encouraged by considerable state investment in nationalized industries and by a 'social partnership'. This involved government, employers and unions being able to settle industrial disputes via consensus; all parties were keen to develop this approach for they had long memories of the difficult political and economic consequences of conflicts in the 1930s.

Switzerland, as a neutral, escaped many of the worst economic effects of the Second World War and made steady, if unspectacular, growth during this period. This was based on highly efficient but protected agricultural production, high-quality competitive industrial exports, tourism and a strong banking sector. Despite the strength of the Swiss franc and high standards of living, exports have remained competitive because of the use of short-term immigrant labour, low inflation, and the development of international networks of branch plants by Switzerland's strong group of multinational companies.

The Scandinavian states have performed strongly both as a group and individually. Sweden has a very strong resource base in terms of forestry and minerals, but the key to its post-war success has been the shift from exports of primary products to manufactured goods. Immigration from Finland (and southern Europe) has helped reduce labour costs, and abundant hydro-electric power cushioned the effects on its economy of the 1970s oil crises. Sweden has also benefited from a strong sense of social consensus between employers, workers and the state over economic policy and the welfare state (Lundberg 1981). Among other policies, this led to a commitment to Keynesian economic management which, while effective in the 1960s, was to prove inadequate to meet the crises of the 1970s and 80s (see pp. 47–8). Norway, too, has a strong resource base, and this has been considerably strengthened by the discovery of vast North Sea oil resources (see pp. 102–6). Exports of primary materials, and of manufactured goods from a few very competitive industries such as shipbuilding and wood products, have been important in its expansion; in the 1960s some 50 per cent of growth could be attributed to exports (Andersen and Akerholm 1982).

Denmark lacked a strong resource base, and has pursued an open economy policy, promoting exports of food to northern Europe (especially dairy and meat products) and of selected manufactured goods (especially to Sweden). In this respect it has successfully exploited the potential advantages of membership of, first, EFTA, and, latterly, the EEC. Modest labour costs and a policy of slightly undervaluing the Danish kroner have also been important in export growth. Finland was one of the least developed European countries in 1945 and it is still an important exporter of labour. However, there has been expansion of some modern sectors in its small manufacturing base, including shipbuilding, paper and pulp, and specialist engineering. These provide important exports alongside primary products, especially timber.

Southern Europe transformed

Western Europe's Mediterranean flank has experienced some of the most rapid growth in the post-war period, especially in the cases of Italy and Spain. Italy was devastated in the war, and so there was a short burst of repair and recovery immediately after 1945. However, the economy really 'took off' between 1958 and 1963, the years of the 'economic miracle'; industrialization was export-led and competitivity was maintained by the availability of labour reserves in the south drawn to work in northern factories. Spain's growth came a little later, not least because the ravages of the Civil War from 1936 to 1939 were followed by years of political isolation and economic autarky. Pre-civil-war levels of production were only restored by 1950, and the remainder of the decade was given over to a steady growth of the indigenous manufacturing sector, sustained by heavy state investment in basic industries. Then, in 1959, the Stabilization Plan committed Spain to a strategy of internationalization. Growth became more export-orientated and foreign investment and technology were imported on a large scale; however, constant visible trade deficits (on goods) could only be covered by receipts from tourism and emigrants' remittances (Williams 1984). Nevertheless, average annual growth rates exceeded 7 per cent in the 1960s and, by the 1980s, Spain was considered by the World Bank to have joined the group of industrial market economies.

Growth came later to Greece and Portugal and both are still classified only as upper-middle-income countries by the World Bank. Portugal, under the Salazar regime, was subject to autarkic economic policies between the 1930s and 50s; industrialization was limited to traditional sectors (such as cork or textiles) and an inefficient agriculture dominated employment and retarded development. In the 1960s the economy was opened to foreign investment and trade, Portugal joined EFTA, and the manufacturing sector expanded rapidly. However, it was tourist receipts and emigrants' remittances which covered the net deficit on exports/imports of goods, and agriculture continued to stagnate. Greece is also a late industrializing economy which experienced sustained growth in the 1960s. This was very much based on export-led industrialization, as is indicated by a rise in the share of manufactured goods in exports from 11 to 50 per cent between 1962 and 1975. However, in common with Portugal, it had a stagnant agriculture, and was heavily reliant on emigrants' remittances and the tourist industry. Politically, if not economically, it remained somewhat isolated from the remainder of western Europe, both because of ostracism of the Colonels' regime (1967–74) and because of the desire to foster links with eastern Mediterranean countries.

Summary
While individual economies expanded at varying rates during this period, the most impressive feature was the collective expansion of the economy of western Europe. One consequence was that, in the course of the 1950s and 60s, the balance of trade between Europe and the USA changed, as deficits accumulated for the latter. Given that fixed parity between gold and the dollar had been agreed at Bretton Woods, the USA was unable to rectify these deficits through a strategy of devaluation so as to strengthen its comparative advantages in exports. Instead, the continuing trade deficits were paid for by drawing on gold reserves, and these diminished by 2.3 billion dollars in the 1950s alone (Ballance and Sinclair 1983).

This relationship benefited western Europe because the dollar surplus facilitated multilateral world trade and boosted demand. However, the gold drain could not last indefinitely and, towards the end of the 1960s, the international monetary system came under intense pressure. At the same time, the nature of the world economy was changing. For the time being the relationship between Europe and the USA remained central. However, Japan – growing at over 9 per cent a year in the 1950s – was already emerging as a major industrial power, at first in Asia but latterly on a global scale. The consequences of these developments are considered later in this chapter; but next we consider the institutional framework of the European economy.

The European economic framework

The framework for economic relationships in western Europe has been transformed during the post-war period. This has involved the emergence of new groupings within Europe, increasing integration within the region, and changing relationships with other economic powers. For convenience, the changes can be categorized into early integrationist movements, formation of the EEC and EFTA, and enlargement of the EEC.

Early integrationist movements

The first step towards economic integration was the founding of the Benelux Customs Union in 1948, a free trade area incorporating the Netherlands, Belgium and Luxembourg. Discussions followed on how to enlarge the union, and there were attempts to organize free trade areas elsewhere in Europe – for example in Scandinavia and between Italy and Austria.

None of these proposals came to fruition and the next major step was the Treaty of Paris, establishing the European Coal and Steel Community (ECSC) in 1951; this included Belgium, France, West Germany, Italy, Luxembourg and the Netherlands. It sought to create a genuinely free trade area for coal and steel products. While the plan was launched to secure the advantages of free trade, it was the Korean War which led to its speedy adoption (Blacksell 1981). The demands of the war made it imperative to expand German iron and steel production, despite lingering fears of a resurgence of militarism in that country. At least the ECSC seemed to offer a framework which would permit some international control over expansion in these key industries. The British Government was hostile to the idea of any external control over its industries and therefore chose to remain outside the ECSC. Already the UK was becoming isolated within Europe, and this was exaggerated by government delusions that it had the ability to maintain a colonial and global role.

Progress was slow in the ECSC (for example, freight rates for coal and steel were only equalized by 1957), because the central authority of the grouping was relatively powerless. It was not constituted, as originally conceived, as a powerful council of neutral experts making technocratic decisions. Instead, it was an administrative body operating complex regulations arising from careful balancing of the interests of member states. If this was to be the model for the European Community the omens were not favourable. As Milward (1984, p. 418) states: 'the future political economy of the Treaty of Rome, the analysis of which by any neo-classical formulations is likely to reduce the analyst only to a state of bewildering despair, had already taken shape.' Nevertheless, formation of the ECSC was a stage towards establishment of the EEC.

The EC and EFTA

In 1955 the Messina Conference was held to discuss the creation of a large economic community. The UK clearly underestimated the importance of this meeting, at which it had only junior representatives; similarly, it was out of step with the other major European powers when the Treaty of Rome was signed in 1957. From 1 January 1958 this created the European Economic Community and the European Atomic Energy Community (EURATOM). The members of the EEC were the same six states that had set up the ECSC; their motives were a mixture of the political (creation of a political union which could incorporate the Federal Republic of Germany) and the economic (establishment of a 'third' economic power block to rival the USA and the USSR). The Treaty of Rome laid down a number of general principles for the operation of the EEC (Robertson 1973):

- free trade among the members
- common external tariffs against third parties
- free movement of capital and labour
- a Common Agricultural Policy
- a Common Transport Policy
- a European Social Fund
- a European Investment Bank
- associations with other countries so as to increase trade and development

Only general principles were agreed in the Treaty of Rome: detailed policies were to be determined later, so as to facilitate rapid establishment of the EEC. Progress in the early years was often difficult as the conflicting interests of individual states emerged in the course of negotiations. This culminated in 1965 in France's 'empty chair' gesture when, for a number of months, it refused to participate in negotiations over finance since it was not prepared to agree to measures giving the EEC self-financing powers. Nevertheless, economic integration did continue and, by 1967, formal steps were taken to unify the three European communities – the ECSC, the EEC and EURATOM – in a single European Community (the EC).

The UK had been wary of joining the EEC, not least because the British Government's principal interest was the creation of a free trade area for industrial goods only. It certainly did not favour loss of sovereignty to a supranational body such as the EEC. Instead, in 1956/7 the UK actively promoted the idea of forming a European industrial free trade area as an alternative to the EEC, but this was rejected by the six who signed the Treaty of Rome. Therefore, the UK was left, with some of the smaller European states, to form a residual free trade area, the European Free Trade Association (EFTA), established by the 1960 Stockholm Convention. The original members were Austria, Denmark, Eire, Norway, Portugal, Sweden, Switzerland and the UK, each of which had its own reasons for accession (Blacksell 1981). Iceland joined in 1970, while Finland, in order to maintain its political neutrality between West and East, became only an associate member (see Figure 2). The aims of EFTA were limited: to set up a single free trade area for manufactured goods. Rapid progress was made towards this end so that, by 1970, all tariff restrictions on industrial goods

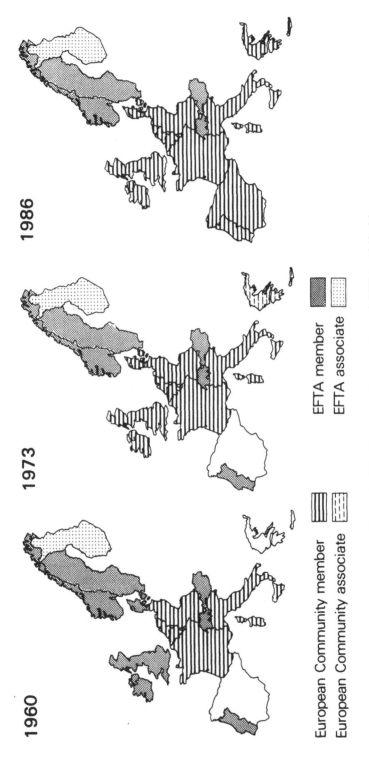

1986

1973

1960

European Community member

European Community associate

EFTA member

EFTA associate

Figure 2 The European Community and EFTA, 1960–86

had been removed in trade between the members; Portugal was an exception, being allowed to defer removal of import duties until 1980 in view of its low level of development. The new institutional framework for Europe was in place and, along with GATT, helped further European economic integration.

Enlargement of the European Community

There were no major changes in these institutional arrangements over the next decade. Associate membership of the EEC was granted to Greece in 1962 (but was frozen between 1967 and 1974 – a period of military rule), and to Turkey in 1964. Spain and Portugal considered full membership of the EEC in the early 1960s but their dictatorial governments made them politically unacceptable as members; instead they signed preferential trade agreements with the EC (Spain in 1970 and Portugal in 1973). The preferential agreements gave improved access to EEC markets for their manufactured goods, and made possible development aid from the European Investment Bank.

It was only in the 1970s that the institutional framework for the European economy was revised (see Figure 2). The critical steps were the accession of Denmark, Ireland and the UK to the EC in 1973, followed by Greece in 1981 and Portugal and Spain in 1986 (see Blacksell 1984; Seers and Vaitsos 1982). EFTA was left in a much reduced form, but the remaining members were promised associate status with the EC. In part, these changes were a political recognition of an economic integration which had already occurred in terms of trade, investment and labour flows (see Chapter 2). They have, however, considerably altered the nature of the EC, both at the organizational level (with decision making being made more complex) and in terms of the economic structure of the Community which now includes substantial Mediterranean interests.

The European institutions in practice

How is the success of these supranational institutions to be evaluated? In terms of its own limited objectives, EFTA has been outstandingly successful. Most tariffs on industrial goods – with some important exceptions – were dismantled by 1967, three years ahead of schedule. This contributed to a strong growth of trade within EFTA itself, by about 15 per cent a year in the 1960s. It is much more difficult to assess the performance of the EC, if only because of its more diverse aims. Although we return to specific policies in later chapters, some general points can be noted here.

Transition to a full customs union was planned by 1969, whereby all inter-EC tariffs and trade barriers would be removed, while a single common external tariff was adopted against imports from non-members. Progress was relatively rapid and most targets were achieved by July 1968, although non-tariff barriers have proven obstinate. Partly in consequence of this, intra-EC trade has increased in importance for the member states

(Commission of the European Community 1983a), which is indicative of a shift to greater economic integration (see Table 2). There has also been clear progress in establishing free movement of labour, at least for manual workers; international entry to many professional posts is still restricted by different national criteria for validating credentials.

The EC has also developed trading agreements with less-developed countries (LDCs). The first stage was the Yaoundé Convention which offered preferential trading terms to some seventeen African states, mostly French ex-colonies. In 1968 the Arusha Convention brought Kenya, Tanzania, Uganda and Nigeria into the scheme. Finally, as a third stage, the Lomé Convention brought a total of 46 African and Caribbean states into special agreements with the EC; many were British ex-colonies. The conventions allowed duty-free importation of industrial products and most (non-competing) agricultural products from these countries into the EC. Nearer home, the EC has signed a number of preferential import agreements with Mediterranean countries; with Israel in 1975, and with the Maghreb and the Mashreq countries of North Africa in 1976/7. When the agreements were signed they were a logical means of securing supplies of Mediterranean, sub-tropical and tropical agricultural products of which the EC had a deficit. However, enlargement in the 1980s has modified the EC's position with respect to Mediterranean produce, so that the agreements are now highly contentious in the face of potential overproduction in the EC of, for example, citrus fruits.

Another objective of the EC was monetary union, but progress has been slow and irregular, not least since the breakdown of the international monetary system after the late 1960s (see pp. 40–1). An initial compromise was that some member states (excluding Ireland, France and the UK) adopted 'the snake in a tunnel' system whereby currencies were allowed to fluctuate within a margin of 2.25 per cent. In 1979 the European Monetary System was launched (without the UK). This again provided for limited currency fluctuations (backed by credit and systems to exchange information) as a step towards a unified currency system. Only the UK remained outside this system in 1986.

The Common Agricultural Policy has been the most contentious element of the EC economic framework. The aims were to establish uniform farm price regimes within the Community, and minimum prices for imports. The fiscal framework is operational but, in practice, there has been a tendency to overproduction at prices well above world levels, while decisively favouring the interests of farmers over consumers. However, a Common Agricultural Policy does exist, unlike the vaunted Common Transport Policy, an area of conspicuous neglect. There has been some success in establishing a European Social Fund, although it still accounts for only about 6 per cent of the Commission's budget (Commission of the European Community 1984a). Its role in the 1980s is illustrated by its priorities, which are to assist unemployed young people (75 per cent of

Table 2 *Analysis of western European trade in 1984 (percentage figures)*

	Origins of imports					Destinations of exports				
	EC10	USA	Japan	Rest of world	Total	EC10	USA	Japan	Rest of world	Total
Austria	60	4	3	33	100	53	4	1	42	100
Belgium	64	6	2	28	100	69	6	1	24	100
Denmark	47	5	3	44	100	44	10	3	44	100
F.R. Germany	50	7	4	39	100	48	10	1	41	100
Finland	36	5	6	54	100	58	8	1	33	100
France	54	6	2	38	100	49	8	1	42	100
Greece	47	3	8	43	100	54	8	1	36	100
Ireland	70	14	3	13	100	69	10	2	20	100
Italy	43	6	2	49	100	45	11	1	43	100
Netherlands	52	9	2	37	100	72	5	1	22	100
Norway	46	9	5	41	100	70	5	1	24	100
Portugal	36	14	2	48	100	58	9	1	33	100
Spain	33	11	3	53	100	49	10	2	40	100
Sweden	53	8	5	34	100	48	11	1	39	100
Switzerland	67	7	4	22	100	50	10	3	37	100
United Kingdom	42	14	5	39	100	44	15	1	40	100

SOURCE: Eurostat (1985b).

expenditure in 1983) and less-developed regions (especially Ireland, Northern Ireland, Greece, southern Italy, Greenland and the French overseas departments).

Taken together, the policies of the EC and of the individual member states have come to dominate western European economic development. The EC does not operate as a fully integrated unit at the economic, let alone the political, level, but it is at the heart of European trade, investment and labour flows. However, while the EC was evolving, the larger international economic context for development was also changing.

Into recession: the 1970s and the 1980s

The unprecedented post-war expansion of the European economy was running into difficulties by the mid-1960s, when growth slowed markedly in France and the UK (and in the USA); this became more generalized at the end of the decade.

In general, demand tends to increase irregularly, with periods when it is static, while production increases constantly. It is inevitable, therefore, that there are occasional crises of overproduction, resulting in economic stagnation. Kondratieff (1935) argues that such crises tend to occur at 40 to 45 year intervals and, according to Wallerstein (1984), were prevalent in western Europe in the 1970s. While the precise causes of the crises are complex, of key importance was the conjuncture of four major features:

- failure of the new international economic order
- greater social militancy and labour-market rigidities, leading to increased production costs, reduced profits and investment
- competition from the newly industrialized economies
- unpredictable major oil price rises.

As a result many of the industries – such as car, steel and chemical production – which had led expansion in the 1950s and 60s went into sharp decline. While the service sector was more resistant to recession, it, too, tended to have lower productivity growth which contributed to the overall limitation of expansion (World Bank 1984).

Collapse of the international monetary system

The international monetary system established at the 1944 Bretton Woods conference was based on two main reserve currencies, sterling and the US dollar. Sterling has never really proven strong enough for this role because of the UK's recurrent balance of payments problems. The pound had been devalued as early as 1949 and, in the late 1960s, it came under pressure, having to be further devalued in 1967. This contributed to a decline in confidence in the other key currency, the dollar.

The US economy had exhibited a negative balance of payments since the late 1950s, and this was partly caused by the fixed parity between the dollar and gold. This meant that the dollar could not be devalued, which would have been one way to ease the country's trading deficit. However, the huge gold reserves held by the USA did allow it to continue financing these deficits. While the gold reserves were sustained, European governments actually welcomed the US deficit for it ensured there was a dollar surplus to finance world trade. However, in the course of the 1960s the deficits continued, with heavy expenditure on the Vietnam War and large-scale international aid programmes, so that US gold reserves were reduced from $18 billion to $11 billion (Scammell 1980). Once sterling was devalued the pressure was truly on the dollar, and this was compounded because Japan and West Germany were both reluctant to revalue their undervalued currencies. Finally, the President of the United States had to act; in 1971 the dollar was devalued by 8 per cent against gold prices, dollar convertibility to gold was ended, and an import surcharge was introduced to reduce imports. The Bretton Woods system had broken down.

Immediate international negotiations followed this crisis. The Smithsonian Agreement in December 1971 led to revaluation of several other currencies against the dollar, and a move to a system of more flexible exchange rates (with ± 2.2 per cent limits). This gave only a temporary respite; loss of confidence still led to large-scale sales of dollars, and US trade deficits continued. By February 1973 the dollar again had to be devalued, this time by 10 per cent. At the international monetary conference in Paris, in March of that year, it was agreed that there was no alternative to adopting a system of flexible exchange rates. This, of course, turned currencies into 'tradeable' commodities and increased the potential for speculative pressures on all currencies (especially the weaker ones), leading to further loss of confidence in the international economic system. It also meant that countries such as Switzerland, which previously had undervalued exchange rates, lost their competitivity as their currencies floated upwards to their 'real' values.

Social militancy

Conditions of production, too, were changing both within and outside Europe. Within Europe, many countries experienced growing social militancy in the late 1960s, following two decades of steady increases in labour costs and rising profits. Formal and informal workers' associations demanded improved conditions of work, holidays and pay levels, leading to a rise in real wages. Employers' social security contributions increased even more rapidly to help sustain state welfare programmes, thereby pushing up total labour costs. As a result, from the late 1960s real wages outstripped productivity increases, reducing profits and contributing to falling investment levels. Most countries attempted to use some form of wages or incomes policy to contain labour costs but, other than in Austria and to a lesser extent in West Germany, these proved unsuccessful (World Bank 1984).

The precise timing of the crisis point varied, but it occurred in the late 1960s in West Germany and Italy and in the early 1970s in France (Mazier 1982). Italy provides a good example of the pressures that were in operation. Mounting trade union demands culminated, during the 'Hot Autumn' of 1969, in some four million workers going on strike, initially over wages and contracts, although this later broadened to a concern with working conditions (Slater 1984). The struggle also became linked with wider social campaigns outside the workplace over, for example, urban services and housing (Marcelloni 1979). The social costs of the economic transformation of Italy – evident in appalling urban living conditions in cities such as Turin (Lagana *et al.* 1982) – were finally catching up with economic growth. In the end the government and employers acceded to many of the workers' demands, adding an estimated 28 per cent to the costs of Italian industry (Giugni 1971). This contributed, at the national level, to exports being less competitive and to lower growth, and, at the regional level, to productive decentralization outside the established industrial area (see pp. 171–2).

The rise of social militancy, and its consequences, was repeated – if less spectacularly – throughout western Europe. This was fuelled by a tightening of labour markets in the 1960s as agricultural labour reserves were depleted, at the same time as early successes in wage negotiations fostered rising expectations of further gains. This contributed to a squeeze on profits and falling investment levels, as well as to price inflation (Cox 1982). Matters were compounded because the narrowing of the USA–Europe technology gap meant that there was little scope for substituting capital for labour so as to reduce costs, at least in the short term. Furthermore, when the economic downturn came in the 1970s, labour-market rigidity resulting from the strength of workers' organizations made it difficult for employers to cut real wages or reduce the labour force, hence retarding restructuring of the labour process.

Competition from the newly industrialized countries

Production has also been reorganized at the global scale as a result of export growth from the Socialist countries, fuelled by 'dumping' of products at arbitrarily low prices, and by European companies having to buy-back from joint ventures they have invested in within these countries. There has also been the emergence of newly industrialized countries (NICs) such as South Korea and Brazil in the Third World. This had consequences both in their import penetration into western European markets and in reduced export opportunities for European companies elsewhere in the world market. In turn, this limited the opportunity to 'expand the basis of the world economic system', one of the ways in which new long waves of expansion had been initiated historically (Wallerstein 1984). There was already rapid economic growth in many NICs in the 1960s, but, significantly, they surpassed the industrialized countries in the 1970s. Whereas the growth of manufacturing in Europe slumped in the 1970s, it held up much better in the NICs: comparing the two decades, value-added manufacturing fell from 6.2 to 3.0 per cent in the advanced market economies, but only from 7.3 to 5.8 per cent in the NICs (Ballance and Sinclair 1983). For the less-developed countries as a whole, Figure 3 shows continual expansion of their share of non-oil world exports between 1970 and 1985.

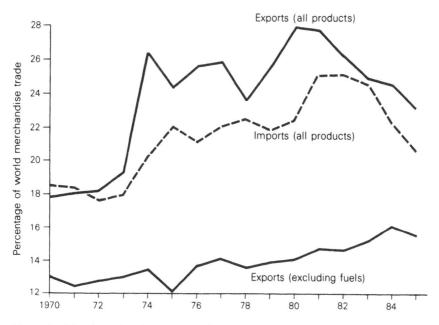

Figure 3 The less developed countries' share of world imports and exports, 1970–85

SOURCE: *Financial Times*, 15 September 1986.

Some of the more important NICs are located in the 'Pacific Ring', including Korea, Taiwan, Hong Kong, Singapore and India; elsewhere, Brazil, Argentina and Mexico are important. This group alone accounted for more than three-quarters of the manufacturing exports of the less-developed countries as a whole in 1976 (Keesing 1979). Manufacturing growth in the NICs usually began as import substitution but later became export-led, sometimes under sub-contract to multinational companies.

Governments encourage this process through under-valued exchange rates and export credits. Above all, however, their competitiveness was based on abundant supplies of cheap and compliant labour, which showed little resistance to often appalling conditions of work. Lack of labour militancy and acceptance of low wages were reinforced by repressive political regimes in most of these countries. The extent of the wage-cost differentials is considerable: whereas in 1981 the hourly costs of labour in manufacturing industries exceeded ten ECUs (European Currency Units, used for accountancy by the EC) in most European states, and were eight ECUs in Italy and six in the UK, they were less than one ECU in Brazil and South Korea.

The impact of the growth of the NICs can be seen in a small but important increase in their share of world manufacturing output, while the share of the advanced industrial centres declined sharply (see Table 3).

Table 3 *Shares of world output* in terms of value-added manufacturing (percentage figures)*

	1953	1970	1980
Industrial centres**	55.2	39.6	33.4
Recently industrialized	10.4	25.8	24.3
Other industrialized	6.4	8.0	7.5
Newly industrialized, Third World	3.2	6.0	7.7
Other Third World	1.6	2.8	3.3

* Non-socialist world only.
** Includes all of western Europe, except Greece, Ireland, Italy, Portugal and Spain, which are in the 'recently industrialized' group.
SOURCE: Ballance and Sinclair (1983, p. 14).

Their exports have, indeed, penetrated western Europe, although to a lesser extent than is sometimes thought. For example, in 1979 the industrialized countries obtained only 19.4 per cent of their manufacturing imports from the NICs (Hughes and Waelbroeck 1983). This is important, but NIC imports are estimated to have been the cause of less unemployment in the developed economies than, for example, have been productivity changes (Greenaway 1983). Nevertheless, the sheer magnitude of growth of NIC exports is impressive: between 1973 and 1979 the exports of manufactured goods from South Korea, Taiwan, Hong Kong, Singapore and Brazil alone increased in value from $18 billion to $53 billion. The major impact of these exports has been in other Third World countries where, increasingly, they have displaced exports from the developed world (Jenkins 1984). This severely reduced the potential for recovery in the advanced economies in the 1970s and 80s.

A double blow: oil crises in the 1970s

Rising production costs, increased competition from the NICs and a breakdown of the international monetary system all combined to produce a reduction in economic growth and in business confidence in the early 1970s. This was to be shattered in the course of the decade by two major oil crises.

In the 1950s and 60s, oil production was dominated by multinational companies known as 'the Seven Sisters': BP, Texaco, Exxon, Gulf, Shell, Mobil and Socal (Sampson 1977). This oligopoly was able to hold prices at around $1.8 a barrel during the 1960s (see Figure 4). However, by the early 1970s many oil production rights had been nationalized and more militant members of the Oil Producing Exporting Countries (OPEC), such as Algeria and Libya, wanted to gain control of world markets. As a result, most of the larger companies were reduced to the role of distributors and the only important remnants of multinational ownership were in the North Sea, Nigeria and Indonesia (Adelman 1983). Oil production and pricing thus became more subject to political considerations.

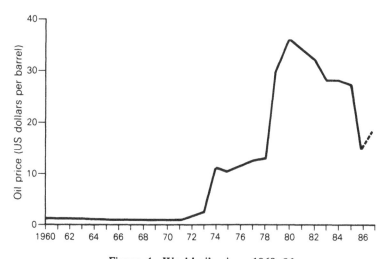

Figure 4   World oil prices, 1960–86

Crisis one

The spur to the first oil crisis was Israeli victory in the 1973 war with its Arab neighbours, leading to an Arab oil embargo on the West. Between October 1973 and January 1974 the price of oil almost quadrupled, and the results were widespread and immediate. Between 1973 and 1974 the oil imports bill of the developed countries increased from $35 billion to $100 billion, and this alone, amounted to about 20 per cent of their GNP (Cox 1982).

The timing could hardly have been worse because, in 1973, several major Western economies were already pursuing deflationary policies to try to reduce price inflation. This was because, in 1972, prices had soared as a result of reflationary policies which, in turn, had attempted to reverse the economic downturn resulting from the collapse of the international monetary system in the previous year. Therefore, the oil crisis gave a further sharp downwards push to economies already deflating. As wages in western Europe did not fall, the result was price inflation. Whereas prices in the developed countries increased by 2.9 per cent a year in the 1960s, the rate was 8.1 per cent in the 1970s. Price inflation increased their balance of payments deficits and led to further compensatory deflationary policies. The European experience compared particularly unfavourably with Japan, where real wages fell so that price inflation was lessened and profits and investment levels were sustained.

The effects of the recession were immediately evident as the growth of GDP slumped. In the advanced industrial economies this had averaged 6.1 per cent in 1973, but was negative (−0.4) in 1975 (World Bank 1984). Unemployment rates in the OECD countries also increased from an average of 2 per cent in the 1960s to 5.5 per cent in 1975 (Cox 1982).

The economic collapse in 1974–5 was rapid but equally striking was the strength of the recovery; between 1976 and 1978 the developed countries returned to moderate growth rates of around 3 per cent a year, with inflation at 7–8 per cent, even though unemployment persisted at the relatively high level of 5–6 per cent. Why did such a rapid recovery occur? One reason was that collapse of the Western economies had global consequences, and this extended to the OPEC countries because of reduced demand for oil. It was therefore more difficult to sustain prices and real oil prices fell between 1974 and 1978 (Cox 1982).

In addition, the OPEC economies were unable to absorb internally the surplus foreign exchange earnings they accumulated after 1973, so large quantities of capital had to be recycled via international banks in the advanced industrial economies. This, in turn, helped recovery in both the less-developed countries and in western Europe. Many governments also took steps to ameliorate the effects of the oil price rises. These ranged from reducing real taxes on retail oil prices to adopting expansionary fiscal policies, even at the risk of higher inflation. Inflation, combined with a depreciating dollar (in which oil prices were fixed), were, in fact, beneficial because they acted to reduce the real cost of oil imports. As a result, the economies of western Europe partially recovered in the mid-1970s – although unemployment remained at an unprecedented level.

Crisis two

In 1979–80 the second oil price crisis broke on the West. The catalyst on this occasion was the Iranian revolution, which resulted in Iranian oil sales being stopped completely by December 1979. Panic followed as the developed countries rushed to buy all the oil available on the 'spot' market. As a result prices were forced up from $12 to $31 a barrel between December 1979 and May 1980, even though there was no real supply shortage. The consequence was a repeat of history: the oil import bill for the OECD states doubled in 1980, and GNP growth rates fell while unemployment and inflation both increased. The effect on GDP growth was particularly severe: an increase of 3.3 per cent in 1979 became a contraction of −0.5 per cent in 1982.

The effects, however, were even more severe this time. For a start, aggregate growth rates were much lower in 1979 than they had been in 1973, so that the economies were less able to absorb such an external shock. In addition, there was a new consideration: the OPEC economies were better able to absorb new investment internally so that there was far less recycling of petro-dollars through European and US markets. To cap matters, western Europe was severely affected because West Germany – the dynamo of the economy – was particularly badly hit this time (Grahl 1983). Europe, and indeed most of the world economy, was in the grip of global economic crisis.

Global economic crisis: western Europe in the 1980s

Over the period 1973–83 economic growth rates in GDP per capita in western Europe were little more than half those recorded in the 1960s (see Table 1). At the same time national differences were pronounced. Ireland (benefiting from EC membership, especially from the Common Agricultural Policy) and Norway (with its major oil and gas resources) performed quite well in this period, but most countries had very low growth rates. The lowest growth rates were in Sweden, Switzerland and the UK, and the latter experienced the greatest difficulties in the early 1980s. Belgium, Denmark, the Netherlands and Spain also performed relatively weakly; a particular problem in the Netherlands was that natural gas exports led to an overvalued foreign exchange rate which made exports less competitive. Among the major western European economies, France and West Germany performed relatively well, although the French economy was to run into growing problems of restructuring and unemployment during 1984–6. Probably the effects of the crisis are most clearly visible in the virtual bankruptcies of some of the major European companies in the 1980s, including the Swedish shipbuilder Kockum and the Austrian metals group Voest Alpine.

Another symptom of the global recession was persistent inflation in the 1970s and early 1980s, which only lessened as the crisis itself deepened. Over the period 1973–84 as a whole, the highest rates of inflation were recorded in southern Europe, followed by Ireland and the UK, all of which

Table 4   *Basic indicators of the western European economies*

	Percentage average annual rate of inflation 1973–84	*Current account balance 1984 (millions of dollars)*	*Unemployment rates 1984*
Austria	5.3	− 633	3.8
Belgium	6.4	205	14.4
Denmark	9.4	− 1634	9.8
Finland	10.7	1	6.1
France	10.7	− 820	9.9
F.R. Germany	4.1	6130	8.4
Greece	17.3	− 2123	1.9
Ireland	14.4	− 916	16.5
Italy	17.2	− 2902	11.9
Luxembourg	—	—	1.7
Netherlands	5.9	4879	14.5
Norway	9.4	3228	3.0
Portugal	20.3	− 502	—
Spain	16.4	2323	20.1
Sweden	9.2	356	3.1
Switzerland	3.9	4019	0.9*
United Kingdom	13.8	1417	11.8

* 1983
SOURCES: World Bank (1986) and Eurostat (1985b).

were over 13 per cent (see Table 4). Austria, West Germany and Switzerland had the lowest inflation rates.

There was also a persistent growth of unemployment in western Europe, particularly compared with the USA and Japan (see Figure 5). The highest rates occurred both in older-established industrial economies such as Belgium, Netherlands and the UK, and in more recently developed economies such as Ireland and Spain (see Table 4). The lowest rates were in the Alpine and some Scandinavian states, and in Luxembourg and Greece (where interpretation of the official statistics is difficult). Unemployment, has, of course, been age-, sex- and ethnic-specific. For example, youth unemployment has been a special problem: in 1982 there were one million people under 25 in the EC who had been unemployed for more than one year (Commission of the European Community 1984a). Female unemployment was also rising significantly, a consequence of greater female participation in the labour force in the 1960s and 70s.

Another consequence of global economic recession has been the growth of protectionism which has seen the rolling back of some of the GATT trade-liberalization gains. This is evident in both trade restrictions and in state aids to particular industries. Trade restriction agreements have mostly been of the non-tariff type. An example is the 1978 Multifibre Agreement to cut EC textile imports from several leading NICs; worst affected was Taiwan with a 25 per cent reduction (Balassa 1981). National governments have also taken such measures and the UK, for example, has signed voluntary agreements with Japanese car producers so as to reduce imports (Turner 1982). Protectionism is greatest in France and least in Austria, Italy and Norway. This has led to increasing friction over trade, particularly between Japan, the USA and western Europe.

Government aids to industry became more widespread after 1974–5 and assumed a variety of forms, including preferential taxation and credit schemes and direct subsidies to producers, especially nationalized industries. Examples abound: the West German Government covered 75–80 per cent of the wage cuts of employees put on part-time work in some industries; the British Government operated a temporary-employment subsidy scheme to encourage firms to keep workers on their payrolls, especially in textiles and clothing; the Swedish Government subsidized its shipyards; and the French Government offered special aids to a number of industries, including cars, electronics and paper pulp.

1986: another turning point?

There were signs in 1986 that the economies of western Europe had reached another major turning point in their histories and, again, oil prices seemed to be at the root of this. From a level of about $28 a barrel in 1985,

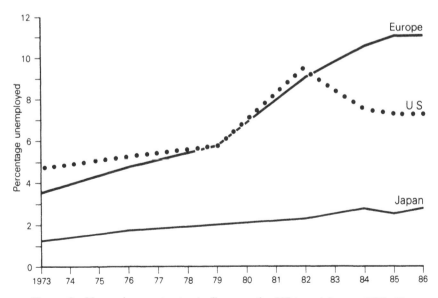

Figure 5 Unemployment rates in Europe, the USA and Japan, 1973–86

prices dropped drastically to $10 a barrel in the summer of 1986 before settling at around $16 a barrel by the end of the year (see Figure 4). The reasons for this lay in reduced demand, owing to improved energy efficiency in the main consumer countries and the effects of recession, and in over-supply as both OPEC and non-OPEC producers failed to cut back output.

Reduced oil prices gave a boost to a European economy which was already showing signs of recovery. This greatly improved the prospects of growth, for the $60 billion saved on imports was equivalent to between 0.75 and 1 per cent of the GNP of the developed countries. Lower oil prices also led to a reduction of about 1.3 per cent in already low inflation rates, which also permitted greater scope for economic expansion. The exceptions to this pattern were those western European countries which had become important energy exporters – Norway, the Netherlands and the UK. In Norway, for example, the fall in oil prices meant a reduction of about 10 per cent in state revenues. In addition, the medium outlook for growth in Europe was made uncertain by the Plaza Agreement of 1985, which aimed to force down the value of the dollar so as to reduce the USA's balance of payment deficits; this made American exports more competitive than those of Europe. Revaluation of the yen, leading to a sharp reduction in Japanese exports, provided some compensation for this development.

Further reading

1 General reviews of European economic development: D. H. Aldcroft (1980), *The European economy 1914–1980*, London: Croom Helm; W.M. Scammell (1980), *The international economy since 1945*, London: Macmillan.
2 Southern European development: A.M. Williams (ed.) (1984), *Southern Europe transformed*, London: Harper and Row.
3 European institutions: M. Blacksell (1981), *Post-war Europe: a political geography*, London: Hutchinson.
4 Crisis in the 1980s: A. Cox (ed.) (1982), *Politics, policy and the European recession*, London: Macmillan.

2 Capital, labour and the state in western Europe

Transformation of the European economy

The major sectors

The aggregate economic growth patterns outlined in Chapter 1 have been based on a broad sectoral redistribution of labour. In every western European economy the relative importance of primary activities has declined while that of tertiary activities has increased. The share of industrial employment tended to increase at first, but, since the mid-1960s (see Table 5), has been in at least relative decline in most countries. Industrial employment is here defined so as to include mining, construction and the utilities, as well as manufacturing, but the latter has been the most influential component of long-term change in most countries.

The declining share of primary activities has been all-pervasive but, not surprisingly, has been most marked where agriculture was still dominant in

Table 5 *Sectoral distribution of the labour force, 1965–81 (percentage figures)*

	Primary		Industry*		Services		Total	
	1965	1981	1965	1981	1965	1981	1965	1981
Austria	19	9	45	37	36	54	100	100
Belgium	6	3	46	41	48	45	100	100
Denmark	14	7	37	35	49	58	100	100
F.R. Germany	10	4	48	46	42	50	100	100
Finland	23	11	36	35	41	54	100	100
France	17	8	39	39	43	53	100	100
Greece	47	37	24	28	29	35	100	100
Ireland	31	18	28	37	41	45	100	100
Italy	24	11	42	45	34	44	100	100
Netherlands	9	6	41	45	50	49	100	100
Norway	15	7	37	37	48	56		
Portugal	38	28	31	35	32	37	100	100
Spain	34	14	35	40	32	46	100	100
Sweden	11	5	43	34	46	61	100	100
Switzerland	19	5	50	46	41	49		
United Kingdom	3	2	47	42	50	56	100	100

* Industry = manufacturing, mining, construction and utilities.
SOURCE: World Bank (1985; 1986).

1945; examples are Iberia, Ireland and especially Greece. Changes in industrial employment have been more diverse. The majority of the more-developed economies have experienced relative decline, although this has been more marked in some of the lower growth countries such as Belgium and the UK than in, say, West Germany and France. At the other extreme, however, there is a group of relatively late-industrializing countries where industrial expansion is still an important element of recent economic growth; this includes the southern Europe fringe of Greece, Italy, Spain and Portugal, as well as Ireland and Finland. The Netherlands and Norway also recorded gains in industrial employment, although the development of energy resources partly accounts for this (see Chapter 3). In contrast, tertiary sector employment has increased in importance in all countries, with particularly large gains in both more-developed countries, such as West Germany, and less-developed ones, such as Spain.

The outcome of these changes in terms of the sectoral division of labour is shown in Figure 6. The variations among the individual countries of western Europe reflect both different stages (early and late) of development, and different forms of development. However, it is clear that primary employment is of relatively limited importance in Europe, being less than 10 per cent in most countries and as low as 2 per cent in the UK. The exceptions are the Mediterranean economies, especially Greece (37 per cent) and Portugal (28 per cent), and, to a lesser extent, Ireland. The secondary sector is more evenly distributed and accounts for between 35 and 45 per cent of employment in all countries (excepting Greece); the highest levels are to be found both in mature industrial economies such as the UK, Belgium and the Netherlands, and more recently industrialized countries such as Spain and Italy. Finally, the tertiary sector has become the largest employer in every country (again excepting Greece), and is particularly important in the UK and Scandinavia. This reflects the maturity of development, the extent of deindustrialization, and the role of state expenditure in these countries.

Spatial divisions of labour
The underlying causes of the sectoral differences will be considered in more detail in Part Two. However, at a general level, it is possible to conceptualize these employment shifts as being the outcome of capital and labour flows between and within both sectors and countries. As the conditions for production and distribution have changed over time, so capital and labour have constantly been channelled to new areas and new activities. Capital has been guided by the need to realize profits, and labour by the need to seek out higher wages and/or job vacancies. The ways in which labour and capital have been combined is critical; their ratios have shifted over time, especially in response to changes in technology and in their relative costs (note the effects of social militancy, pp. 41–2), and over space, according to local conditions for production and distribution.

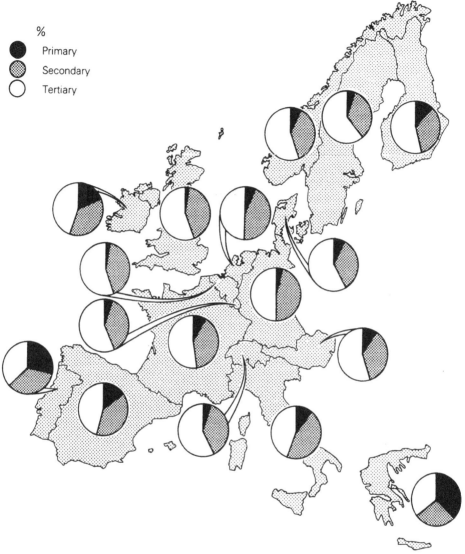

Figure 6   Sectoral distribution of the labour force, 1980

SOURCE: World Bank (1986).

Usually – but not always – this has meant substitution of capital for labour. Differences in the rate of such substitution partly account for the redistribution of employment between sectors and countries. For example, there are greater opportunities to increase the ratio of capital to labour in industry than in services, hence contributing to the greater relative increase in employment compared with output in the latter.

The concept of the spatial division of labour offers a useful descriptive device for comprehending the evolution of different combinations of economic activities over space. According to Massey (1978, p. 114), the term refers to:

> [the way in which] economic activity responds to geographical inequality in the conditions of accumulation – the particular kind of use made by capital of such inequality. This will differ both between sectors and, for any given sector, with changing conditions of production.

The historical process of development may be viewed as a series of rounds of new investment, each of which establishes a new spatial division of labour, involving new combinations of labour and capital in different locations. Once created, this new spatial division of labour (and associated infrastructure such as transport and housing) conditions any further rounds of investment. This process has an underlying logic, related to the nature of capital accumulation and profit realization. In addition, the state may act to modify or reinforce this process according to its own dictates (see pp. 79–82.

Internationalization of the European economy

A major feature of post-war Europe has been growing internationalization of the context within which spatial divisions of labour are created, especially – but not exclusively – for manufacturing activities. Internationalization has proceeded through three main stages (Cohen 1981; Palloix 1975) but, it should be emphasized, these are not clearly demarcated and are not temporally exclusive; the stages have succeeded one another, but more than one type of internationalization process may operate at any particular time. The processes have also operated at different scales, both between Europe and the Third World, and between more- and less-developed European economies.

The first stage is the internationalization of commodities, or the building-up of international trade. This was important within Europe from an early date whilst, later, it was one of the major motives behind colonization, a system which created protected enclaves for the trade of imperial powers. The second stage is the internationalization of capital which, although it has long historical roots, became particularly important towards the end of the nineteenth century. Initially, some of the more-developed economies – such as the UK, Germany and France – exported capital to other developed economies for investments in infrastructure (especially the railways) or to less-developed countries for the exploitation of raw materials. This form of internationalization was epitomized by the establishment of major mining, and agricultural plantation, companies. Only later – and, at a significant scale, only in the twentieth century – was capital exported in order to establish manufacturing bases in foreign countries. Typically, small-scale manufacturing plants were set up to produce for local markets. It is at this stage that the characteristic modern multinational company (MNC) emerged.

The third stage of internationalization relates to production itself, so that the division of labour within a firm is extended over international boundaries. In this way, only elements of, rather than the whole process of, production is located in any one country. This is especially characteristic of some branches of manufacturing industry such as clothing or electronics. Technological changes associated with neo-Fordism (see pp. 163–4) permit the sub-division of production into different stages, such as headquarters functions, research and development, skilled labour tasks and unskilled labour tasks. Each stage (in terms of labour costs, market access, etc.) has different production requirements. As conditions of production are differentiated between countries and regions, the outcome is a spatial division of labour. Typically, headquarters and research functions remain in the country of origin of the multinational company, while more labour-intensive functions are located in a low-wage country, whether this be Portugal, Brazil or Taiwan. Depending on the specific nature of the product, the final assembly stage may be located in either developed or less-developed countries. For example, motor vehicle assembly is often undertaken at or near the potential market, wherever that is, because of the greater cost of transporting the final product compared with the components. By contrast, assembly of more advanced micro-electronic goods, such as computers (for which transport costs are relatively insignificant), is usually located in the developed countries where the small but essential reserves of highly skilled labour are available.

One tendency in this complex organizational pattern has been decentralization of more labour-intensive stages of production to less-developed economies, whether in Europe or elsewhere, where wages are lower and the labour force is more compliant in accepting new work practices (Fröbel *et al.* 1980). The attraction of such international decentralization strategies for multinational companies has increased in the post-war period, especially following the exhaustion of large-scale indigenous labour reserves and the rise of social militancy in the developed countries since the mid-1960s. Even so, as Jenkins (1984) emphasizes, the precise form of decentralization varies by product. It is most common in industries such as clothing, where economic and technological considerations limit the possibilities of relying on increased mechanization as an alternative response to rising domestic labour costs.

Lipietz (1984) suggests that there have been two very different ways in which less-developed countries have been tied into the internationalization of production. One of these he labels 'bloody Taylorization' (see pp. 163–4), which can be typified as cheap-labour, sweatshop production such as occurred in textiles or electronics in Hong Kong or Singapore during the early phase of their industrialization. Ownership is mainly indigenous although production is tied to exporting, often under sub-contract. The second type is 'peripheral Fordism', whereby multinational companies locate branches of their operations in particular countries so as to exploit both labour reserves and markets, locally and in adjoining less-developed countries.

Although Lipietz referred specifically to the division of labour between developed and less-developed countries, his distinction between the two forms of decentralized production also has some relevance for Europe. Production has been internationalized within Europe by multinational companies setting up branch plants in different countries, responding both to marketing and production needs. Southern European countries and, increasingly, the UK have tended to attract some of the production stages requiring low-cost labour (see pp. 177–9, 187–9 on the textiles and motor vehicles). Production has also been internationalized through indigenously owned small firms (partly representing 'bloody Taylorist' sweatshop conditions) undertaking international sub-contracting. This applies particularly to some areas in northern Italy, northern Portugal and Catalonia where conditions of diffuse internationalization exist (see Hudson 1983a; Murray 1983; also pp. 171–2).

While internationalization of production has mostly applied to manufacturing industries, it is of growing importance in parts of the tertiary sector. One obvious example is the tourist industry, for many tourist facilities (such as golf courses) and much of the tourist accommodation in Mediterranean Europe are owned by companies whose headquarters and clients are located in northern Europe; examples are Thomson in the UK, Scharnow Reisen in West Germany, and Spies and Tjaereborg in Scandinavia. Retailing, too, is becoming internationalized and British companies such as Marks and Spencer or Laura Ashley have networks of European retail outlets. Other examples are the Italian clothing retailer Benetton and American fast-food specialists such as Macdonald's, both of which have chains of outlets throughout Europe. Perhaps of greater significance, however, has been the internationalization of producer services. Banks and other financial services such as insurance and accountancy companies have international networks, and these range from the small-scale, such as Finland's Postipankki, to the truly global, such as Credit Suisse. Some property development/management agencies operate at an international scale, an example being the UK's Wooton, Jones and Laing which is active in several European capitals. Such services also support multinational manufacturing companies, helping them to adjust to the legal, economic and political frameworks of foreign countries (Cohen 1981).

The creation of new spatial divisions of labour and the process of internationalization have involved reorganization of both capital and labour in post-war Europe. The next two sections examine some of the major features of these, and especially their international dimensions.

The supply of labour

Lack of labour supply constraints was important in early post-war expansion, and some theorists (for example, Kindleberger 1967) consider this to have been the key element in a continuous circle of growth. Availability of labour ensured that wage demands were moderated, labour costs were kept low and high profits and investment rates were maintained. Only in the late 1960s were these conditions challenged while, in the late 1970s and the 1980s, global crisis has restored the conditions of labour surplus. Even so, labour supply remains important to the needs of firms since the increasingly rapid process of restructuring requires a high turnover of labour. For example, between 1982 and 1985 there was a 25 per cent turnover in the labour force of firms in West Germany (*Financial Times*, 4 September 1986).

Establishment of international branch plants is one means of securing a suitable labour supply, as was emphasized in the previous section. Within a particular country, various options are available. For example, in Spain (1939–75), Portugal (1926–74) and Greece (1967–74) there were dictatorial governments which used repressive means – such as Spain's 1939 Basic Labour Law (Mérigo 1982) – to ensure that labour costs were kept relatively low. However, in most of western Europe democratic regimes prevailed and, in practice, the labour supply was secured by two processes: internal reorganization of labour and international migration. Government policies have supported the reorganization of labour supplies because 'the intervention of the State has become central to the articulation of migration and capital accumulation' (Miles 1986, p. 50).

National sources

Internal (national) reorganization of labour has been achieved in three main ways: agricultural restructuring, restructuring in existing industrial areas, and increased female participation in the workforce. Agricultural restructuring has been important in releasing labour in every country in western Europe (see Table 5). Production increased in agriculture several-fold, despite large-scale shedding of labour, because there had been considerable hidden unemployment in many agricultural regions, while mechanization also reduced labour requirements. The release of labour was only one aspect of agricultural restructuring, however. This was a process which created mass consumer markets in many rural areas which had been pre-capitalist (and more self-sufficient) prior to 1945, as labour was drawn from peasant production to waged employment in services and industry (Carney 1980). This process has been particularly important in regions such as southern and western France, most of Italy, and southern Germany.

Restructuring of existing industrial areas is also a major source of labour, and classic cases have been such regions as the Ruhr, North-East England, Alsace and Limburg (see Chapter 8). Many were coalfield areas experiencing a decline in production as a result of changes in European energy sources. However, many coalfield-based industries, such as steel production, have also been restructured, as have other traditional industries such as textiles, leading to large-scale redundancies. These are industries affected by intensified international competition or by new products (such as plastics) replacing traditional products (such as rubber or some metal alloys). Redeployment of labour released from such declining industries was one means of easing the supply conditions of expanding firms, but it was also a means whereby incomes and, therefore, demand could be sustained in these regions.

Increased female participation in the workforce is another source of labour. Women entered paid employment in many countries during the Second World War to meet strategic needs, but this had longer-term implications in the acceptance of women in the labour force. For example, in the UK many women left the factories after 1945 but, as labour shortages emerged at the end of that decade, there were active campaigns to re-recruit them for industrial work. As a result, in many countries there was a social re-evaluation of the role of women, especially married ones, in the labour force. It became more acceptable for women to continue in paid employment after marriage and after childbirth, and this was facilitated by state provision of pre-school child-care services. The nature of the job market has also changed and the shift from industry to service employment has created more demand for women workers. There is nothing intrinsic to most office or factory jobs which makes them naturally more suited to males or females; rather, certain types of jobs are socially defined as being for males or for females. A major post-war trend has been the expansion of jobs socially defined as suitable for women workers. Female workers may be preferred by employers because they are less well paid, tend to be less unionized and are more compliant in accepting new working conditions and technology.

As a result of these changes the proportion of women in the working population has increased throughout the post-war period, and in the EC, for example, had reached 38 per cent by 1980. There are, however, still variations between countries, especially with respect to married women, reflecting both differences in job opportunities and cultural norms concerning the role of women. Nearly 50 per cent of all married women in Denmark and the UK, but only 18 per cent in the Netherlands and 14 per cent in Ireland, are in waged employment.

In spatial terms, these domestic sources of labour supply can be drawn into production in one of two main ways; by migration to areas of employment growth, and by the in-movement of investment to create jobs in the labour reserve regions. Until the 1970s, at least, the former was dominant and, indeed, the two main national migration flows of the post-war period have been from rural to urban regions and from declining to growing industrial regions (Salt and Clout 1976; Clout *et al.* 1985). According to Fielding (1975), many rural areas had annual net migration losses greater than 17.5 per 1000 in the 1960s. Classic examples of internal migration flows are from the south to the north of Italy, from Andalucia to Barcelona, from the Alentejo to Lisbon, from the south and west of France to Paris, and from the north to the south of Scandinavia.

There has also been substantial net out-migration from such traditional industrial regions as Lorraine and Nord in France, the Ruhr (West Germany), Wallonia (Belgium) and the North-East in the UK. However, these migration flows have tended to be on a lesser scale than the rural–urban shifts because, at least until the 1970s, there was a tendency for capital to move into such areas so as to reindustrialize them. This was often encouraged by state policies (Damette 1980).

The second locational strategy for utilizing labour – investing in labour reserve regions – has generally become more important in the 1970s and 80s, as conditions of production in many existing metropolitan regions have become less attractive. As a result, the rural–urban migration flow has been reversed in many countries. Capital has moved into rural areas because firms have sought to use labour reserves in what are considered to be less-militant local communities (see pp. 168–9).

International sources

Two stages

Traditionally, Europe has been a labour exporting region: between 1815 and 1913 some 45 million Europeans are estimated to have emigrated, mainly to the Americas, South Africa and Australia (Ilbery 1981). Emigration has continued since 1945, especially to North America and Australia, and, more recently, on short-term contracts to the Middle East. Some states have actively encouraged emigration; for example, early post-war governments in the Netherlands were concerned with the potential problem of over-population and, between 1946 and 1963, some half a million persons received assistance to emigrate. The Francoist state in Spain also encouraged emigration, in order to acquire foreign exchange, as a deliberate part of macro-economic strategy.

Since 1945, however, Europe has experienced net in-migration, with workers being attracted to the labour-hungry northern economies. It has been estimated that, in the EC9 alone, there were 1.78 million immigrants in the 1950s, 3 million in the 1960s and 3.1 million in the 1970s (Haselen and Molle 1981), although substantial illegal immigration makes all such estimates highly uncertain (King 1984). Nevertheless, by the early 1970s there were probably some 12–13 million foreigners resident in western

European countries (Kane 1978; Kirk 1981), although not all were active in the labour force. A large proportion of the estimated world total of 16 million emigrant workers (Lewis and Williams 1984) is to be found in western Europe.

Emigration in western Europe has proceeded through two main stages since 1945, divided approximately by the first oil crisis in the 1970s (Castles *et al.* 1984; Tapinos 1982). In the early stage (1945–73) the characteristic *gastarbeiter* (guest worker) system of single, adult migration prevailed. Faced with potential labour shortages at home, many northern European governments encouraged immigration. Initially the UK had recruited 90,000 workers from among the refugees of Europe, under the European Voluntary Workers Scheme. France, as early as 1945, established the Office National de Immigration to recruit temporary labour, mainly from southern Europe. Belgium also had an organized system of labour recruitment, the *contingenten systeem*, whereby bilateral agreements were signed with southern European countries (especially Italy), mainly to secure workers for the coal and steel industries. The colonies were also a ready source of labour. Governments either took a passive role in this, presenting no obstruction to immigration of colonial citizens, or actively encouraged recruitment.

Emigration was also encouraged at the institutional level; for example, the Scandinavian countries formed the Nordic Common Labour Market in 1954, which largely facilitated Finnish migration to Sweden. Free movement of labour was a basic tenet of the EC, as laid down in the Treaty of Rome. In practice this initially provided a framework for Italian emigration to other EC member states, but the EC has also been involved in signing labour migration contracts with other countries, such as Turkey.

After the 1973–4 oil crisis there were lower levels of 'new' immigration and more family reunification as dependents followed. This was encouraged by increasingly restrictive immigration laws in many countries which, while permitting the arrival of dependents, limited the entry of new single migrants. Subsequently, the children of some of the earlier wave of immigrants have come on to the labour markets in the host countries. By 1980 there were an estimated 4.5 million foreigners aged under 25 in the main European immigration countries (Lebon 1981).

A number of factors contributed to the dramatic turn-around in international migration. Most obviously, the growth of unemployment from the late 1960s reduced the demand for labour, especially following the 1973–4 oil crisis. In addition, as unemployment rose among the indigenous population, there were serious outbreaks of racial tension in several countries, notably the UK and France. This increased the political pressure on governments to implement anti-immigration measures. However, immigration was already lessening in importance for the host countries before these events. As immigrant populations age and pass through the family life cycle, so their costs to the state in terms of social investment (health, housing, education, etc.) increases sharply (Castles *et al.* 1984). At the same time, immigrants become less compliant and deferential in the workplace, as they become more socially integrated in the host country, and learn its language. Symbolically, immigrants were heavily involved in the national wave of strikes in France in 1968 and in the prolonged stoppage at Ford's Cologne works in 1973 (Paine 1977). Consequently, the real cost of immigrant labour was rising, before the demand for it fell.

As a result of these new conditions, several European countries took measures to restrict immigration. Given the weakness of its economy, it was the UK which took the lead: the 1962 and 1968 Commonwealth Acts introduced work vouchers for immigrants and restrictions on overseas passport holders, while the 1971 Immigration Act restricted the entry of dependents. Later, Belgium, Switzerland and West Germany introduced bans on immigrant recruitment in the course of 1973–4. France, too, banned new recruitment in 1974 and, in 1977, introduced a programme of financial incentives (at 10,000 francs per head) to 'assist' immigrants to return to their countries of origin (Lawless and Findlay 1984). These measures did have an impact, and Böhning (1979) estimates that between about 1.5 and 2.0 million foreign workers lost their jobs in western Europe between 1974 and 1978. However, total numbers of immigrants hardly fell at all as the process of family reunification was allowed to continue unhindered in most countries, except the UK. The case of Switzerland is not untypical; it had some 593,000 foreign workers in 1974 but this fell gradually to 489,400 in 1978, before slowly recovering to 526,200 by 1982. Immigrants, therefore, continue to be important in the labour forces of many European countries in the 1980s.

The geographical pattern
The broad pattern of emigration flows that contributed to the stock of emigrants in western Europe in 1980 is shown in Figure 7. At that date the main recipients were West Germany, France and the UK, followed by Switzerland, Belgium, Sweden, the Netherlands, Austria and Luxembourg (see Table 6). Some two million emigrants had come from beyond Europe and the Mediterranean region, while the largest absolute flows from within Europe were from Italy, Yugoslavia, Portugal, Spain, Greece and Finland. In relative terms there is a somewhat different picture, and immigrants are most important in Switzerland (18 per cent of the labour force), followed by West Germany, France and Belgium (Vilaça 1984). Among the sender nations, emigration has been most important in Ireland (30 per cent of the labour force) and Portugal (11 per cent).

By about 1980, Castles *et al.* (1984) estimated that approximately 38 per cent of immigrants in Europe were from Ireland, Finland and southern Europe, while some 54 per cent were from outside Europe. There were four types of emigrants. Large numbers of returnees from ex-colonies had settled in France, the UK, the Netherlands and Belgium. These were

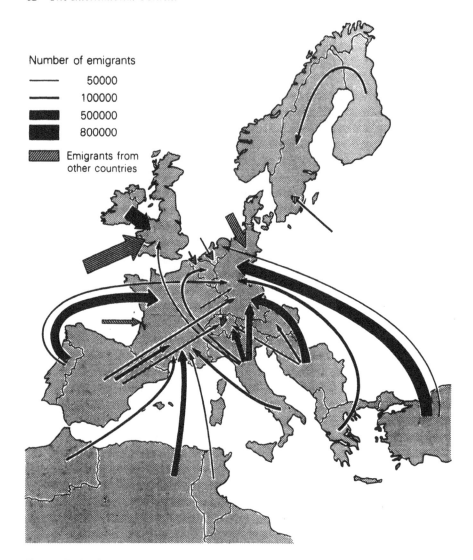

Figure 7   Major international migration movements contributing to the stock of emigrants in western Europe in 1980

SOURCE: ILO (1984).

mainly emigrants from Europe and their children, who chose or had to return to their country of origin after decolonization; examples include flows from Algeria to France, from the Indian sub-continent to the UK and from Indonesia to the Netherlands. The Federal Republic of Germany provides a special case of this type of immigration, having received some 3.5 million people from the German Democratic Republic since the initial partitioning of Germany. A second group of immigrants is from the

Table 6 *Numbers of international migrant workers in 1980 in selected western European countries*

Countries of origin		Countries of destination	
Algeria	327,500	Austria	174,700
Finland	112,700	Belgium	332,600
France	117,200	France	1,487,000
Greece	166,900	F.R. Germany	2,168,800
Italy	958,400	Luxembourg	51,900
Morocco	203,708	Netherlands	194,600
Portugal	519,700	Sweden	234,100
Spain	394,400	Switzerland	706,600
Tunisia	71,100	UK	929,000
Turkey	773,000		
Yugoslavia	616,200		
Other	2,018,400		

SOURCE: ILO (1984).

indigenous populations of colonies and ex-colonies, attracted by job opportunities in the UK, France, the Netherlands and Belgium in the 1950s and 60s. Although ethnically and culturally distinctive, they were fairly easily absorbed in these decades. However, in the 1970s, as many attempted the transition from temporary guest-worker to settler, they did so against a rising tide of unemployment and increasing racial tensions.

A third group of immigrants are the classic *gastarbeiters* or temporary migrants from southern Europe and North Africa. Some of the more significant of these flows are from Greece, Turkey, Yugoslavia and Italy to West Germany; from Italy to Switzerland; and from Portugal, Spain and Algeria to France. With the passage of time, the onset of recession (making re-immigration more difficult in the face of new entry laws) and the enlargement of the EC, many have become settlers. Finally, in recent years, there has been a growing number of skilled labour migrants – both manual and non-manual – who have been mobile within Europe, mainly among the more industrialized countries. Precise data on this is fragmentary, but West Germany seems to have been the major recipient.

Most countries are both importers and exporters of labour. The UK, for example, has experienced emigration on a permanent basis to Australia and the USA, and on a temporary basis to West Germany, while also being a major recipient of immigration. In general, labour is exported to more-developed or more-rapidly expanding economies while it is imported from less-developed or stagnant ones. The same basic rule applies to southern Europe. Greece draws labour from southern Mediterranean countries, especially to work in the Piraeus dockyards and in the building industry. Portugal continues to draw labour from Cabo Verde, again for the construction industry. The outstanding example in this category, however, is Italy. It is estimated that while there were almost a million Italians working abroad in 1980, at the same time there were some half a million immigrants (mostly African) in Italy, mainly employed in menial jobs (King 1984).

The stocks of migrant workers in the main destination countries reflect diverse and changing histories of immigration. For example, Sweden traditionally relied on Finnish workers and, even in the 1980s, this still constitutes over 40 per cent of immigrant labour. However, in the 1960s and 70s it had to attract labour from further afield, partly because of the expansion of jobs in Finland itself. Hence, in the 1980s, some 17 per cent of the immigrant labour force had originated in Greece, Turkey and Yugoslavia (Castles *et al.* 1984).

The economic role of international labour migration

There is considerable debate regarding the economic implications of international migration for both the origin and destination countries (see King 1984; Lewis 1986). There certainly are some benefits for the sender nations. Emigration is a source of foreign exchange earnings and contributes to the balance of payments: for example, in 1978–9, emigrant remittances as a proportion of merchandise exports were 69 per cent in Portugal, 30 per cent in Greece and 13 per cent in Spain (World Bank 1981). Emigrants may use their savings to invest in new forms of production and, on return, may transfer job skills or innovations from abroad. Emigration may also be useful in absorbing pools of unemployment or hidden unemployment. Additionally, it can facilitate agricultural reorganization, through reducing population pressure on limited land supplies, especially in areas of peasant production.

All of these benefits exist, but they tend to be sporadic and there is little evidence of any systematic advantages for the sender nations, apart from the acquisition of foreign exchange earnings (King 1984). Indeed most case study evidence suggests that there is little reorganization of land and scant transfer of skills. Most returnees use their savings for consumption rather than investment, and on establishing cafes and bars rather than factories (for example, Lewis and Williams 1984; Toepfer 1985). Even where emigrants seem to have preferred industrial employment after their return, there are usually few opportunities to undertake this, especially in rural regions (King *et al.* 1985).

In contrast, there are considerable benefits for the host countries, and Castles and Kosack (1973, p. 428) consider that 'labour migration is a form of development aid given by poor countries to the rich countries'. There are a number of reasons for this (Paine 1977): immigration prevents bottlenecks emerging because of labour shortages; it permits the realization of scale economies; it helps maintain investment levels by reducing labour costs and sustaining high rates of profit; it contributes to export-led growth in output by helping to keep down export prices; and it helps contain inflationary pressures and consequent government deflationary policies. It is not simply that immigrant labour has low direct costs, although this is certainly true – especially for women – given acceptance of low wages and poor working conditions. The indirect social costs are also lower than for indigenous workers: the costs of early education and socialization, and of health care in later life, are borne by the sending rather than the receiving country. There are advantages both for individual companies hiring immigrant labour and, as a result of wage levels being depressed, for all other firms in the economy. Where the immigrants are illegal, then the advantages for employers are even greater: living in fear of repatriation, illegal migrants will accept even poorer wages and working conditions. Switzerland has maximized the economic benefits by only issuing short-term work visas to immigrants, hence maintaining a high degree of flexibility in the face of fluctuating demands for labour.

Capital: the international context

Capital may be organized in several different forms, although a few types are usually dominant. One classification (Fahrenkrog 1984) identifies six main variants:

- individual capital
- simple associations of capitals (two or more independent capitalists)
- stockholding capital (complete separation of ownership from the means of production)
- cooperatives of capitalists
- cooperatives of direct producers
- state enterprises.

There are a mass of individual capitals in most European countries, mostly small businesses with owner-managers. However, the 'standard' form of larger business organization is the stockholding company, which ranges from large trusts with considerable assets to single factories or offices owned through shareholding. Each of these different organizations of capital has its own distinctive features, in terms of management style, reliance on producer services, and the spatial division of labour, as is made clear in subsequent chapters.

While ownership remains mostly national – whether public or private – there has been a growing internationalization of capital. This can take the form of portfolio (non-specific) investments or Eurobond loans to governments, but most significant – and certainly most controversial – are the multinational companies (MNCs). These have assumed a key role in economic development in many western European countries and most of this section will concentrate on this particular form of capital.

There are a number of definitions of MNCs, and some of the more stringent emphasize that they own capital drawn in several countries whilst their organization and planning is conceived on a global scale. Companies such as the UK's ICI or France's Pechiney Ugine Kuhlmann certainly meet such criteria, but this does exclude many smaller multinationals which, characteristically, may have holdings in only two or three European countries. Therefore a more practical approach is to adopt Fishwick's (1982, p. 18) definition of MNCs as having 'significant production outside the country of origin', while accepting that they vary enormously in their ownership forms, size and degree of transnationality.

The evolution of multinational companies

Multinational companies have a relatively long history. Between 1820 and 1913 Europe is estimated to have exported some \$37 billion of capital, mainly to the 'New World' (Arnell and Nygren 1980); about 43 per cent originated from the UK, 20 per cent from France and 13 per cent from Germany. Most of this capital was portfolio investment (Dicken 1982) but, even at this stage, European and American companies were engaged in direct overseas investments, creating international networks of branch plants. By 1914 US firms had 122 foreign subsidiaries, UK firms had 60, and other European firms had 167. The MNC had arrived on the European scene.

Several European companies had MNCs at an early stage. Bayer, the German chemical group, opened its first international branch plants in the USA in 1865 and, by 1908, had extended into Russia, France and Belgium (Hörnell and Vahlne 1986). In Sweden, the chemical firm Alfred Nobel was the pioneer, having branch plants in 10 countries as early as the 1870s. By the 1890s, Ericsson and Alfa Laval had followed it abroad. Switzerland, too, had its MNCs, led by Bally who established several shoe factories in South America in the 1870s. By 1914 Ciba Geigy (chemicals), Brown Boveri (electrical goods), Hoffman La Roche (medical goods) and Nestlé (dairy products) had also become MNCs.

In the inter-war period internationalization of capital slackened and the next major phase of expansion was after 1945. American companies took the lead in exporting capital but, as Europe recovered, they were joined first by the UK and later by West Germany, France and the Netherlands (see Table 7). Formation of the EC and EFTA encouraged internationalization of production within Europe. The pace of international investment quickened in the 1960s and early 1970s and the world stock of such investment increased from 108 to 287 billion dollars between 1967 and 1976 (Sauvray 1984). By the early 1970s the MNCs already controlled 20 per cent of world production (Blackbourn 1982). For historic reasons, the

Table 7 *Growth of international investment by country of origin, 1914–76*

| | Value of assets as a percentage of the world total | | | | | |
	1914	1930	1960	1967	1971	1976
USA	6.3	35.3	59.1	55.0	52.0	47.6
UK	50.4	43.8	24.5	16.2	14.5	11.2
F.R. Germany	17.3	2.6	1.1	2.8	4.4	6.9
Japan				1.3	2.7	6.7
Switzerland				3.9	4.1	6.5
France	22.2	8.4	4.7	5.5	5.8	4.1
Canada	0.5	3.1	5.5	3.4	3.6	3.9
Netherlands	3.1	5.5	4.2	2.1	2.2	3.4
Sweden	0.3	1.3	0.9	1.4	2.1	1.7
Benelux				1.9	2.0	1.2
Italy				1.9	2.0	1.0

SOURCE: Sauvray (1984) and Shepherd *et al.* (1985).

USA still dominates the stock of international investment, although Japan has been catching up rapidly (Table 8*a*).

Multinational companies have become an increasingly important element in the European economy, although there are considerable differences between countries. Some European countries have their own MNCs and Europe's share of the world's largest companies has increased in the post-war period (Table 8*b*). By 1983 France, West Germany and the UK were the European leaders, followed by the Netherlands and Switzerland. Italy is the most recent entrant into this 'super league', with such giants as Fiat and Montedison. It is also represented in advanced electronics by Olivetti, which was strong enough to have bought out Germany's Triumph Adler office equipment producer, and the UK's Acorn computer company in 1986.

The relative strengths of European- and non-European-based MNCs are different. In terms of turnover in 1978, the largest six MNCs were all Japanese, and they were closely followed by two giant US corporations, General Motors and Exxon. Whereas the world's largest MNC, Mitsubishi, had a turnover of $106 billion, the largest European MNC – the Royal Dutch–Shell group, based in the Netherlands and the UK – had a turnover of only $44 billion (Nakase 1981). The only other European MNCs with turnover in excess of $12 billion were British Petroleum (UK), Unilever (Netherlands/UK), Philips (Netherlands) and Hoescht (West Germany). These are the European giants and, in some ways, the exceptions, for direct foreign investment is also increasingly important to medium-sized companies operating in only a small number of countries.

In recent decades the process of internationalization has increased. For example, the proportion of Swedish MNC production located abroad

Table 8   *The world's largest multinational companies by country of origin*

a   Percentage distribution of sales

	1962	1982
USA	67.3	47.8
Europe	26.9	31.9
Japan	3.6	12.8
Rest of world	2.1	7.5

b Sales ($) 1983	Number of companies	
	over 10 billion	*5–10 billion*
Austria	0	1
Belgium	0	1
F.R. Germany	7	5
France	4	5
Italy	2	1
Netherlands	1	2
United Kingdom	3	7
UK/Netherlands	2	0
Sweden	1	0
Switzerland	1	2
Europe total	21	24
USA	29	36
Japan	7	12
Rest of world	4	11

SOURCE: based on Dunning and Pearce (1985).

increased from 26 per cent in 1965 to 39 per cent in 1978 (Hörnell and Vahlne 1986). The reasons for this are considered in the following section.

Reasons for the expansion of MNC activity

The underlying reasons for the post-war expansion of the MNCs are not hard to find; companies have had to internationalize so as to maintain profit levels in the face of changing economic circumstances. Chief among these are the needs to secure economies of scale in production, diversify their production bases, secure markets and lower labour costs. For example, BP obtain 58 per cent of turnover from their overseas operations, but 73 per cent of their trading profits from these (Shepherd *et al.* 1985). In addition it can be argued that, as ownership is divorced from management in larger companies, the aims of management may become expansion *per se* rather than profit realization, which adds to the imperative of internationalization (Scammell 1980).

Technology and scale economies
The primary reason for international expansion has probably been the need to achieve economies of scale, given changing production and technological requirements. MNCs characteristically operate in 'new' industrial sectors, either involving bulk inputs (such as oil) or high-level technology. This is reflected in the dominance of oil, chemicals, vehicles and electrical interests in the list of the top 25 European companies (see Table 9). Research in modern high-technology industries is very expensive and must be recouped with large-scale production runs. A monopoly is especially preferable in such cases so as to protect huge initial outlays on research and development.

Table 9 *Europe's top 25 companies by turnover, 1985*

Company	Turnover ($ million)	Country	Major sector
1 Royal Dutch Shell	110,293	Neths/UK	Oils
2 British Petroleum	61,838	UK	Oils
3 IRI	33,368	Italy	General indust.
4 ENI	30,504	Italy	Oils
5 Elf Aquitaine	25,378	France	Oils
6 Unilever	25,186	Neths/UK	Food mfc.
7 Siemens	24,557	F.R. Germany	Electricals
8 Total	24,320	France	Oils
9 Philips	23,909	Neths	Electricals
10 Volkswagen	23,610	F.R. Germany	Motor vehicles
11 Daimler-Benz	23,564	F.R. Germany	Motor vehicles
12 Nestlé	22,953	Switzerland	Food processing
13 Veba	21,850	F.R. Germany	Diversified indust.
14 BAT Industries	21,836	UK	Tobacco
15 Deutsche Bundespost	21,526	F.R. Germany	Utility
16 BASF	21,442	F.R. Germany	Chemicals
17 Bayer	20,649	F.R. Germany	Chemicals
18 Hoechst	19,209	F.R. Germany	Chemicals
19 Fiat	17,699	Italy	Motor vehicles
20 Renault	17,158	France	Motor vehicles
21 Electricité de France	16,743	France	Utility
22 Electricity Council	16,208	UK	Utility
23 ICI	16,181	UK	Chemicals
24 Thyssen	15,640	F.R. Germany	Metals
25 Petrofina	14,740	Belgium	Oils

SOURCE: *Financial Times* 'Top 500', 1986.

Classic examples of industries with such technology-related scale economies are computers and aerospace engineering. The market for non-specialist computers is dominated by the US firm IBM, which has about two-thirds of the EC market (Williams 1979). In the aerospace industry, American companies such as Boeing, McDonnell Douglas and Lockheed again dominate the main market for passenger aircraft. European companies can only survive through specialization (for example, France in military aircraft; Rolls Royce in engines; British Aerospace in airframes), or through joint ventures, such as the European Airbus, whereby the development costs are spread among several countries and companies. By and large, European companies have not performed especially well in such high-technology sectors and the main challenge to US supremacy – especially in electronics – has come from Japan.

Securing market access

Another reason for MNC expansion is to secure market access, which is linked to the need to obtain scale economies in production. Given increasing international competition, both from the developed countries and the NICs, it is imperative for MNCs that they have secure markets. This applies to both final consumer markets and intermediate industrial customers. While markets could be supplied from the MNC's factories in the home country, relocation of production abroad offers two distinctive advantages. First, for some products – such as motor vehicles – the relative costs of production and of transport make it cheaper to locate at the market. Secondly, as many countries operate tariff duties against imports, this may lead MNCs to locate production abroad.

This is considered to be one of the prime reasons for US investment in the EC (Kemper and Smidt 1980), and a major reason for joint production agreements such as that between Honda and British Leyland. In reverse, given the smallness of French markets, the only way for Pechiney Aluminium to survive was (Savey 1981, p. 318):

> to endeavour simultaneously to organise the national market, to reach a controlling position in some sectors of the European market, to gain access to the United States and to open up some prospective outlets in developing countries where aluminium consumption was increasing.

Such internationalization strategies are contagious for they may prompt similar investments by other MNCs for defensive reasons (Fishwick 1982). In addition, firms may find it advantageous to locate abroad those production processes involving mature technologies.

Reducing costs

The growth of MNCs has also been encouraged by the need to reduce costs. This became particularly important in the 1970s as a combination of rising energy prices, soaring labour costs (see pp. 41–2) and intensified competition from the NICs affected the developed countries. One strategy adopted by the European companies to counteract the resulting tendency for profits to fall was to decentralize at least some stages of production to lower-cost locations in southern Europe or the Third World. In addition, internationalization allows companies to spread the risks of disruption resulting from social militancy in any one country: this was a major reason for car companies adopting European, and even global, production and assembly strategies. At its crudest, this means that if Ford UK is on strike, then the company can still import cars from other European factories. Furthermore, the existence of alternative production facilities can be used by the MNCs as a veiled threat in wage negotiations with the trade unions.

Given that there are strong tendencies for the expansion of MNCs, their growth has been facilitated by two important conditions: the strength of finance capital and new managerial methods. Finance capital – the fusion of different types of industrial and commercial capitals – is an important feature of the world economy. For example, by 1978 some 125 of the 487 leading companies of the world were owned by banks (Grou 1983). The availability of capital resources on a global scale greatly eases the financial requirements of companies seeking to expand abroad. At the same time, a tendency for diseconomies to occur with increasing scale – as close control is lost over a company's activities – has been counteracted by revolutions in management and communications methods. These permit the evolution of decentralized managerial hierarchies which are held together by modern means of communications.

Sectoral variations
Internationalization is essentially a means of raising or maintaining a company's profit levels. Given that operating conditions vary between different types of economic activities, foreign investment has been unevenly distributed sectorally. To begin with, there are broad differences according to the nationality of the MNCs, related to the timing of internationalization. US and UK multinationals expanded relatively early, at a time when securing raw materials was important; hence, over a quarter of the MNCs in these countries were involved in extractive activities (Shepherd *et al.* 1985). West German MNCs had strong expansion in the 1960s and 70s and now tend to be more involved in manufacturing and other activities (75 per cent). Japanese MNCs are different from all these; about 25 per cent of companies have been involved in extractive industries, which is not surprising given the lack of indigenous raw materials. Furthermore, the early phase of internationalization was based on trading companies, so that in the mid-1970s more Japanese MNCs were involved in service than in manufacturing activities (Dicken 1982).

There are detailed variations in the precise sectors that have attracted foreign investment (see Table 10). Oil and natural gas have attracted high levels of MNC activity in most countries, and this is reflected in the data for

Table 10   *Foreign participation in manufacturing sectors in France, West Germany, Italy and the UK in the late 1970s*

	Shares of sales turnover (per cent)			
	France	*Germany*	*Italy*	*UK*
Food and drink	16	18	21	13
Chemicals	30	26	23	30
Iron and steel	17	40	9 }	11 }
Non-ferrous metals	10	28		
Mechanical engineering	24	20	25	20
Electrical goods	22	30	*	25
Motor vehicles and parts	17	24	5	36†
Textiles and clothing	9	7	21	6
Paper and products	17	14	19	17
Oil and natural gas	55	93	*	*

* not available.
† estimated.
SOURCE: Fishwick (1982)

France and Germany. These are industries with very high entry costs, long realization periods, and advanced technological requirements. Similarly, high research and development costs in industries like vehicles manufacture, chemicals and electrical goods are an important barrier to national, as opposed to international, operations. Services have tended to attract less MNC investment and, for example, in 1983 only 23 per cent of US investment in the UK was in banking, finance and trade.

MNCs within western Europe

US companies

US investment is still dominant in western Europe but is unevenly distributed within the region. The UK was originally the major recipient and, in 1950, still had a half of all US investment in Europe. However, this had fallen to 28 per cent by 1975, as US investment shifted to Belgium (7 per cent), the Netherlands (7 per cent), West Germany (18 per cent) and Switzerland (Kemper and Smidt 1980). Yet, in 1985, the UK was still the dominant destination of US cumulative investment, followed at some distance by Switzerland and West Germany (Table 10). Even taking into account population sizes, the UK had US investment of $127 per capita by the mid-1970s, while its nearest rival, Belgium, had only $75 per capita (Hamilton 1976; 1978).

   US companies were originally attracted to Europe by expanding consumer markets and, at least in the 1950s, by cheap labour; at this stage US wages were four times those in the Netherlands. Location within EC and EFTA tariff barriers was also important, as was availability of aid from European governments. The UK was an obvious initial base because it had a stable political environment, was a trade and shipping centre, offered access to Europe as well as the British Commonwealth, and had no linguistic barrier. Later, in the 1960s, many American companies (for example Ford and ITT) sought to diversify their European production and established plants on the continent, usually in West Germany. Subsequently, the Netherlands and Belgium benefited as investment shifted to the coast to 'the Golden Delta', especially Rotterdam and Antwerp.

The Europeans

Traditionally, European MNCs concentrated on colonies or ex-colonies in the Third World. UK companies have a long history of exporting capital, with firms such as ICI being among the pioneer MNCs. Their investments initially were directed both at Old and New Commonwealth countries, with Lonrho's African interests being an excellent example of the latter. Only more recently has there been a relative shift to Europe, especially to West Germany. In contrast, German MNCs have concentrated more investment within Europe (56 per cent of the total in 1972–6; Dicken 1980), especially southern Europe. In the early 1970s Spain received 8 per cent of German investment – a larger share than, for example, the UK (Dicken 1980). Indeed, the formation of the EC seems to have given a considerable boost to MNC activity within Europe. Between 1959 and 1970 the 69 largest firms in the 'EC Six' established no fewer than 1436 foreign sales and manufacturing subsidiaries within the EC (Franks 1976), confirming the evidence of increased integration that was also evident in trade statistics (see p. 39).

In the 1970s the strategies of the European-based MNCs have changed significantly. Faced with higher production costs at home, the need to compete with lower-cost MNCs from other countries (whether the USA, Japan or India), and uncertain demand in Europe following the 1973–4 oil crisis, many companies have shifted relatively more investment outside Europe. This is new in so far as branches in these LDCs are being used for outward processing, that is to make goods for re-export rather than for local markets. An example is Philips who make electrical goods in southern Asia so as to be able to compete with Japanese exporters. It is part of the creation of a new international division of labour in which European capital still has a central role. It is impossible to generalize about this disinvestment in Europe, as it varies according to sector and timing. Nevertheless, two examples serve to illustrate the processes involved. There was a 29 per cent reduction in ICI's employment in the UK between 1971 and 1979/80, while the company continued to expand in the Far East and Australia (Clarke 1982). Similarly, a study of fourteen of the largest

French MNCs (such as Michelin and Thomson) has shown that, between 1973 and 1977, their employment at home fell by 4 per cent while employment abroad increased by 18 per cent (reported in Sauvray 1984).

European companies adopted another strategy in the 1970s and 80s, that of locating in the USA so as to break into the world's largest single consumer market, and/or gain access to advanced technology. Some companies sought to achieve this by setting up new plant, but acquisitions are usually favoured. For example, Holderbank of Switzerland purchased the Denver-based Ideal Basic Industries, the third-largest US cement producer, while the UK's Saatchi and Saatchi purchased the US's Ted Bates, one of the world's largest advertising agencies. Investment in the USA has become particularly important for European companies (Dicken 1980) but more recently European companies have become increasingly concerned with trying to break into Japan's highly protected markets.

Japan and the Middle East
Japan is a new but significant source of international investment; but it has only achieved global significance since the late 1960s when government restrictions on capital exports were eased, partly because of domestic labour shortages. In the course of the 1970s, overseas investment increased nine-fold (Dicken 1983). Japanese investment is very distinctive and, in Europe, has mainly been concentrated in commercial and service activities. Thus Europe received only about 7 per cent of Japanese investment in manufacturing, compared with 35 per cent allocated to other Asian countries and 23 per cent to North America (Loeve *et al*. 1985). Within Europe the UK, followed by West Germany and the Netherlands, have attracted most Japanese investment (Table 11). Nevertheless, in

Table 11   *Japanese and US investment in Europe, 1985*

| | Cumulative investment ($ billion) | |
	By Japan	By the USA
UK	3.14	33.96
Netherlands	1.69	7.06
F.R. Germany	1.34	16.75
Luxembourg	1.22	0.46
France	0.82	7.83
Belgium	0.74	5.10
Switzerland	0.66	16.23
Spain	0.51	2.60
Ireland	0.26	3.75
Italy	0.18	5.64
European total	11.00	106.8

SOURCE: *Financial Times*, 10 November 1986.

global terms the UK, with $1823 million of accumulated Japanese investment in 1980, was still only the fourth ranked recipient (after the USA, Brazil and Indonesia). No other European country figured among the top ten ranked host countries. Japanese companies mainly undertake green-field site investments; large-scale outright acquisitions such as that by Ashaki Glass of Belgium's Glavarbel are exceptional.

Europe has also been the recipient of Middle Eastern investment, especially given the need to recycle petro-dollars following the 1970s oil crises. Much of this investment has been channelled into property markets, especially in London, but there have also been important direct investments in manufacturing companies; for example, in West Germany the Iranian government has a 25 per cent holding in Fredrich Krupp, while Kuwaiti interests own 14 per cent of Daimler-Benz (Owen Smith 1983).

Centralization/decentralization of MNC activity
There has also been a debate on whether MNCs have been subject to spatial centralization tendencies within particular countries. Control is certainly highly centralized in a few cities and, for example, the headquarters of the 198 largest non-US corporations are highly concentrated; 30 in Tokyo, 28 in London, 13 in Osaka, 12 in Paris, 10.5 in Rhein–Ruhr and 3.5 in the Randstad (Cohen 1981). Concentration of headquarters is to be expected given their reliance on producer services, efficient communications and face-to-face contacts with other corporations.

There is less clear evidence on whether MNC production is subject to spatial centralization (see Massey 1978 on the general issues in this debate). Holland (1976) believed that, until the 1970s, there was a tendency to spatial concentration but this was offset by state power. However, with the growing strength of the MNCs – which he terms the meso-economic sector – they gained the upper hand in bargaining with the state, and thereafter could locate virtually wherever they chose. Holland's thesis is long on concepts and short on concrete examples and, in fact, the empirical evidence on this question is mixed. Hamilton (1976; 1978) found that MNCs were subject to centralization tendencies and that, for example, foreign investment in France was concentrated in the Paris region and, in Belgium, in the north (especially around Antwerp). This he accounts for by the need to be in or near major cities and/or ports so as to facilitate links with international suppliers and markets.

Hamilton also found important differences according to the origin of the MNCs. American companies were subject to stronger centralization tendencies than were western European MNCs. US companies were more likely to have European-wide strategies, with branches in the core regions of several countries, while western European companies were more likely to expand intranationally and, therefore, to be better represented in peripheral regions. Until recently this was illustrated by the car industry:

US companies had international production strategies but European companies' operations were almost entirely nationally based.

In the Netherlands, contrary to Holland's assertion, there seems to have been a deconcentration of foreign investment (Smidt 1983). Up to 1960 the Randstadt dominated foreign investment but, thereafter, a significant proportion of new investment was at the coast or (to utilize domestic labour reserves) in the north (Pinder 1976). There is also evidence that American multinationals in the UK tend to be disproportionately located in Assisted Areas (McDermott 1977), and that foreign capital in Ireland is generally more likely than indigenous capital to locate in Development Areas (O'Farrell 1980).

The role of MNCs

The role of MNCs in national economies is contentious, sometimes being regarded with outright hostility. The French politician, Servan-Schreiber (1968) stated that '. . . if present tendencies continue, the third industrial power in the world, after the US and USSR, would not be European but American industry in Europe.' There were certainly some grains of truth in this provocative statement, but both advantages and disadvantages are likely to result from MNC activities.

On the positive side, MNCs are a source of capital, frequently linked to advanced technology and managerial skills. Unlike most less-developed countries, western Europe has actually benefited from technological transfer; in 1975, the EC and Canada were the recipients of more than four-fifths of American companies' overseas research expenditure, with West Germany being especially favoured. MNCs are also an important source of employment – about 13 per cent in the UK in 1975 (Dicken and Lloyd 1980) – and may contribute to a country's exports, while taxation of its profits may contribute to government revenue. Finally, there is at least some evidence that MNCs contribute to diminishing regional inequalities in some European countries.

On the debit side, some negative aspects of MNC activities can be highlighted. Capital inputs into the economy are counterbalanced by outflows, such as the repatriation of profit and the purchase of inputs from abroad, hence reducing the net financial benefits to the economy. MNCs also wield enormous corporate strength because of their size, and can challenge the power of the state over such things as pollution controls and the location of plant (Watts 1979; 1980). The MNCs have an advantage in that, being locationally flexible, they can always threaten uncooperative governments with closure of local plant while diverting expansion abroad. This is increasingly likely to occur anyway, given the rapidity of modern production and investment cycles (so that plant may have a life of only 20 years), a process termed the hypermobility of capital (Damette 1980). MNCs also control the quantity and quality of new technology made available in particular countries; as part of their strategies, this often means that only mature technologies and products are devolved from the company's overseas headquarters. This is part of a broader process which Hayter (1982) terms 'truncation', whereby foreign investment replaces and pre-empts the growth of indigenous firms. It takes over domestic markets and replaces local producers, thereby weakening the potential of indigenous firms in such countries to generate growth.

The state and the economies of western Europe

Theoretical issues

There is considerable debate concerning the nature of the state in capitalist society, but this will only be pursued here in so far as it has a direct bearing on economic development. Weberian theories have concentrated on the organization of decision-making (for example Albrow 1970) while Marxist theories of the state are usually categorized as instrumentalist (for example, Miliband 1973) or as structuralist relative autonomy (for example, Poulantzas 1969, 1978). The former sees the ruling class (representing different capitalist interests) occupying a privileged position both inside and outside the state, and using this to achieve its own ends. The structuralist view argues that the state reflects the balance of power among classes (mainly the fractions of capital but also labour at any given time). Both theoretical approaches have been criticized as over-simplifications (Clark and Dear 1981; Cooke 1983; Holloway and Picciotto 1978; Jessop 1982; Offe 1975).

Instead, it can be argued (Amin 1983b, p. 142) that '. . . the necessity of the capitalist state arises when the circulation (distribution, exchange and consumption) of commodities is impeded.' Individual groups of capital (industrial, financial, property, etc.) cannot guarantee their own long-term survival because they cannot guarantee the general requirements of production and investment, such as the needs for provision of infrastructure, education and housing. This therefore requires the existence of a mechanism which is separate from capital – that is, the state – to guarantee the reproduction of these conditions. As such the state is partly a reflection of the balance of social classes and of class conflict, although it is partly independent from and mediates these. However, herein lies a contradiction for, being an external mechanism, the state does not have the power to guarantee to resolve the crises in the economic system. In general, these theories have been based on liberal capitalist democracies and do not, therefore, apply to Greece, Portugal and Spain while under dictatorships (Herz 1982).

While there is dispute over the precise relationship between capital and labour and the state, there is broader agreement over the view that the state has three main types of functions (O'Connor 1973): social investment, social consumption and social expenditure. Social investment involves expenditure on trade protection (export subsidies) and on fixed items of constant capital which are necessary to assist and ensure production. Examples include transport and other infrastructure as well as – in many countries – investments in key industries such as energy. This may require nationalization to ensure the provision of such investment, although in the 1980s there has been a counter-move to privatization in some European countries, notably the UK and France. These investments may be linked to macro-economic policies – such as Keynesian regulation – to ensure smooth functioning of the economy.

The state both invests directly in infrastructure and is the overall coordinator of other agencies involved in this field (Lapple and Van Hoogstraten 1980). Its role is particularly influential in macro-spatial, economic planning, notably in regional planning, or in developing industrial growth-centre complexes such as Fos-sur-Mer in France or Sines in Portugal (see pp. 257–9). According to Lipietz (1980a), such state intervention may be necessary in order to make up for the lack of a law of value in space, that is, that individual capitals cannot guarantee a geographical organization of the economy which is rational in terms of the needs of the economy as a whole. It is also a means by which capital has managed to socialize some of the costs of production (Mandel 1975), transferring part of the burden of new investment costs to the state.

Social consumption involves investment in social capital for collective consumption, mainly to ensure the reproduction of the labour force. This includes housing, health and educational provision which, left to the private sector, could not be guaranteed to reproduce the quality of labour required in particular locations. The precise form of such expenditure varies and may include direct state provision and building programmes, subsidies to consumers and/or subsidies to the private sector. Finally, social expenditure is undertaken to consolidate social control and the state's ideological apparatus. This may involve expenditure, at one extreme, on unemployment relief and, at another, on the police and repressive measures.

It is in response to the need to fulfil social consumption and social expenditure functions that the welfare state has developed. By the 1930s this existed in most western European states, in so far as some provision was made for old-age pensions, health and unemployment insurance for most occupational groups, although agricultural workers were often excluded. However, by the end of the 1970s, with very few exceptions, the population of western Europe was covered by universal old-age pensions, unemployment, sickness and industrial injury schemes (Sawyer 1982). The very provision of such services acted to legitimize the role of the state, while socializing the costs of the reproduction of labour which, otherwise, would have to be borne by particular employers. The state does not simply develop a role in social consumption and expenditure in response to the needs of capital. As the state is an area for class struggle, it is open to working-class pressure for social reform. Indeed, many forms of state intervention – unemployment benefits, nationalization, housing provision, etc. – have only been secured through historical struggles. Despite this, there has been some 'rolling back of the state' in several European countries in the 1980s. This has been in response to the crisis of accumulation, the heavy demands of social expenditure on unemployment payments and in some countries, including the UK, a political will to promote individual property ownership.

The state in western Europe

Although all states have some involvement in all three types of functions described in the previous section, there are international variations which are the outcome of interrelated social, political and economic considerations. This is crudely reflected in the percentage of GNP which is accounted for by central government expenditure. The range in western Europe is from 19 per cent in Switzerland to 58 per cent in Ireland, with Belgium, the Netherlands, the Scandinavian states and Italy also recording high levels (World Bank 1986, Table 22). These differences are partly accounted for by variations in local government expenditure (which is particularly high in federalist Switzerland). However, there are also real differences in the extent of state involvement in social and economic life. These variations can be illustrated by some examples of western European states, indicating differences both in the origin of and in the current operation of policies.

The state in Sweden encourages operation of the free market while, at the same time, having a strongly developed role in terms of all three major state functions. It owns some important industrial holdings (managed by Statsfönetag), including shipbuilding, railway and forestry interests; these account for 6 per cent of all Swedish industry and are supposed to operate according to market principles. Nationalization was extended in the 1970s as one response to recession in key industrial sectors. The state also exercises control over the private sector through a series of investment funds. Companies are allowed to allocate up to 40 per cent of pretax profits to capital investment in any year on a tax-free basis. However, the timing and location of such investment has to be approved by the Labour Market Board, which aims to even-out cyclical and regional inequalities. In the fields of social consumption and social legitimation, Sweden has one of the world's most advanced welfare systems, based on seeking out those in need rather than just responding to expressed need. There are also strong mechanisms for income redistribution.

There is a comprehensive system of state intervention in economic management in the UK and some 25 per cent of employment is in the public sector (Maunder 1979). Except for attempts by the Thatcher governments to privatize nationalized companies and parts of the welfare state, this approach has largely been accorded bipartisan political status (Hudson and Williams 1986). The range of state involvement has been considerable. A number of important economic sectors have been nationalized, including the Bank of England, coal mining, steel production, civil aviation, shipbuilding, British Leyland, gas, electricity and water. Recently, there has been partial or complete denationalization of British Aerospace, British Telecom and several other state-owned companies. Faced with a ·long-term decline of manufacturing, the state has been particularly active in this sector; the Industrial Reorganization Corporation (1966–71) was empowered to encourage mergers (as between Leyland and BMC in 1968) and the National Enterprise Board acted as a holding company for both 'lame duck' industries, such as Rolls Royce, and modern high-technology companies such as INMOS, the silicon-chip producer. In addition there is a loose system of national economic planning based on sector working parties, which have sought to achieve greater industrial efficiency through a corporatist strategy. The social consumption function is fulfilled through a variety of housing, health and welfare policies.

France is a highly planned capitalist economy (Owen-Smith 1979). Nationalized companies date back to the establishment of the Compagnie Française des Pétroles (CFP) in the 1920s (Holmes and Fawcett 1983). This was considerably extended in the years immediately following the Second World War when munitions, aircraft production, electricity, gas, coal, the Banque de France, several insurance companies, Renault, and Gnome et Rhone (aeroengine production) were nationalized. By the late 1970s the state was a major shareholder in about 500 companies, and had minor interests in a further 600 (Hough 1979). There was a further round of nationalization in 1982, under the socialist government. This included the five major private manufacturing companies (Saint Gobain, Compagnie Générale d'Electricité, Rhône-Poulenc, Thomson-Brandt and Péchiney-Ugine-Kuhlmann), 36 banks, two finance houses (Suez and Paribas), and the iron and steel groups USINOR and SACILOR. Together these added half a million people to the state's payroll. With the election of a conservative government, steps were taken to denationalize some of these companies, beginning with Saint Gobain, Paribas and Assurance Générales de France.

State planning is highly organized within a framework of long-term national plans which analyse change, forecast trends and coordinate investment plans. Early plans concentrated on particular sectors, but more recent ones – such as the 8th National Plan, 1981–5 – cover most economic activities. Policy is particularly strong in the field of industrial production, and the state (via the Institut de Développement Industriel) has sought to encourage mergers among domestic companies so as to establish French companies capable of competing on a global scale. The merger of Pechiney and Ugine Kuhlmann, encouraged by tax concessions, is the most spectacular example of this policy.

Italy represents a different form of state economic intervention, being based on control of investment (via ownership of the national and the commercial banks) and a series of public corporations. The largest of these is the Instituto per la Riconstruzione Industriale (IRI), established in 1933; it has three-quarters of all employment in state holding companies (King 1985). The IRI encourages mergers, takes over key companies facing bankruptcy, and has important holdings such as Finsider (steel), Alfa Romeo (cars) and STET (telecommunications). Its interests are so extensive that Bethemont and Pelletier (1983, p. 75) comment that '. . . given the size of IRI and the fact that it could be considered the largest investment and service company in Europe, one may question whether the Italian state controls it or vice versa.' State holding companies own over two-thirds of iron, steel and chemical production in Italy. In addition employment in the service sector has been swelled – especially in the south – as this has been used for political clientelism, in order to create and maintain support for the governing parties (Slater 1984).

The post-war economy of West Germany has been modelled on the lines of a 'social market economy', with the state maintaining a relatively low profile. Market principles dominate and the state's role is largely limited to obviating poverty and policing business competition (Owen-Smith 1979). Nevertheless, there are important state holdings, and some 8 per cent of total employment is in companies where the state is the majority shareholder. Other than the railways and the post office (both nationalized in the nineteenth century), the most important of these are VEBA (energy, oil, chemicals and transport), Volkswagen (cars), Salzgitter (steel and shipbuilding), Saarbergwerke (coal) and VIAG (electricity, aluminium and electrochemicals). Industry is also favoured by subsidies, tax concessions, and exchange policies which undervalue the deutschmark so as to encourage exports. Major banks play a key role in the economy, holding some 70 per cent of the shares of the 400 largest firms (Hall 1983), and this ensures a flow of investment funds to industry, as well as overall coordination of the market. The banks are also involved in administering state aid to industry, largely on commercial lines, although social and strategic needs can be considered, as in the cases of coal, shipbuilding and the railways. A high degree of social consensus has facilitated the implementation of this strategy.

Although there are variations among these different national strategies, they all involve state intervention to sustain capital accumulation (profits and reinvestments). The major economic policies have been trade protection, export assistance, incentives to attract foreign investment, and financial and institutional aids to help firms to modernize and reorganize in the face of international competition. State ownership is prevalent in the utilities, energy, transport and key sectors of manufacturing (see Table

12). Agriculture and manufacturing have been substantially influenced by state policies, but less so services.

However, growing internationalization of the economy, and especially the complex interlinking of MNCs with transnational banks (Susman 1984), has weakened the already limited ability of the state to manage the national economy. The state has 'begun to lose control' (Damette 1980) over a number of important economic features, including currency exchanges (see pp. 40–1), investment, and production itself. The most visible signs of these failures have been a series of crises, including dramatic currency speculation (against the pound in 1976–7 and the dollar in 1985), rapid deindustrialization (as in the British car and petrochemical industries), or the eruption of severe regional crises (as in Nord and Lorraine in France during 1978–9). Regional crises are, in fact, only one particular spatial and social manifestation of the economic, social and political consequences of uneven development, and these are considered in more detail later (see pp. 243–52). Creation of the EC has also reduced the responsibility of individual states for some areas of activity, notably farming and fishing.

The European economy: core and periphery

The European economy has become increasingly locked into an international system of production and distribution, in which individual countries have very different roles both within Europe and globally. Essentially there is an unequal exchange of goods and service between countries (Wallerstein 1984). This is evident in the dominant trade patterns, as well as in the capital and labour flows which have been examined in this chapter. These unequal exchanges constitute core–periphery relationships. According to particular countries' roles, they can be classified as parts of the core, the semi-periphery or the periphery of the world economy. While the model has been criticized on the grounds of oversimplification and functionalism (Cooke 1983; Johnston 1982), it nevertheless provides a useful classification which facilitates interpretation of variations among western European economies.

The 'core' is constituted by three types of economies, each of which is represented in Europe. There are old colonial powers (such as the UK, France and the Netherlands), 'new' industrial powers (such as West Germany), and closely linked modern economies (such as Sweden, Switzerland and Belgium). These core economies are the 'homes' of the larger MNCs and recipients of major immigration flows and, in a few cases, are centres of international finance capital. The periphery, essentially the less-developed countries, is dominated by primary producers who have suffered from poor export prices in the post-war era, relative to the manufactured goods they import. These countries also import capital and

Table 12 *Levels of public ownership in selected European countries in the mid-1980s (percentage figures)*

	Posts	Telecoms.	Electricity	Gas	Oil Production	Coal	Railways	Airlines	Motor industry	Steel	Shipbuilding
Austria	>75	>75	>75	>75	>75	>75	>75	>75	>75	>75	N/A
France	>75	>75	>75	>75	N/A	>75	>75	75	50	75	<25
F.R. Germany	>75	>75	75	50	25	50	>75	>75	25	<25	25
Netherlands	>75	>75	75	75	N/A	N/A	>75	75	50	25	<25
Italy	>75	>75	75	>75	N/A	N/A	>75	>75	25	75	75
Spain	>75	50	<25	75	N/A	50	>75	>75	<25	50	75
Sweden	>75	>75	50	>75	N/A	N/A	>75	50	<25	75	75
UK	>75	<25	>75	<25	<25	>75	>75	75	50	25	>75

SOURCE: AMEX Bank Review.

are sometimes the sources of large-scale out-migration. Many of these countries – such as Nigeria, Kenya and the Congo – are intimately related to particular European economies by virtue of previous colonial ties. Finally, between core and periphery stands the semi-periphery, which includes much of southern Europe and Ireland. These countries stand in relation to northern Europe as 'a periphery' (see Seers *et al.* 1979) but also act as 'cores' in relation to many less-developed countries. (An example is Portugal, which is a periphery in relation to, say, West Germany, but a

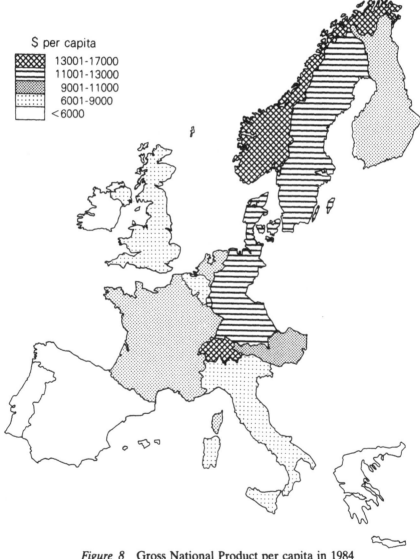

$ per capita

- 13001-17000
- 11001-13000
- 9001-11000
- 6001-9000
- <6000

Figure 8   Gross National Product per capita in 1984

SOURCE: World Bank (1986).

core in relation to its ex-colonies such as Angola or Mozambique.) These countries are partly industrialized and may have small-scale MNCs, but they may also export large quantities of labour and are net recipients of foreign investment.

There is nothing fixed and immutable in this classification. In the past some countries have risen spectacularly from periphery to semi-periphery positions (for example Brazil or Argentina) while others have climbed from semi-periphery to core. Japan provides the most spectacular example of the latter, although Italy is a notable European case. More recently, Spain has been in the process of shifting from semi-periphery to core status, as a result of rapid industrialization and modernization of agriculture. Movements need not necessarily be upwards, and it is quite conceivable that long-term decline could lead to a core economy being reduced to semi-periphery status – a fate sometimes predicted for the UK and Belgium. State policy is often directed at preserving or modifying the role of a particular economy within this hierarchy.

GNP per capita provides a crude indicator of the state of development in western Europe in the mid-1980s (Figure 8). The highest levels are in Switzerland, Scandinavia and West Germany, followed by France, the Netherlands and Austria. This broadly accords with a larger core group, with Belgium and the UK hovering at the edge of it, although the latter continues to be an important centre of international finance. Southern Europe still lags behind, although Italy and Spain have closed the gap significantly.

However, the concept of core and periphery is necessarily a generalized one and, in practice, there may be considerable variations between industries and regions. Thus, in Spain, Catalonia may conceivably have reached core status, but Andalucia (as a major source of emigrants and with a weak industrial base) is still firmly semi-peripheral. Differences between industries are also evident, and in Greece, for example, world leadership in shipping and dominance of eastern Mediterranean banking coexists with many weak and traditional manufacturing activities. The next part of this volume seeks to analyse these sectoral differences within western Europe.

Further reading

1   The international division of labour: R. Jenkins (1984), 'Divisions over the international division of labour', *Capital and Class*, **22**, pp. 28–57.
2   Emigration and return: S. Castles, H. Booth and T. Wallace (1984), *Here for good: Western Europe's new ethnic minorities*, London: Pluto Press; R. King (1984), 'Population mobility: emigration, return migration and internal migration', in A. Williams (ed.), *Southern Europe transformed*, London: Harper and Row.
3   Multinationals: P. Dicken (1980), 'Foreign direct investment in European

manufacturing industry', *Geoforum*, **11**, pp. 289–313; J. H. Dunning and R. D. Pearce (1985), *The world's largest industrial enterprises, 1962–1985*, Farnborough: Gower.

4   The state: S. Bornstein, D. Held and J. Krieger (1984), *The State in capitalist Europe*, London: George Allen and Unwin.

Part Two

The Sectors

3 Energy

Energy in western Europe

The first two post-war decades were characterized by perceived – and sometimes real – energy shortfalls in western Europe, and by concerted state actions to ameliorate these. However, the 1970s were years of acrimonious debate about the future of energy, encouraged by a succession of oil price crises, awareness of the advantages and disadvantages of North Sea resources, and rising conservationist opposition to some forms of energy production. These developments occurred alongside the long-term decline of traditional coal resources, linked to major social and regional crises in most of the major European production centres. By the mid-1980s, the spectre of energy shortages was receding and there was global overproduction of many types of energy.

Despite changes in the form of the energy problem in western Europe, this has remained a central concern of post-war policy making. Economic development is dependent on the availability and the cost of energy supplies, and this applies both to agriculture and services as well as to, more obviously, energy-consuming industries such as steel furnaces and petrochemical plants. The price of energy is an important element in the constant capital costs of industry and, consequently, has a direct bearing on rates of profit and capital accumulation (Carney 1980). In addition, energy is an important element in international trade. Most western European countries are net importers of energy. For example, fuels already constituted 11 per cent of the merchandise imports of the industrial market economies in 1960 while, by 1981 (immediately after the 1970s oil crises), this share had risen to 28 per cent (World Bank 1984). Only the Netherlands, Norway and (to a very limited extent) the UK were net energy exporters in the early 1980s.

Energy has also been important as a focus of political conflicts, and these can be divided into two types. The first were regional crises arising from the selective run-down of coalfields after the late 1950s and the 1960s, especially in France, Belgium and the UK. As many of the affected coalfields were in politically sensitive regions – such as South Wales, Nord and Wallonia (see pp. 250–2) – local opposition to pit closures was often linked to wider regionalist political movements and demands for greater influence over the state policies that had led to the run-down of previously booming mining areas (Carney 1980). Although these regionalist issues are still important – as was sharply illustrated during the 1984–5 coal industry conflict in the UK – energy questions have also become important as conservationist issues in the 1970s and 80s. 'Green' politics have assumed an important place in the political life of many European countries, especially in the opposition to nuclear energy. This is evident, for example, in West Germany, where the Greens are represented in the federal parliament and in many regional assemblies, and in Sweden where a national referendum was taken in 1980 to determine the country's nuclear future.

Energy questions have been an obvious target for political conflict because the state itself has played such a dominant role in this field in most European countries, notably through nationalization of coal, and some oil and gas production. Macro-economic policies – such as reflation/deflation – obviously influence energy consumption, as do exchange-rate fluctuations, especially as oil prices are fixed in dollars. However, historically, there have been few attempts to develop specific demand-management policies, and this has only changed since the 1970s when conservationist measures were widely promoted in the wake of the oil crisis. Since then, energy resources have been revalued and many governments have sought to introduce more energy-conscious building designs and more fuel-efficient transport technologies.

In contrast, the state has been far more active in energy production. The immediate post-war priority was provision of a secure and relatively cheap supply of energy. As it became clear that the private sector would not be capable of guaranteeing this (as it sought above-average profits which would push up prices), the state took a direct or indirect guiding role in production. The initial aim was to boost output of existing forms of energy (mainly coal) and, later, to seek out and encourage the use of new energy sources. This often meant cooperation with multinational companies to ensure cheap imports of oil, a strategy which subsequently shifted to development of indigenous oil and gas reserves. Long preparation times and enormous investments are required in developing most energy sources. This has ensured that where the private sector is active, as in the British sector of the North Sea, then large-scale and multinational capital is involved rather than medium-scale and national capital. Another state response was to develop nuclear energy sources, but this, and other specific sources of energy, will be discussed later.

Energy production has been influenced by a combination of the activities of the state and of multinational companies; within Europe supranational bodies have only a limited role. The European Coal and Steel Community (ECSC) did influence the restructuring of coalfields in continental Europe in the 1950s and 60s, but the EC has tended to be little involved (Odell 1976). Since the 1973–4 oil crisis, the EC has sought a more active role, particularly in respect of energy conservation. It set a target for reducing the link between increases in energy consumption and GDP to 0.7 but, by the late 1970s, this was still operating at a ratio of about 0.84 (Ray 1983), although it has probably fallen in the 1980s. The EC has also encouraged plans to increase coal and nuclear energy production, to integrate national electricity grids, and to build up oil stocks as a buffer against the panic effects of oil shortages such as occurred in 1979 (see pp. 46–7), so as to reduce overall dependency on international oil suppliers (Commission of the European Community 1984c). However, both the consumption and production of energy remain conditioned essentially by state policies and international market changes.

Energy consumption: two phases

Energy consumption is best considered in two phases: prior to and after 1974. The first phase was characterized by sustained growth in the consumption of energy which was the result of the long post-war economic boom in western Europe. There were energy shortfalls in the early years because coal production had either been disrupted by the war and its aftermath (note West Germany's loss of the Upper Silesia coalfield to Poland) or had been hampered by the neglect of repairs and of new investment. Nevertheless, the coalfields were soon brought back into full production and, backed by growing imports of oil, this facilitated the growth of consumption. By the 1960s, energy consumption in the industrial market economies was increasing at a rate of 5.3 per cent a year and, prior to the 1973–4 oil crisis, there was little reason to expect that this would not continue in the foreseeable future.

Phase one: the long boom

Expansion of demand was based on both the general economic boom and the precise nature of this. Post-war expansion saw an increase in capital–labour ratios as new technology was introduced to reduce the cost of production; inevitably this tended to increase the consumption of energy relative to each unit of output. A notable example is the petrochemicals industry, for it both consumes oil as a raw material and uses energy-intensive production processes, such as oil-crackers. At the same time, energy demand was bolstered by the consumer boom, especially in countries such as France which had low levels of ownership of consumer goods in the 1940s (see pp. 30–1). This has involved both the extension of car ownership between and within households, and mass-acquisition of such consumer products as washing-machines and refrigerators.

As a result of these changes, western Europe became a highly energy-intensive economy. By the 1970s, for example, the EC10 consumed 15.9 per cent of world energy even though it had only 6.2 per cent of population. Although it was surpassed by the US which, with a similar share of population, accounted for 29.2 per cent of world energy consumption, per capita energy consumption was well above the world average. In contrast, the less-developed countries, with 79.9 per cent of population, only consumed 31.1 per cent of energy (Commission of the European Community 1983b). However, energy consumption also varied within Europe, as can be seen in the per capita changes for 1960–74 (see Figure 9). The largest increases in this period of overall growth were in Greece, Spain, Italy, Finland and the Netherlands, all of which had rates in

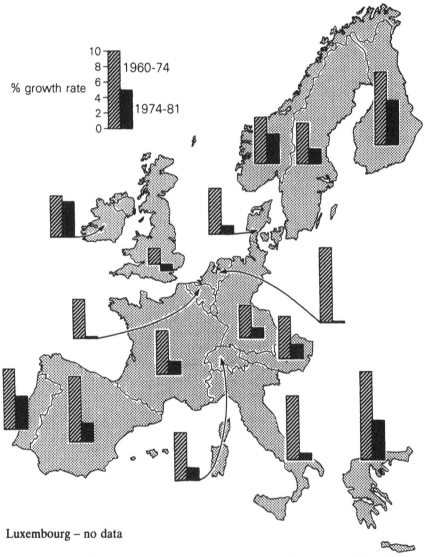

Luxembourg – no data

Figure 9 Per capita consumption of energy in 1960–74 and 1974–81
SOURCE: World Bank (1985).

excess of 7.5 per cent. The first four of these were among Europe's least-developed economies and were experiencing a period of 'catching up' in terms of both domestic and industrial consumption. The Netherlands, although a relatively advanced economy, was industrializing rapidly in this period (see p. 31) and many of the new industries were energy-intensive. In contrast, the lowest increases were in the UK (2.2 per cent) and West Germany (4.4 per cent) which were relatively mature industrial economies, experiencing relative shifts to less energy-intensive service activities.

Phase two: crisis and conservation

The 1973–4 oil crisis was a catalyst which transformed energy consumption. All the European economies were affected by the four-fold increase in oil prices (see Figure 4); even those, such as Spain, who had particularly friendly links with the Arab oil producers were only shielded from a supply boycott and not from the price rises. As a result of these price increases, the recession which followed and, later, the introduction of energy conservation measures, energy consumption fell dramatically. Whereas per capita consumption increased at 5.3 per cent a year between 1960 and 1974, between 1974 and 1981 the increase was only 1.1 per cent. Indeed, the EC10 consumed less gross energy in 1982 than in 1973 (Commission of the European Community 1983b). Energy forecasts had to be drastically revised downwards: whereas in the early 1970s it had been predicted that western Europe would consume 5000 million tons coal equivalent per annum at the end of the century, by 1980 this had been revised to only 2700 million tons (Odell 1981).

Again, there were considerable national variations within Europe (Figure 9). The largest increases (over 5 per cent a year) between 1974 and 1981 were in Finland, Greece, Ireland and Portugal, being less-developed economies still in the process of catching up. Most of the remaining economies – including Italy and the Netherlands this time – had smaller increases of 2 per cent or less. In the UK, per capita energy consumption actually fell (by 0.8 points) and in Belgium it was static; both these economies were severely affected by recession. Since 1981, it is likely that energy consumption has fallen even further owing to a combination of deepening recession and more effective conservation measures. Over the period 1973–83 as a whole, energy efficiency in western Europe increased by 18 per cent (Odell 1986). This trend has occurred even in Portugal – which had rapid increases in consumption in the 1970s – as macro-economic deflationary policies aimed at reducing imports (including oil) took effect. At the time of writing it is too early to assess the impact of the 1985–6 fall in oil prices on consumption.

While some of the less-developed economies have experienced relatively large increases in energy consumption, there are still considerable absolute differences within Europe. The highest levels of consumption in 1981 were determined by two factors: level of development and environmental conditions, especially the different needs in northern and southern Europe for domestic heating. As would be expected, consumption is greatest in the Scandinavian states, which are characterized both by relatively advanced economies and by long cold winters. Equally predictably, southern Europe brings up the rear in this comparison and, at the extreme, consumption per capita in Portugal is only one-seventh of that in Norway. In the case of Norway, the availability of indigenous energy resources may contribute to its relatively high consumption level. The general question of energy production is considered in the following section.

Energy production: changing sources

Post-war recovery

In 1945 there was a critical need to expand coal output in order to avoid an escalation of energy costs which, ultimately, might have threatened profits and investment generally. Production did recover remarkably well initially, and by 1947 had regained 1939 levels. This was essential for European recovery as coal provided about 80 per cent of primary energy requirements. In the 1950s it proved more difficult to expand production for large-scale new investment was required to bring new coal mines into production. The shift from coal to oil as an energy source and the greater demand for electricity (which could be generated from diverse sources) also meant that the demand for coal, but not for energy, lessened in the late 1950s and 60s. In addition, the availability of imported coal threatened domestic production. Energy-hungry Europe was, therefore, transformed into a multi-energy economy as relatively cheap oil was imported, first to supplement and later to substitute for coal. The share of coal in energy provision therefore started to decline, although, prior to exploitation of North Sea resources, this still accounted for some 70 per cent of indigenous energy production.

Electricity became increasingly important in energy supply as it supplanted the direct use of primary energy (oil, coal, etc.) in many industries. Compared with these other fuels – and especially coal – electricity represented a cleaner and more flexible source of energy, better suited to the requirements of many industries. Electricity also proved more suitable for domestic heating systems in many countries and was crucial in facilitating the post-war boom in household consumer goods.

Expansion of electricity production at first helped cushion the decline in direct demand for coal, because most power stations in the 1950s were still coal-fired. However, alternative primary sources of electricity became more important over time; at first oil, but later gas and nuclear reactors.

Power stations also became more fuel-efficient as a result of technological changes and, for example, between 1958 and the late 1970s there was a 50 per cent increase in the output of electricity from each unit of primary energy input (Minshull 1980). By 1978, some 64 per cent of western Europe's electricity was generated in conventional thermal (oil or coal) stations, while 26 and 10 per cent came from hydro and nuclear sources, respectively. These proportions obviously varied between countries, with, for example, almost all of Norway's electricity being derived from hydro sources, and almost all of Denmark's from thermal stations.

The 1960s: multi-sourced energy production

The 1960s was the crucial decade in the substitution of oil for coal. Before this oil did not account for more than a half of the energy inputs of any western European economy but, by 1972, it exceeded this in all cases except the Netherlands (where gas was important). In western Europe in aggregate, between 1950 and 1981, the share of coal in energy production fell from 83 to 23 per cent, while the share accounted for by oil rose from 10 to 49 per cent (Ilbery 1981). Natural gas accounted for 15 per cent and primary electricity (from nuclear or hydro sources) for 13 per cent by the latter date. The share of oil actually declined from a peak of 57 per cent in 1975.

There are, of course, considerable variations between countries (see Figure 10), and these are accounted for by differences in indigenous resource endowments, and by state policies. In 1981 coal accounted for over 30 per cent of primary energy requirements in only four countries: the UK, West Germany, Denmark and Luxembourg. The first two of these are western Europe's major coal producers, while the latter two, lacking hydro-electric potential and indigenous carbon fuels, are heavily dependent on imported coal and oil. Gas was generally of limited importance, the exceptions being the Netherlands, the major European producer (48 per cent), and the UK (21 per cent) which has effectively harnessed North Sea gas for domestic consumption. The only other countries where natural gas provided more than 15 per cent of total primary energy requirements were West Germany, Belgium and Italy which rely on imports, by tanker or pipeline, mainly from the Netherlands, Norway and North Africa.

Nuclear energy was of least importance in 1981 and accounted for more than 10 per cent of energy requirements in only four countries: Sweden (where it held the largest share, 18 per cent), Finland, France and Switzerland. In all these cases there have been deliberate state strategies to develop nuclear energy so as to lessen reliance on conventional energy sources. France has been especially active in building nuclear power stations in the 1980s. Many countries – Denmark, Greece, Portugal, Norway, Ireland, Luxembourg and Austria – did not produce any nuclear

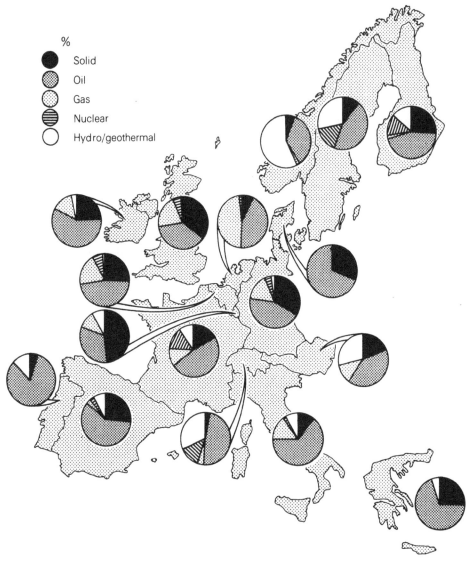

Figure 10   Primary energy requirements by fuel type in 1981

SOURCE: International Energy Agency (1983).

energy in 1981. The capacity for producing hydro-electricity is also unevenly distributed, although, unlike nuclear energy, this is partly accounted for by resource endowment. Not surprisingly, hydro-electricity was most important in the Alps and Scandinavia (excepting Denmark) – but even so, only in Norway (57 per cent) did it account for more than a half of national energy needs. However, it is oil, despite the setbacks of the 1970s and the development of alternative energy sources, which is still the dominant primary energy source. It accounted for more than 40 per cent of total requirements in all countries excepting Norway, the UK and Luxembourg. Oil is most important in Denmark, Greece and, above all, Portugal (82 per cent), all of which lack any substantial indigenous energy resources.

Energy dependency

The changing sources of primary energy in western Europe have led to increases over time in the dependency of both individual countries and of the region as a whole on external supplies. Whereas, in 1950, the EC imported only 10 per cent of its energy, by 1973, on the eve of the first oil crisis, this had risen to about 60 per cent. This was a direct consequence of changing relative production costs and of state energy policies, notably the drive to expand and diversify energy supplies through importing oil (and, to a lesser extent, coal) at the expense of indigenous coal production. By the 1970s all western European countries had become net importers of energy. Even after the first oil crisis the level of dependency continued to increase, and it only peaked in 1979, after many of the major North Sea oil and gas fields had come fully on-stream.

The level of dependency varies between countries according to their resource endowments and state policies towards different energy forms. Denmark has the highest level of dependency, importing 100 per cent of its energy; but several other countries, including Ireland, France, Greece, Belgium and Italy, import over 70 per cent. Particular countries have also been able to absorb the shocks of the oil crises in very different ways. Given the availability of indigenous resources in the UK, and state policies for their exploitation, it is not surprising that, between 1960 and 1983, the share of fuel in merchandise imports remained largely static (World Bank 1984; 1986). During the same period, their share in Italy doubled. The most extreme example of increased energy dependency, however, is Spain (Harrison 1985), where the share of energy in merchandise imports rose from 13 per cent in 1973 to 40 per cent in 1983. This can be explained by reference to Spain's export-led economic expansion in the 1960s which was based on extremely energy-intensive industries such as petrochemicals and cement production. At the other extreme there are some countries for which fuels are an important element in exports; these are fairly predictable, being the UK (26 per cent in 1983), the Netherlands (26 per cent), and especially Norway (62 per cent).

Energy is also largely in a dependent relationship with other sectors of the economy; put simply, a rise in economic output will usually lead to an increase in energy consumption, although the exact nature of the relationship depends on the precise form of development. Nevertheless, energy can also play a leading role in encouraging economic development. Some examples serve to illustrate this. It has been argued that industrial growth in both the UK and the Netherlands has been hampered by their strong energy-production sectors which has given them unduly strong currency exchange rates, hampering industrial exports. However, the Netherlands Goverment has also used cheap energy to subsidise industry. Expansion of the energy-production industry also has spin-offs for manufacturing and service firms; this is evident, for example, in the forward linkages between North Sea oil development and platform construction and financial services in the UK. It is clear that all such linkages and, indeed, many features of energy production cannot be generalized; instead, the particular features of each energy sector must be considered separately.

Specific forms of energy

Coal production: boom and long slump

In 1945 western Europe was a one-fuel economy, and that fuel was coal. It was therefore an immediate priority of all governments to secure a supply of this essential energy source. There were frantic attempts to reactivate the capacity of existing coalfields and to develop new ones as, for example, in the case of Belgium's Kempenland. Industry required low-cost energy which the coal mining industry could not always ensure. Therefore, as Carney (1980) emphasizes, there was a need to divorce investment from the rate of return on production, and this could only be achieved via state intervention. This could involve nationalization, as occurred in the UK with establishment of the National Coal Board in 1947, and in France with the setting up of Charbonnage de France in 1946. However, in West Germany the industry remained in private ownership which, in the face of recurrent crises, became increasingly concentrated. This led in 1969, with state encouragement, to formation of the consolidated grouping Ruhrkole AG, which took over the 26 major coal mines in the Ruhr.

By 1950 the immediate post-war shortages had been overcome and, in that year, the OECD countries as a whole produced 86 per cent of their own coal requirements. The leading European producers were the UK (220 million tonnes) and West Germany (126 million tonnes), although there was also significant production, exceeding 10 million tonnes per annum, in France, Belgium, the Netherlands and Spain. Coal output continued to rise over the next few years and, indeed, was fostered within the six members of the ECSC (formed in 1951) by a free-trade framework. However, output peaked in most countries by the mid-1950s. The reasons for this included rising costs, because coal mining remained a highly labour-intensive process in which it was difficult to introduce more capital-intensive methods, while well-organized trades unions were able to maintain relatively high wages. There was also slow, but steady, growth of import penetration as supplies arrived from non-West European producers such as Australia, Poland and the USA. At the same time, markets for coal were gradually weakened by a shift to oil and gas in both domestic and industrial consumption, although this blow was partly softened by coal's initial dominance of electricity production.

There were signs of overproduction by the late 1950s, and matters were brought to a head by several mild winters which reduced the demand for coal. As a result, stocks of coal in western Europe rose from 16 million tonnes in 1957 to 67 million tonnes in 1959 (Clout 1981b). The resulting cutbacks in coal production brought about a mini-recession in the late 1950s in many European coalfields. This signalled a new phase in coal production – one of long, sustained decline, as can be seen from Table 13. Production in the Netherlands was exhausted by the mid-1970s, not least because of the low-cost competition from gas, and there were substantial falls in output in all the main producer countries.

The major coal producers
By 1984 production in Belgium had fallen to less than one-third of its 1961 level, and was a mere 6 million tonnes. The Sambre–Meuse coalfield was almost completely abandoned and most of the output came from Kempenland. Production is relatively high-cost in many Belgian mines, and the industry suffered from the greater international competition introduced by the ECSC. In France, the Monnet Plan had aimed for a major expansion of coal production, but this peaked at 60 million tonnes as early as 1952 (Holmes and Fawcett 1983). Output had fallen to only 17 million tonnes by 1984. Nord du Pays suffered most, where output fell from 20 million tonnes in 1961 to only 2.5 million tonnes by 1984, not least because poor geological conditions contributed to very low labour productivity. The Centre/Midi coalfield also declined sharply, but production in the fairly modern mines of Lorraine has been relatively stable since 1971.

West Germany has consistently been the second largest producer in Europe, but its output also peaked early, with 152 million tonnes in 1956 (Mellor 1978). Between 1961 and 1984 output fell by 43 per cent, a relatively modest amount in comparison with other countries, but this still reduced production to only 84 million tonnes by the latter date. The Lower Saxony field ceased production in the mid-1970s, so that output of hard coal comes from three major areas: the Ruhr (dominant, although output has been halved since 1961), the Aachen and Saar coalfields. In addition, Germany possesses major reserves of lignite; in the late 1970s these yielded about 120 million tonnes a year, mostly from the Cologne region (Mellor 1978).

Finally, the UK has been and continues to be the major coal producer in western Europe, despite a 40 per cent reduction in output between 1961 and 1983. Although the state has encouraged the use of coal in electricity

Table 13   *Coal production (in million tonnes) in major western European coalfields,* 1961–84*

	1961	1971	1977	1984
F.R. Germany				
Ruhr				66.3
Aachen				5.4
Saar				10.8
Totals	148	117	91	84
France				
Nord/Pas de Calais				2.5
Lorraine				10.9
Centre/Midi				3.2
Totals	52	33	21	17
Belgium				
Kempenland				6.0
Totals	21	11	7	6
UK†				
Scotland				0.4
North				2.1
Yorkshire				4.2
North-West				4.5
Midlands/Kent				21.5
South Wales				1.3
NCB Open-cast				13.5
Totals	193	147	121	50

* Only the major coalfields in each country are listed.

† 1984 was an abnormal year, affected by the 1984–5 NCB–NUM conflict, when annual production fell by more than 50 per cent. Output in 1983 was 116 million tonnes.

SOURCES: Eurostat (1985a) and Minshull (1980).

generation, it has lost many of its major markets in the UK; there are also significant imports from lower cost producers, especially Australia. The state's policy of reducing production – which has seen employment in the industry fall from 900,000 in 1947 to little more than 100,000 in 1985 – has been vigorously opposed by the miners' unions, and the prolonged stoppage of 1984–5 was only one phase in a long history of industrial conflict. The impacts of closures have been regionally uneven, with Scotland, the North, Lancashire and South Wales losing out mainly to the Midlands and Yorkshire, but also to open-cast production. Investment to open up new coalfields is still being undertaken (boosted by the oil crises of the 1970s) but is mostly concentrated in eastern England, notable examples being the Selby and Vale of Belvoir coalfields.

The future for the coal industry is difficult to predict. In the 1970s, largely as a result of the oil crises, several governments adopted policies to increase coal production. For example, the Spanish Government announced new investment in the Asturias coalfields as part of a strategy to reduce energy dependence and, in the UK, the 1974 Plan for Coal also gave renewed commitment to the industry. Much depends on how the decline of North Sea oil and gas production in the 1990s is to be compensated for. There are large reserves of coal in western Europe, with the UK and West Germany each having an estimated 14,000 million tonnes, although the next largest deposits – in Belgium – are only 800 million tonnes (International Energy Agency 1983). A resurgence of the coal industry is feasible, but much will depend on the relative costs of other energy sources (and the 1985–6 fall in oil prices is a poor omen), the competitivity of European producers against other countries such as Australia, and the state policies for energy.

Oil: long boom and slump

In the post-war period changes in both consumption and production patterns have made western Europe increasingly reliant on oil as a primary energy source. Initially, this was encouraged by many states in order to diversify from excessive dependence on coal production (and hence weaken the power of coal mining unions), and to bridge the gap between coal output and energy demand in the immediate post-war years. In the 1960s, the comparative cost advantages of oil gave it ascendancy in energy markets, and this was only challenged after the oil price rises of the 1970s. In understanding the supply of oil, the post-war period needs to be divided into two phases: dependency on external (to Europe) production and exploitation of indigenous resources, especially in the North Sea.

External dependency

At first, oil production was dominated by multinational companies, working mainly in the Middle East. As demand for oil increased in the post-war period, the profitability of the oil industry increased rapidly. Given largely fixed capital costs, there were low marginal costs of pumping additional barrels of oil from the ground, so that increasing output continued to offer high rates of profit. Indeed, in the course of the 1950s the major concern of the oil industry was not how to expand production but how to limit expansion. There was constant apprehension of fluctuations in output and overproduction, both of which, potentially, could have destabilized prices (Nore 1978). Downward pressures on oil prices were only resisted because the oligopolistic position of the major producers, 'the

Seven Sisters' (see p. 45), 'protected the international market from the massive disruption inherent in the discontinuity between the marginal cost of the vast reserves of Middle East Oil and the establishment of prospective fields elsewhere' (Auty 1983, p. 3). These included the European 'sisters', BP and Shell, which had major distribution interests in Europe as well as global production interests. However, the position of the oil majors was always difficult and became significantly worse with the entry of the USSR into world markets as a supplier in the 1960s (Nore 1978). OPEC was formed in 1960 specifically to counteract falling prices, but these continued to decline in real terms throughout the decade.

At this stage oil production rights were nationalized in several countries so that control of the industry began to pass from the oil majors to such hawkish states as Libya; by 1979 the oil majors' share of 'non-communist' oil reserves had fallen to only 25 per cent (Auty 1983). This led to two major price rises in 1973–4 and 1979 which multiplied the costs of European energy several times during the 1970s. In the face of the recession which followed, and of new energy conservation measures, western Europe's demand for oil fell from a peak of 750 million tons in 1973 to only 585 million tons in 1983. There followed a glut in oil production – which was accentuated by the expansion of North Sea output – and, in the 1980s, the real price of oil fell sharply, especially in 1985–6. One indicator of the crisis is the fact that, in 1985, oil refineries in Europe were only operating at 60 per cent of their potential, even though a third of capacity had disappeared since 1977; plants have been mothballed, demolished or, as in the case of Esso's Milford Haven refinery, dismantled and shipped to the Middle East.

The North Sea bonanza
In part, the overproduction crisis of the 1980s stemmed from the development of indigenous European oil reserves. There had been sporadic oil prospecting in on-shore Europe since the 1930s when, for example, there were small-scale finds in the Trent Valley; but the major oil discoveries have been off-shore. There are some reserves in the Mediterranean and, in 1980, Spain produced some 31,200 barrels, mostly from the Casablanca field near the Balearic Islands, while Italy produced 6200 barrels, mostly from the Gela field off Sicily (Luciani 1984). Greece, too, has proven reserves in the north-west Aegean, although political tensions with Turkey make it difficult to exploit these. However, the dominant sphere of exploration and production has been the North Sea.

While it had long been predicted that there were major oil reserves in the North Sea, exploration was held back until the question of sovereignty was finally resolved by the 1958 Geneva Continental Shelf Convention. This apportioned the North Sea to the surrounding coastal states in accordance with the principle of median lines – despite the wishes of the UK to use the Norwegian trench as a boundary (Mackay and Mackay 1975). The use of median lines apportioned the North Sea in the following way (percentage figures): the UK, 46.7; Norway, 25.1; the Netherlands, 10.7; Denmark, 9.2; West Germany, 6.8. However, oil (and gas) reserves were not equally distributed among these sectors (see Figure 11). The search for oil began in earnest in 1962 and the first major discovery, in 1968, was the Ekofisk field in Norwegian waters, followed in 1969 by the Montrose field in UK waters. This established the basic pattern of

Figure 11 Major oil, gas and coalfields in western Europe and the North Sea, 1986

discoveries, and there have been only minor finds in the Danish, Dutch and German areas. Production began in 1971, and has since risen rapidly, reaching 22 million tonnes in 1976 and 150 million tonnes in 1979, before levelling-off in the mid-1980s. Expansion has eased, and exploration has plummeted, however, following the decline in oil prices in 1985–6. In large part, production has been determined by the policies of the British and Norwegian Governments and of the oil companies.

In the UK the main government aim was rapid exploitation, which is hardly surprising in the context of the 1960s situation when energy shortages were still conceivable (Odell 1978). The state considered that the two UK oil majors, BP and Shell, did not have the resources to develop the North Sea at a sufficiently rapid rate; the participation of non-British oil companies was therefore welcomed. Operators are licensed to prospect for oil in a particular block, while abiding by certain conditions. They pay a small fee for their licence but have to pay 12 per cent of wellhead oil revenues to the government and are subject to UK taxation laws. The government sought to establish a UK presence in the industry and set up a state holding company, British National Oil Company (BNOC), in 1976 to compete with the majors, while also encouraging smaller UK companies such as Tricentral and Ultramar. Over time the share of blocks allocated to UK companies has risen, but the British sector of the North Sea has been opened up to MNC activity at relatively favourable terms. By the mid-1980s US companies controlled almost 50 per cent of potential production in the North Sea. Even though the UK Secretary of State for Energy has the power to control the rate of output, this has only been used once so far (to delay production in the Clyde Field), and the dominant role has been taken by the private – often multinational – sector. This tendency has been strengthened by the privatization of Britoil (part of the BNOC) in 1982.

UK policy has been successful in that it has encouraged rapid exploitation of the North Sea. Oil production in the British sector only began in 1974 but, by 1982, with 103 million tonnes, the UK had become the fifth largest producer in the world, and a net exporter. At this time the oil industry already produced 5 per cent of the UK's GDP, yielded significant government revenue (of about £7 billion), and had virtually eliminated energy imports (Manners 1984). The industry has had less impact on employment – estimated at only about 10,000 direct jobs – which is highly concentrated regionally. Most of the oil comes from about twenty major fields, of which Forties and Brent are the largest (Figure 11). The development of North Sea oil has also had negative implications for the UK economy. In particular, it has strengthened sterling and (arguably) overvalued it, so that export-led manufacturing firms may have been disadvantaged.

Production in the Norwegian sector in the mid-1980s was roughly equivalent to that in the UK sector. The first, and still the major, discovery was the Ekofisk field, which came on stream in 1971. By 1980 three-quarters of Norway's oil still came from this field and, indicative of the way that exploration has moved northwards over time, the next two most important fields were Frigg and Statfjord (Hansen 1983a). Given that Norwegian oil consumption is a modest 7 million tonnes per annum (with much of its electricity coming from HEP), this has yielded a large surplus for export.

State policies in Norway were, initially, broadly similar to those in the UK. There was no government participation in the early prospecting, and the return to the state was subject only to 10 per cent royalties and 42 per cent company tax. The oil companies were supposed to favour Norwegian suppliers but only if these were competitive (Nore 1978). Given the absence of Norwegian oil majors, domestic participation was very limited at this stage. The Norwegian conglomerate NOCO had a 15 per cent share in nine blocks, but the MNCs were dominant: Phillips Petroleum in the Skofisk field and Elf Aquitaine in the Frigg field. This approach was in keeping with the Norwegian tradition of non-state ownership of production.

However, in the course of the 1970s it became clear that Norwegian capitalists were unable to operate effectively in the oil industry, and so, however reluctantly, the state became more interventionist. STATOIL, the Norwegian state oil company, was established in 1972 and was given preference in the allocation of new blocks for exploration; this bore fruit with its discovery of the Statfjord field in the late 1970s. Less successful was the policy of encouraging several large Norwegian interests to form the SAGA company, so that there would be a major national private capital interest in the North Sea. SAGA has only acquired a small share of exploration, so that the private sector is still dominated by the MNCs. Nevertheless, the state has tightened its controls over these; it reserves the option to acquire up to 40 per cent participation in any new fields, and has pressured for more sub-contracting to be allocated to Norwegian firms. Through SAGA and NOROL, the state has also secured a share for Norwegian capital in the refinery and petrochemical industries.

In general terms, then, the Norwegian Government has played a more active role than the British Government in influencing the oil industry. Nowhere is this more clearly illustrated than in their different attitudes to the rate of exploration. Whereas the British Government has encouraged rapid development, so as to ameliorate its recurrent balance of payments crises, the Norwegian Government preferred a more gradual, longer-term development strategy. No exploration at all was permitted north of latitude 62° until the late 1970s. There was considerable domestic political debate concerning this question but, in the end, fears about the effects of an overvalued exchange rate on the competitivity of Norwegian industry was a crucial argument for gradual exploration. In the light of UK experience (and that of the Netherlands with gas production), this seems to have been an appropriate strategy.

Oil forms an even more significant element in the Norwegian economy than in the UK economy. In 1982, the oil industry accounted for some 15 per cent of GNP, provided £2.5 billion in revenue and accounted for one-third of export earnings (Hansen 1983a). About 45,000 people (2 per cent of the workforce) were employed in oil and oil-related industries. Among the latter, Norway has developed some important onshore services and platform construction capacity, but very little refining or petrochemical capacity; construction of a small petrochemical plant at Posgrann is likely to prove an exception rather than the rule in this respect.

Gas: from LNG to Groningen and the North Sea

Before about 1960 gas had a relatively limited role in energy provision in Europe, and was mostly supplied from coke ovens. However, after that date the costs of gas were significantly reduced by two major developments: imports of liquified natural gas and the development of indigenous natural gas reserves. At first, natural gas was imported in liquified form by tankers, and came mainly from Algeria. The first terminals for receiving liquidified natural gas (LNG) were opened at Canvey Island in the UK in 1964 and at Le Havre in France in 1965. After the Trans-Mediterranean link between Algeria and Italy had been constructed, gas began to arrive by pipeline. More recently, western Europe has been connected by pipeline to the vast reserves of natural gas in the USSR. The 1960s were an important phase in the development of gas sources in Europe. There has always been a large potential market for gas as an industrial input and as a source of domestic heating because it is relatively clean and flexible; but it was natural gas imports which allowed demand to be met by relatively cheap supplies, and hence opened up the market.

The second phase in the expansion of gas supplies came with the development of indigenous resources. There were long-established gas reserves on the European mainland but, until the late 1950s, only about 11 million tons equivalent (mtoe) was produced, mostly from south-west France and Italy's Po Valley. The turning point came in 1959 with the discovery by Shell and Esso of the Groningen gas field in the Netherlands, with recoverable reserves estimated at 2400 mtoe. This is still the largest known field in western Europe, and it had the advantages of being on-shore and near an enormous potential market. Soon afterwards, surveys of the North Sea began and the first commercial offshore gas find was by BP near Humberside in 1969. Most of the earlier exploration was in the southern part of the North Sea but, latterly, attention has switched to the northern part and to collecting gas from the oilfields. The major producers have been the Netherlands and the UK (especially with its Leman field), but most sectors have yielded useful reserves. The greatest reserves may well lie in the Norwegian sector, with the Frigg field in production, and the Sleipner and the gigantic Troll field in development in the mid-1980s.

Norway consumes relatively little of its own gas production, which is mostly exported by pipeline to the UK and West Germany.

As with oil, state policies have exercised a formative influence on the development of gas reserves. The British Government (unlike with oil production) has taken a strongly nationalistic line over gas production, which is not surprising given that domestic reserves are barely sufficient to meet demand. Companies given exploration rights in the UK sector are legally bound to offer to sell to British Gas at 'a reasonable price'; the final arbiter of this is the government. The monopoly position of British Gas was complete for, under the Gas Act 1965, it has control over the purchasing and final prices of, as well as the distribution of, gas. At times there has been a tense political struggle between the state and MNC interests in the British sector. In 1971 the government refused to sanction the sale of gas from the Viking field to continental Europe, even though the landing costs for this were less than those for bringing it to the UK. Then, in 1974, the UK contracted to buy gas from Norway's Frigg field, a move which enabled it to hold down the price paid to producers in the UK sector. Consequently, prices in the UK have consistently been lower than those in world markets (Russell 1983). More recently, controversy has surrounded supplies from Norway's Sleipner field which the British Government considered purchasing, so as to reduce the prices paid in the UK sector. In the end, this was abandoned because of the demands from companies operating in Britain for compensatory export rights, and because of their threat that depressing UK prices would result in reduced exploration in future. Instead, it has been agreed that, after privatization of British Gas, gas exports will be permitted so as to break the company's purchasing monopoly. Meanwhile, Norway in 1986 was discussing sales from the Troll and Sleipner fields to Germany, France and Belgium, in a deal potentially valued at $60 billion.

The Netherlands Government has a policy of limiting production from its enormous reserves so as to keep the price of gas relatively high; this will conserve reserves for domestic needs into the next century. While gas provides about 50 per cent of domestic energy needs, there is still a sizeable surplus available for sale to Switzerland, France, Luxembourg and West Germany (Clout 1981b). At their 1976 peak, exports amounted to 51 trillion cu. mm. (Russell 1983). This has been sufficient to keep the guilder at an exchange rate high enough to have damaged the exports of the manufacturing sector. The nationalized corporation, Gasunie, has a monopoly over the purchase and supply of gas and the government has, with some success, used subsidized gas prices to encourage industrial decentralization from Rotterdam to Gronigen and other less-developed provinces (Odell 1978). The availability of cheap gas also encouraged the expansion of capital-intensive industries such as oil refining and chemicals. Natural gas production has also been an important source of state revenue – some 10 per cent in 1985 – which has sustained ambitious welfare and social policies. Consequently, the crash in gas prices in 1986 has severe problems.

There are large known reserves of gas remaining in the North Sea (over 2000 mtoe) and the onshore Netherlands (over 1500 mtoe), while future exploration may yield new additions, especially in the north of the Norwegian sector (Odell 1981). In addition there are small, known reserves in the northern Mediterranean, for example in Spain's Gulf of Cadiz. However, these reserves are not expected to last beyond the early part of the twenty-first century. Thereafter, Europe will again have to resort to imports from Nigeria, the USSR or the Middle East where, for example, Iran alone has proven reserves of 13,700 mtoe (Luciani 1984). The greatest problems of adaptation will occur for those countries which are currently most heavily reliant on gas, namely the UK and (especially) the Netherlands.

The nuclear option: promise and problems

In the 1970s the nuclear option seemed to offer great potential as a future source of energy, but its development has been brought into question by two important considerations. Costs have proven higher than anticipated so that this is still far from being a source of cheap energy. In addition, the nuclear option and its potential hazards has become the focus of intense political conflict in many countries, including Austria, West Germany and Sweden. These themes can be examined by tracing the history of nuclear energy in western Europe, which can be categorized into three main phases (Ilbery 1981).

The first phase lasted until 1965, and was initiated by the opening of western Europe's first nuclear power station in the UK, at Calder Hall, in 1956. Over the following nine years the UK established a formidable technological and production lead over its European neighbours and, by 1965, had nine nuclear power stations. There had been a ban on nuclear research in West Germany in the immediate post-war years and, by the end of this phase, only an experimental station had been established. The only other countries to have established nuclear power stations by 1965 – and on a very limited scale – were France and Italy.

In the second phase (1965–80) there was a major expansion of nuclear capacity to six new countries: Switzerland, Sweden, Belgium, Spain, Finland and the Netherlands. The reasons why these countries opted for a nuclear strategy are complex, but current or anticipated lack of indigenous resources was usually paramount. For example, given conservationist pressures, both Switzerland and Sweden had approached the limits for expanding hydro-electric power but lacked carbon-based domestic fuel reserves. Among the early nuclear producers, the UK kept its production lead and had 33 units by 1979. However, it was rapidly being overtaken by West Germany and France. Germany already had sixteen units by this date, and was also experimenting with a fast-breeder reactor at Kalkar. However, France had the most comprehensive nuclear strategy, not least because the centralized power of the state facilitated its development. In 1974 it had ten units, but the Messmer programme planned eighteen units by 1979 and construction of another 50 by 1985, while also developing Phoenix fast-breeder reactors. Costs were reduced by adopting standardized design and construction methods. This policy was pursued as a direct response to the oil crisis of the 1980s in an attempt to make France more energy self-sufficient (Ardagh 1982). The aim was to generate 72 per cent of electricity from nuclear sources in 1985. In practice, progress has been retarded by conservationist concerns and technical problems (Holmes and Fawcett 1983), yet by 1986 France had a surplus of cheap electricity to export. Italy and Spain, too, had ambitious plans for expanding nuclear energy which Ilbery (1981) considered could make them the third and fourth largest producers in western Europe. Given that Luxembourg and Austria also had plans for nuclear stations, this left only Denmark, Greece, Ireland, Norway and Portugal without actual or planned nuclear capacity. The distribution of nuclear reactors in 1986 is summarized in Table 14.

Table 14 *Nuclear power in western Europe in 1986*

Country	Number of reactors in operation	Number of reactors under construction	Percentage of total electricity production, 1985
Belgium	8	0	60
F.R. Germany	20	5	31
Finland	5	0	38
France	44	17	65
Italy	3	2	4
Netherlands	2	0	6
Spain	8	3	24
Sweden	12	0	43
Switzerland	5	0	40
UK	38	4	19

SOURCE: Forum Atomique Européen.

The final phase has seen an adverse reaction to nuclear energy, as a result of publicity given to both the possibility of accidents in nuclear stations (especially following the Three Mile Island incident and the Chernobyl disaster) and to the problems of safely disposing of plutonium waste, which has a life of 200,000 years. The nuclear issue has become important in Green politics, and many countries have either slowed down or abandoned their plans for nuclear energy. Denmark, despite its total lack of indigenous energy sources, has never embarked on a nuclear programme, precisely because there is a strongly organized anti-nuclear movement. Austria actually built a nuclear station at Zwentendorf but, as a result of the anti-nuclear movement's success in a national referendum in 1978, this remained idle and the decision was made in 1986 to dismantle it. Meanwhile, in West Germany the anti-nuclear movement has been able to use the legal system successfully, persuading the courts to revoke licences for new nuclear stations on the grounds of there being inadequate means of waste-disposal. In Sweden, as a result of a national referendum in 1980, the state has committed itself to completing the current large-scale programme of investments in nuclear capacity but, at the same time, these are not to be replaced after their 'normal' lives, so that nuclear electricity generation will end in 2010. Denmark has also protested at the siting of some Swedish nuclear stations in close proximity to Copenhagen. After Chernobyl, the Spanish Government also decided to suspend further construction of a new reactor at Valdecabelleros.

Hydro-electricity: 'white coal'

Hydro-electric power (HEP) has a relatively long history but, in recent years, has been subject to political opposition by conservationist groups. The environmental requirements of HEP are a mountainous relief, and substantial and regular rainfall. Not surprisingly, therefore, it is limited mostly to the Alpine and Scandinavian regions, with important outliers in the Pyrenees, Northern Iberia, France's Central Massif, and northern Britain. Only in Austria, Norway, Sweden and Switzerland does HEP produce more than a quarter of national energy requirements.

Some of the earliest HEP installations were constructed in Sweden and Norway in the 1980s, but the major phase of expansion was in the 1950s and 60s (John 1984). The state in Norway has been massively committed to HEP and that country now obtains over 50 per cent of its energy from this source. HEP has been less important in Sweden because the generating potential lies in the north while most of the population lies in the south. Therefore, the distribution costs of HEP electricity are relatively high, especially given the potential to import cheap energy from its Scandinavian neighbours via the integrated NORDEC electricity network. France was also an early pioneer of HEP production but, since about 1950, the potential for further expansion of this source has seemed limited. Given the potential conflict which exists between tourism and HEP production, and widespread opposition to further dam building and flooding of valleys in what are often beautiful landscape regions, few major extensions of HEP are likely in western Europe in the near future.

Further reading

General reviews: P. R. Odell (1986), 'Energy and regional development: a European perspective', *Built Environment*, **11**, pp. 9–21; H. D. Clout (1981), 'Energy and regional problems', in H. D. Clout (ed.), *Regional development in Western Europe*, Chichester: Wiley.

4 The primary sector

Introduction

The larger part of this chapter concentrates on only one type of primary activity – agriculture – but this accounts for 44 per cent of land use in western Europe (Clout 1981a). While the importance of agriculture in some respects (for example in national employment or output) has declined over time, it is still a key sector in interpreting economic development. The final section, however, considers developments in fisheries.

The conditions of agricultural production in Europe were fundamentally changed in the late nineteenth century when developments in transport technology opened up the possibilities for large-scale imports of foods from low-cost New World producers. Initial reponses to this were varied. The UK opted for a free-trade strategy so as to reduce the costs of urban food supplies, while Denmark and the Netherlands encouraged imports of cheap feeds to develop livestock production. However, most western European governments preferred protectionism, which tended to foster relatively inefficient agricultural practices. After 1945 conditions changed again as demand expanded rapidly. This was due to rising standards of living which required not only more foodstuffs, but also changes in the types of foods consumed. For example, in Finland between 1960 and 1980, per capita consumption of cereals, potatoes and butter fell, while that of meat and vegetables more than doubled (Cabouret 1982); similar changes were evident in other countries. Even though the aggregate elasticity of demand for food was less than unity (that is, increased less rapidly than incomes), Europe could not meet the increased demand solely from its own resources.

There was certainly scope for increased domestic production and, for example, productivity in the UK improved sharply after 1945. In Denmark and the Netherlands there were gains in agricultural productivity and output, not least because the diversion of investment to manufacturing was still modest (Parker 1981). However, in the early post-war years improved food supplies were more readily secured via international trade. This involved both greater participation in international markets and – for the colonial powers – attempts to expand food production in overseas possessions. One implication of this within Europe was the virtual disappearance of regional food markets, which were replaced by international ones, even in such traditionally remote areas as the south of Italy (Bethemont and Pelletier 1983). This did not necessarily imply a greater commitment to free trade. Significantly, agricultural products were excluded at an early stage from the GATT deliberations on trade liberalization (see p. 24), and from EFTA's guidelines on inter-member trade. Even more importantly, the EC from its foundation moved towards a policy of import controls and of guaranteed prices for farmers. By the 1980s Europe had become self-sufficient in a number of major food products but it had also retreated into protectionism.

The most emphatic change in western European agriculture was its declining share in GDP and employment in most countries between 1960 and the 1980s (Table 15). Some of the more exceptional changes were in France, where the shares of agriculture in GDP and employment fell by a third in this period. The UK and Greece experienced the smallest relative changes, although for very different reasons; agriculture in Greece is still relatively unmodernized while, in the UK, the quantitative importance of agriculture in the economy was already minimal by 1960. Despite a tendency to greater uniformity, there were still important international differences in farming in the 1980s: in terms of share of GDP, the UK and West Germany (2 per cent or less) represent one set of extremes contrasting with Ireland, Portugal and Greece (over 8 per cent).

A comparison of the contribution of agriculture to GDP and to

Table 15 *The role of agriculture in 1960 and the early 1980s*

| | 1960 (Percentage of national) | | Early 1980s (Percentage of national) | |
	GDP	Employment	GDP*	Employment†
Austria	11	24	4	9
Belgium	6	8	3	3
Denmark	11	18	5	7
F.R. Germany	6	14	2	4
Finland	17	36	7	11
France	11	22	4	8
Greece	23	56	18	37
Ireland	22	36	11	18
Italy	12	31	5	11
Netherlands	9	11	4	6
Norway	9	20	4	7
Portugal	25	44	9	21
Spain	—	42	6	14
Sweden	7	14	3	5
Switzerland	—	11	—	5
UK	3	4	2	2

* 1984 data except for Spain and the Netherlands, for which 1983 statistics are given.
† 1981 data.
SOURCE: World Bank (1984; 1986).

employment illustrates relatively low labour productivity in farming. In the 1960s, for example, GDP in the world's industrial market economies expanded by 5.9 per cent in manufacturing and by 4.5 per cent in services, but only by 1.4 per cent in agriculture. Even in the 1980s – in the face of global industrial crises – the gap between agriculture and the other sectors narrowed rather than closed (World Bank 1984). Only in the UK does the share of agriculture in GDP match that in employment. Elsewhere, the proportion of employment is up to twice the proportion of GDP, particularly in Austria, Greece and Portugal. This illustrates not only varying levels of labour productivity, but also fundamental differences in the organization of agriculture within western Europe. These differences have persisted despite the influence of the EC's Common Agricultural Policy.

Land-use variations

General variations in agricultural production in Europe are summarized in terms of land use in Figure 12. Environmental conditions, investment levels and organizational features all differ between countries. At one extreme is Scandinavia (excepting Denmark), where less than 10 per cent of the land is used for arable or permanent grassland owing to climatic (a short growing season) and relief conditions. In southern Europe (hot summers with low rainfall) and in Denmark, arable is dominant, accounting for 40 per cent or more of land use. In contrast, permanent pasture is dominant in Ireland, the UK, the Netherlands and Switzerland, reflecting high levels of specialization to utilize comparative advantages offered by climatic conditions (cool summers and high precipitation). In most other countries – including the Benelux, France and West Germany – there is an approximate balance between the major land uses. National comparisons necessarily are broad generalizations and there are many interregional variations. Among these are the specialization of eastern (dryer) Britain in arable production and concentrations of irrigated horticulture in parts of southern Europe. In addition, upland farmers throughout Europe generally work in poorer environmental conditions and tend to have lower incomes; this applies equally well to farming areas in the Massif Central, the Scottish Highlands or southern Italy.

Growth and restructuring in agriculture

The Second World War had a devastating effect on agriculture in most of Europe, and especially on the combatants (excepting the UK). In 1945 food production was below 1939 levels in most countries, and in Austria, West Germany and Greece was less than two-thirds of pre-war output (Clout 1971). There were particular shortages of meat and dairy products because cereals had been consumed directly rather than as livestock feed.

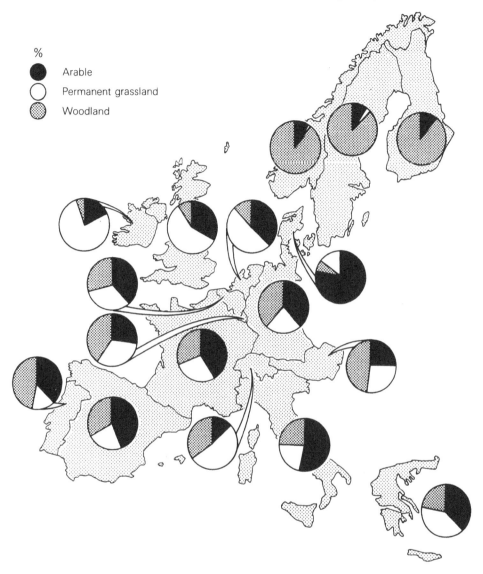

%
● Arable
○ Permanent grassland
◉ Woodland

Figure 12 Rural land use in western Europe, 1983.

SOURCE: Eurostat (1985b)

In the face of supply shortages, long time lags in realizing returns on new investments, and severe winters, recovery was initially very slow; but, by 1949–50, pre-war production levels had been re-established in western Europe, except in Austria, Germany and Greece.

After about 1950 recovery was more sustained but differences in performance soon became evident. The highest levels of output were in the UK, Scandinavia and the Benelux countries; in 1955, output per active male worker in these countries surpassed $1800, peaking at $2310 in the UK (Clout 1971, p. 13). This represented relatively high labour productivity both on the small farms of the Benelux countries and the large farms of Britain (Parker 1981). All other northern European countries, including France and West Germany, had output in the range $1000–1800. Southern Europe, however, consistently lagged behind, with levels of $820 in Italy, $570 in Greece, $435 in Spain and only $385 in Portugal. This represents a north–south divide which had only partially been closed by the 1980s.

The intensification of farming

At first the expansion of output was based on resuming normal farm practices and adopting improved methods which had already been successfully pioneered in North America. In the longer term, however, expansion was to be based on a radical reorganization of agriculture, founded on the introduction of capitalist relationships. This is discussed in more depth in the following section, but here some of the elements which contributed to increasing output are noted. These include technological developments, greater commercialization and specialization, all of which are closely interrelated. While these have changed the nature of production in most countries, their impact has varied, both internationally and between regions.

Changes in agricultural technology include more productive and disease-resistant crop varieties, better weight-gaining livestock breeds, more use of fertilizers, and mechanization of many farm tasks. Mechanization has been applied to many tasks in both arable and livestock farming, but the most commonly used indicator of this is the number of tractors. Whereas tractors were still rare in 1945, they were to be found on most farms in the 1980s. In Europe as a whole, the number of tractors increased from 1.8 million in 1955 to 4.6 million in 1965 (Clout 1971), while individual countries recorded remarkable gains; for example, the number of tractors in Denmark increased ten-fold between 1950 and 1977 (John 1984). Levels of mechanization, however, vary considerably between countries, as can be seen in the case of milking machines and tractors in the EC (see Table 16). The highest levels of mechanization are in West Germany, followed by the Benelux countries and France, with Greece, Ireland and the UK trailing. This does not always reflect efficiency levels because, in the UK at least, lower mechanization levels per hectare/unit of livestock are due to greater efficiency in the use of capital on relatively large farms. Similar reasons account for the apparently small numbers of milking machines in the efficient Netherlands dairy industry. The converse is evident in Greece, where more capital-intensive production methods have been adopted without corresponding increases in economies of scale to reduce unit costs. As a result, 'the growth of capital tended to exceed the growth of output

Table 16 *Indicators of farm performance in the European Community in the early 1980s*

	Tractors per 100 ha, 1983	Milking machines per 100 dairy cows, 1983	Nitrogen fertilizers (kg/ha) 1983	Wheat yields (10^2 kg/ha) 1961	Wheat yields (10^2 kg/ha) 1981
Belgium	7.1	4.4	130	34.8	52.1
Denmark	6.3	5.9	137	41.2	55.6
F.R. Germany	12.3	8.5	121	28.9	51.0
France	4.7	5.3	70	23.9	48.1
Greece	2.1	—	45	13.0	26.2
Ireland	2.5	1.4	52	33.7	56.6
Italy	6.2	3.6	50	19.1	27.1
Luxembourg	6.7	3.8	116	22.2	34.3
Netherlands	8.5	2.7	227	39.3	67.0
UK	2.8	1.5	83	35.4	58.4

SOURCE: Eurostat (1985b).

and the "capital–output" ratio failed to counteract the increase in the capital–labour ratio' (Manitea-Tsapatsaris 1986, p. 120).

A broadly similar pattern exists in the application of fertilizers. Over time there have been considerable increases in the application of fertilizers to improve soil fertility, and this has partly closed the international gap between northern and southern Europe. In 1950 only 8 kg of nitrogenous fertilizers were used per hectare in Italy, which was only one-eighth of the level in the Netherlands (Parker 1981). By 1981 the application levels had risen to 51 kg in Italy and the ratio to that in the Netherlands had narrowed to less than one-fifth. Considerable variations still exist between countries (see Table 16), although in some cases, such as the UK, this is partly due to inherent differences in soil composition and fertility. Nevertheless, the north–south gap in Europe still exists. The number of grams of all plant nutrients applied per hectare of arable land in 1981 was only 767 in Portugal and 672 in Spain, compared with 6094 in Ireland and 7674 in the Netherlands.

Farm yields
As a result of changes in agricultural technology and farm organization, there have been considerable improvements in the productivity of both land and livestock. For example, in the two decades after 1961, substantial gains of more than 50 per cent were recorded in wheat yields throughout the EC (Table 16). Yet, international differences persist, and Greece and Italy trailed with yields which were less than half those of Denmark and the UK, and barely more than a third of those in the Netherlands. Improved

yields have meant substantial increases in food production throughout Europe. Between 1969/71 and 1980/82, food production per capita rose in every country except Portugal, where it fell by 30 per cent owing to low investment and climatic fluctuations. Elsewhere, gains of 10–20 per cent were common (World Bank 1984), while output increased by 20–30 per cent in Greece, Spain, France, Denmark, Norway, Switzerland and the UK. In southern Europe this reflected a late 'catching up' on agricultural practices, but elsewhere it reflected continuing high levels of investment in farming.

These developments were linked to greater commercialization and specialization of farming as more regions concentrated on products in which they held some comparative advantage in international trade. Examples include the UK, where the mixed farms of the early post-war years have been almost completely replaced by specialized farms (Britton 1974). More spectacularly, regions such as southern Almeria in Spain have passed in little more than a decade from traditional peasant production to commercialized and specialized horticultural farms exporting to northern Europe. Not surprisingly, therefore, despite the tendency to protectionism in Europe and despite production improvements, specialization has meant that foodstuffs still represent an important element in the imports of many countries. This ranges from about 7 per cent in Scandinavia to 15 per cent in the Netherlands (owing to a limited supply of land and to specialization in livestock), to 16 per cent in Portugal (World Bank 1984).

In the course of the post-war era, western Europe has moved from a position of acute shortages to – at least in the EC – gross overproduction of some foods. Partly in consequence of this, there has been growing conservationist opposition to continued investment in increasingly capital-intensive agriculture. The reaction has been focused on the implications for the landscape (with the disappearance of traditional features such as hedgerows and marshland), on the quality of food (polluted by chemical fertilizers and insecticides), and on rural social structure (see Phillips and Williams 1984). Public consciousness of these issues has been raised by the publication of such texts as Carson's (1969) *Silent spring*, and has often been channelled into the general growth of Green politics (Lowe and Goyder 1983). In some cases this has also been linked to general reactions against the modern capitalist economy. For example, in France one reaction to the urban social unrest of 1968 was 'antiproductionism' (Trilling 1981), and this became linked to the causes of rural environmentalist groups. The next section examines how the balance between capitalist and non-capitalist organization of agriculture has evolved in post-war Europe.

Transition to capitalist agriculture

Capitalist relationships in agriculture

A theoretical note
Writing in the mid-nineteenth century, Marx predicted that peasant agriculture in Europe would undergo a long-term decline owing to inherent centralization trends in capital accumulation. In other words, he suggested that small-scale subsistence farming would be replaced by large-scale commercial farming. This would be encouraged by the relationship between agriculture and the larger capitalist society, for agriculture is a source of material inputs for manufacturing, and of labour for non-agricultural activities. Arguably, the capitalist penetration of agriculture was not only the likely consequence of capitalist development, but also an important pre-condition for this (Mottura and Pugliese 1980). Cheap food and the release of labour to work in industry were essential for industrial expansion.

The capitalist relationships of agriculture are, of course, more complex than has been suggested. They operate in different combinations by country, let alone by region, while state policies must also be taken into account. Changes in agricultural production methods are the outcome of both changes in market demand and state intervention to support markets and producers. In turn, changes in production methods lead to increased output and may also require changes in land ownership and farm-business organization which also contributes to increased output. However, greater output is likely to increase profits, especially for larger-scale producers, hence leading to further changes in production methods. Changes in farm organization can include greater segmentation of production involving more reliance on off-farm production (say factory- rather than farm-produced feed). This tendency also contributes to further changes in methods of production (see Marsden 1984 for a fuller exposition of these arguments).

Tenure, commercialization and specialization
To what extent have these theoretical expectations been borne out by the post-war development of agriculture? There have certainly been increases in the size of farms because of a number of pressures, including mechanization (requiring increased economies of scale and the need for capital investment). For example, in Greece between 1961 and 1977 there was a 20 per cent decrease in the number of farms smaller than 30 ha, but a 114 per cent increase in the number of units over 500 ha (Manitea-Tsapatsaris 1986). Similarly, in Germany in the 1960s there was a 34 per cent decline in farms of less than 10 ha compared with an 18 per cent gain in farms of over 50 ha (Calmès 1981). Emigration, too, has been important

in some countries, especially in southern Europe, in reducing pressure to sub-divide farm land, although it is also true that many emigrants do not sell their land, expecting eventually to return and work it (King 1984). There are obviously differences between regions and between sectors, with arable farming usually offering larger economies of scale than livestock farming. Nevertheless, an increasing scale of production is a well-established and nearly universal trend throughout western Europe (see pp. 138–41).

Changes in tenure have also occurred, although these are less uniform. In the UK there has been a growth of owner-occupied farms in the twentieth century (Newby 1979), with the major exception being large-scale investments by financial institutions in farm-managed estates in eastern England (Munton 1977). Elsewhere, there has been a shift to the landlord–tenant system. In Italy, this has partly been due to legislative changes to end share-cropping (King and Took 1983). Land reform in Italy, which has broken up many of the latifundia estates of the south, has had a mixed effect, with both owner-occupied and rented farms being established. Changes in tenure have usually been encouraged by state action to increase fluidity in the land market, by replacing traditional inheritance rights with more commercial relationships. Italy again provides an example, for many of the share-cropping farms of Veneto and Emilia Romagna were taken over as commercial farms in the 1960s and 70s (King and Took 1983). It is not surprising that there have been international variations in the precise form of tenure changes, for these have depended on prevailing legal and financial systems (for loans, etc.). However, the modifications to tenure are also indicative of greater underlying fluidity in land markets arising from commercialization.

Commercialization involves the breakdown of traditional subsistence production and entry into market relationships for both the inputs and outputs of the farm. It is usually associated with greater specialization, owing to the need to use expensive machinery fully with large economies of scale. In its most extreme form, specialization involves factory farming in which the 'farmer', with little or no land, concentrates on only one phase of production, such as rearing young chicks or fattening heifers. More commonly, specialization involves product changes within existing farm practices. Not untypical is Denmark where, in the last 30 years, many farms have abandoned dairy cows and pigs to concentrate, instead, on beef cattle or barley (John 1984). Specialization tends to be associated with increased scale of production; but there are exceptions, as in parts of Spain and Italy, where latifundia estates have been broken-up to produce smaller and specialized farms (see Garcia-Ramon 1985). In its most extreme form, commercialized farming is probably best represented by the agribusinesses of the UK which, in terms of their capital–labour structure and their managerial organization, differ little from modern manufacturing firms (Newby 1978).

On-farm and off-farm production

Although the nature of the farm has changed owing to greater penetration by capitalist relationships, the major transformation has not been at the point of production itself but in the processing of inputs and outputs (Murray 1978). With greater specialization, farms have become less autonomous and more part of 'an integrated corporate system of food processing and distribution' (Mottura and Pugliese 1980). In the 1980s relatively little food processing is undertaken on the farms themselves. Farm cheese making or bacon curing are the exceptions and food processing factories are the norm. Many farms have become locked into marketing arrangements whereby the factory arranges for advanced purchasing of crops or livestock; classic examples occur in the milk industry, whether the purchasing agent is the state as in the UK or, as in the case of Nestlé's operations in Spain, a private company. The monopolistic position of such companies often allows them to exercise strong pressure on the prices paid to producers. Individual farmers have little scope to counter this, but agricultural cooperatives are one possible collective response (particularly common in wine production), and direct marketing (via farm shops or 'pick-your-own') offers a possible individual response. In the UK, Bowler (1982) estimates that direct marketing can add 19–32 per cent to a farmer's profit margins, compared with indirect sales to processing/distribution companies.

Processing of agricultural inputs also tends to have been removed from the farm. Bought-in feeds have replaced farm-grown feeds, tractor factories have replaced farm blacksmiths, and chemical fertilizers have replaced or complement organic farmyard manure. The broad outlines of these changes were indicated earlier (pp. 116–8) in the statistics relating to the mechanization of farms and the use of fertilizers. Not-untypical is France (Fel 1984) where, between 1960 and 1980, there was a 450 per cent increase in insecticides and a 223 per cent increase in bought animal feed. Capitalism, then, has appropriated part of the process of production from the farm and relocated it in the factory system. This is a process which has been aided and abetted by the state in many cases (see pp. 127–33). It is rarely a process which has operated in the interests of the farmers themselves. The small number of suppliers is in a strong position *vis-à-vis* the large number of farmers who tend, therefore, to be price-takers usually of relatively high-priced inputs (Newby 1978). Farm labourers, too, have lost out with the removal of jobs from the farmyard to the factory floor. In France, out of 5 million people in agriculturally related employment in 1975, only 2 million were in farming; a further 1.5 million were in trade and commerce, 1 million in food processing, and 400,000 in the manufacture of agricultural machinery (Holmes and Fawcett 1983).

Small farms

In reality, then, penetration of agriculture by capitalist relationships has not assumed the form predicted by Marx. Farms have increased in scale and commercialization has replaced subsistence, but small and medium-sized farms have not disappeared. This is partly because the state has protected small-scale farming for a variety of social and political reasons (see pp. 128–9); but, more fundamentally, it can be related to the nature of land and of organic production. Agricultural production is very risky since it is subject to climatic fluctuations, and it involves a slow circulation time owing to the lag between investment and profit realization. This, itself, is due to long cultivation or animal rearing periods (Mann and Dickinson 1978; 1980). Technological advances have made only minor inroads into these lags, so that agricultural production has not generally been attractive to capital. This has created conditions which have permitted small-scale (family) farms to survive.

While varying definitions make it difficult to estimate precisely the number of family farms in western Europe, some indication of their overall importance is given by the data in Table 17. Non-farm employees are relatively most important in the UK, Netherlands and Denmark, which have already been identified as having the most advanced capitalist agricultural sectors. Elsewhere, more than 90 per cent of farm labour is still provided by farm occupiers (owners or tenants) and their families. One aspect not revealed by these data is the growing feminization of the labour force, especially in southern Europe, where the male household hand may often be working away from home, possibly as an emigrant. In Italy, for example, the proportion of women in the agricultural labour force increased from 24 to 33 per cent between 1951 and 1976, but even this may be a considerable underestimation as women's farm work is not always recognized in the official censuses.

Table 17   *The agricultural labour force in the European Community in 1983 (percentage figures)*

	Farm occupiers	Family members	Non-family workers
Belgium	73	21	5
Denmark	84	4	11
F.R. Germany	53	38	10
France	51	36	13
Greece	52	48	0
Ireland	61	28	11
Italy	49	42	9
Luxembourg	37	50	13
Netherlands	58	26	15
UK	55	10	35
European Community	53	36	11

SOURCE: Eurostat (1985b).

The family farm remains viable because it relies on family labour, which is not usually costed at market rates. Small farms also have a general advantage in that they have few employees and these work in a paternalistic relationship with the owner, contributing to what Newby (1977) has termed 'the deferential worker' – one who is non-militant and usually low-paid. In addition, some types of farming, especially in the livestock sector, do not offer considerable economies of scale. Not least, any gains in fuller utilization of machinery on large farms may be outweighed by the extra time spent travelling to outlying fields (Winter 1984). This is a characteristic of agriculture, stemming from the nature of land, which differentiates it from manufacturing. Furthermore, the supply of land for agriculture is fixed in absolute terms, excepting marginal reclamations from the sea (as in the polders in the Netherlands) or of marshy terrain. Normally, one farm can only expand in size by taking over other farms, but this is often difficult. Small-scale and relatively inefficient farmers may be unwilling to sell land, even if offered an attractive price, because of their attachment to it. As Banaji (1976) emphasizes, the very survival of the peasantry (with its attachment to the land) may, therefore, retard the two basic processes which actually distinguish agrarian capitalism – that is, accumulation and centralization.

The peasantry: extinction or transformation?

A definition
While the peasantry has already been referred to, no attempt has yet been made to define this term, although it is clear that it should be distinguished from small-scale farming *per se*. Drawing on the considerable literature on this subject (Archetti and Aass 1978; Franklin 1969, 1971; Shanin 1979), the peasant farm is best characterized by 'total labour commitment'. The peasant family work extremely long hours for far lower returns than they would earn in a shorter time in a factory. Often their return is little more than food, clothing and shelter for, in its purest form, peasant farming is involved in subsistence and does not enter into market relationships. Peasant farmers are willing to undergo such deprivation because their aims are genealogical rather than commercial, and they are involved in 'land husbandry' not 'land exploitation'; that is, a major aim is to conserve the farm and its resources for subsequent generations so that the family retains its 'inheritance' and 'keeps its name on the land'. In the peasant farm, economic and social life are interlinked and are centred on the family. This is evident in the attitude to labour: employees are family members and so cannot be fired and hired according to market dictates. On the contrary, in times of recession in other industries, the peasant farm may have a social function to perform in offering security and work to members of the family.

The pressures

Over time, peasant production has been subject to economic pressures to enter into capitalist relationships and, in its pure form, it probably no longer exists in western Europe. This has occurred at different times in Europe, and was relatively early in Denmark, the Netherlands and the UK, and relatively late in southern Europe. Indeed, to some extent the dictatorial governments of Italy, Portugal and Spain all encouraged maintenance of the peasantry as late as the inter-war or early post-war periods, as part of the 'rural' national images which they tried to create for their corporate state ideologies (Williams 1984). It has been argued that in 1939, despite commercialization of farming in some countries there was in southern and eastern Europe 'a substantial class of small and dwarf peasant holdings barely supported an undernourished, poverty-stricken and oppressed group of people' (Franklin 1971, p. 9).

Since 1945, however, the pressures on the peasantry have intensified, partly because of the growing capitalist penetration of agriculture, but partly because of the demand for labour (and higher wages) in other industries (Mintz 1973; Shanin 1979). The state may also have been instrumental in this, acting upon 'depeasantization ideology' (Giner and Sevilla-Guzman 1980, p. 14) or the notion that, in the process of modernization, the peasantry is bound to be removed so as to ensure a supply of labour for manufacturing industry:

> The peasants ultimately will have to be sacrificed upon the altars of industrialization to the thirsty gods of modernity and progress. The question is only of how kindly or how cruelly the sacrifice is to be made.

In the face of these economic pressures the peasant family is faced with a fundamental choice – to leave agriculture, as so many have done in the post-war period (see p. 113), or to accept changes on the farm, including commercialization and at least partial replacement of the traditional interlocking of social and economic aims. The changes include a shift from subsistence farming to specialization which, necessarily, implies greater commercialization. Technological changes may also lead to a substitution of capital for labour, hence weakening the traditional role of the peasant farm in offering job security in the face of crises in the capitalist economy. This last change has been facilitated by the expansion of welfare state provision (see p. 78), so that the state has now taken partial responsibility for unemployment, sickness and old-age benefits in most western European countries.

The peasant farm can react in a number of ways to these pressures. Friedmann (1980) stresses that it is important to consider the balance between the internal and external relationships of the household – between divisions of labour within the family and its involvement in larger divisions of labour beyond the farm. In economic terms, she suggests that several different types of survival strategies are available for the family farm in capitalist society: some members may go out to work, while others become involved in on-farm enterprises (such as shops, tourist accommodation, etc.) so as to supplement household income. The peasant farm, then, is faced with a wide range of options in the process of being integrated into the capitalist economy, and the modern peasant farm can no longer be considered as a single category.

A typology of modern peasant farms
Franklin (1969; 1971) considers that there are three main types of peasant farm, although the boundaries between these were somewhat blurred: full-time, part-time and marginal. Each of these is characterized by some elements of the traditional definition of the peasantry. The full-time units have become more akin to family farms as they have shifted from subsistence to commercial farming, and most, if not all, of their income comes from agriculture. Family farms are present in large numbers in most countries, exhibiting varying degrees of commercialization, but the element of subsistence is still relatively most important in southern Europe, the Alps and parts of West Germany. Contrary to the general decline in the number of farms in western Europe, this group is still increasing in many countries, drawing new members from two very different sources. One source is peasant producers who previously operated part-time but who have had to accept greater commercialization. Sometimes the families have been able to undertake the necessary changes through their own initiatives, perhaps using local cooperatives to market their produce. However, the state has also been active in encouraging the formation of viable, modernized family farms as for example in Spain and Italy. The other route into family farming has been from commercial (often larger) farms which have shed non-family employees, in the face of rising labour costs and the substitution of capital for labour; these farms are operated by the family, with some reliance on hired part-time or seasonal labour. This has been particularly important in the UK in the twentieth century; although most farms are heavily commercialized, some 60 per cent of their labour is drawn from the family (Winter 1984).

The second of Franklin's categories are marginal farmers but this is a problematic term. Hirsch and Maunder (1978) consider that even in the UK about 60 per cent of farms are commercially marginal. The term 'marginal farm' is best understood as meaning full-time units with very low incomes (Frank 1983). Such farms are unable to provide a full-time living comparable to that obtainable in non-farm activities for a number of reasons, including excessive land fragmentation and/or operating in very poor environmental conditions. Often the farmer is ageing but is unable to, or chooses not to, give up the farm because it offers a home and security, or because of strong sentimental attachment. It is difficult to estimate the number of such farms, but an EC survey in 1975 showed that only 50 per cent of part-time farmers had off-the-farm jobs, while the remainder, presumably, were underemployed and probably aged (quoted in Frank 1983). Marginal farming survives in a number of regions, including parts of the Scottish Highlands and Islands (crofting), the Alps and Scandinavia. Writing about Scandinavia, John (1984, p. 134) states that '. . . in many remote areas isolation and harsh physical conditions make large-scale commercial farming operations quite impossible and small farmers persist in the use of techniques which may not have changed for centuries.' Arguably, such farms may represent the last vestiges of a true peasantry. Whether it is in its final phase, as Franklin asserted, is more difficult to assess.

Part-time farms are Franklin's third category. Some commentators (for example, OECD 1972a) have viewed part-time agriculture as only a transitional stage for individuals moving from full-time farming to full-time non-farm employment, but this is dubious. To begin with, part-time farming is still widespread (see Table 18), especially in the Alps and Scandinavia. Furthermore, part-time farming, far from being transitional, may represent one very logical survival strategy in the modern capitalist economy. This can take one of two forms, associated with very different types of part-time farms: those with second businesses and those involving worker-peasants.

Table 18 *Full-time and dual-jobholders in selected countries in western Europe (percentage figures)*

| | *Full-time* | *Dual-jobholders* | |
Country		*Mainly dependent on farm jobs*	*Mainly dependent on non-farm jobs*
Austria (1976)	39.5	4.3	56.2
Norway (1972)	34.4	21.5	44.1
Switzerland(1975)	48.6	9.1	42.3
Belgium (1970)	56.7	9.1	34.2
Finland (1969)	63.1	20.1	16.8
France (1975)	80.2	4.4	15.4
Ireland (1972)	77.8	22.2	
Italy (1970)	62.4	5.0	32.6
Netherlands (1975)	74.1	6.3	18.4
F.R. Germany (1978)	48.0	13.2	38.8

SOURCE: OECD (1980).

Part-time farming with a second business is a traditional element of the rural economy, whereby agriculture was combined with brewing, cooperage or other trades. This was effectively destroyed by factory production in the nineteenth century (Mottura and Pugliese 1980) but it is now reappearing in new forms. For example, in the UK many farms also run such businesses as bed and breakfast, farmhouse teas or farm shops (Gasson 1977a). In extreme form, there is growth of 'hobby farming', whereby wealthy professional and business persons run small farms for weekend recreation rather than as serious farming enterprises (Gasson 1966).

In the worker-peasant farm the family's principal employment is in factories or other non-agricultural activities, with the land being worked in the evenings or at weekends. This has necessarily been important in Scandinavia (see Table 18) because climatic conditions limit farm work to the summer so that, during the winter, jobs are taken in rural industries such as sawmills (Hackman 1976). Worker peasants are also important elsewhere in Europe, representing (of all farmers) 15 per cent in France and 30 per cent in West Germany. A recent survey (Moss 1980) has also shown the importance of this type of farm in Northern Ireland; almost a half of the farmers obtained more than 50 per cent of their income from off-the-farm activities, mainly as full-time employees in the building industry, transport or factories. Similar processes are also evident in southern Europe, with industrialization providing more off-the-farm jobs (Jones 1984). For example, Agostini (1977) has shown the importance of part-time farming in rural–urban fringe areas in Italy, while Cavazzani (1976) has recorded the growth of female participation in agricultural work in Marche, in consequence of growing off-the-farm male employment.

Worker-peasants occupy a special role in the relationships between the agricultural and non-agricultural sectors of the economy in some regions. On the one hand, they use earnings from off-the-farm jobs to maintain the peasant-family on the land; this represents a form of subsidization of agriculture by industry and service activities. The non-agricultural earnings may be channelled into investment to modernize the farm, but they can also be used for consumption and raising living standards. On the other hand, the relationship can be viewed as part of a system which sustains industry. In northern and central Italy a system of dual job-holding in farming and industry acts to the advantage of manufacturing firms, because it enables relatively low wages to be paid (see pp. 291–4). Employees laid-off at times of recession also have some farm income to fall back on.

The state and agriculture

The framework of national and supranational policies

Agriculture is a key area of state intervention both because of its strategic importance and because of the ability of farmers to exert pressures on government. Its economic role includes providing a reserve of cheap labour for industry, cheap food, and raw materials for agro-processing firms. As a result, agriculture in most western European countries is subject to measures to regulate prices, subsidize producers, modernize farming methods and reform land ownership. These are mostly organized at state level, but they are complemented by EC policies.

While the agricultural community exerts pressure on policy making, this is often diffuse because of the varied interests of different types of farming. Small and large producers, arable and livestock farmers, suppliers of home markets and exporters all have different requirements (Buttel 1980), and the difficulty for the state is to balance these against each other and against non-farming interests such as food consumers, agro-processing industries and conservationist groups.

Farmers and their organizations influence policy making in a number of ways, including the use of political parties and direct action. In terms of political party affiliation, the two extremes are represented by the UK and Sweden (Wilson 1978). The National Farmers Union (NFU) in the UK has adopted an apparently apolitical stance, preferring instead to develop very strong institutional links with the Ministry of Agriculture, to the exclusion of other interested parties such as farm workers and conservationists (Newby 1979). In practice, this is supplemented by strong informal ties between land-owning interests and the Conservative Party (Roth 1973). Sweden offers a contrast, for the Swedish Centre Party actually started life as a political party to advance farmers' interests; only later was its appeal broadened. Between these two extremes are France and Italy where farmers' groups seek to use both institutional and party channels of influence; this may involve shifting allegiances between parties – as Philipponneau (1975) has demonstrated in the case of Brittany – in pursuit of self-interest. The effectiveness of the farmers' lobby depends, among other considerations, on how united they are. In the UK the NFU, despite criticisms from smaller farmers, has gained recognition as the representative association of farmers; but in Italy the main organization, Coltivatori Diretti, has been strongly opposed by the Communist Party-backed National Peasants Alliance. Farmers have also been able to exert pressure on the state, in such countries as France and West Germany, because of their electoral significance, nationally and regionally.

When electoral pressure or party political contacts have failed to achieve the desired ends, farmers have not been loath to resort to direct action so as to embarrass their governments. Opposition to imports has been the main focus of such action; French farmers have destroyed imports of Italian wine and Spanish fruit, Welsh farmers have (in 1975) prevented imports of Irish cattle, while the EC has been subject to some form of direct action by farmers from virtually every member country. By and large such pressure has been effective, and perhaps no other group has had such a disproportionate influence on policy making in Europe.

State policies have sought to secure a number of objectives (Buttel 1980; Grigg 1984; Mann and Dickinson 1980; Marsh and Swanney 1980), in terms of the interests of farmers and consumers (although the former have usually held precedence). The aims include increasing output and improving national self-sufficiency for strategic and balance-of-trade reasons. Stabilizing prices for farmers faced with climatic fluctuations (see Rees 1978) and securing reasonably cheap food for consumers have also been common objectives. Other aims have focused on employment – either to help maintain the population on the land, or to secure the release of labour for expanding sectors of the economy. Finally, most governments have sought to secure decent standards of living for farmers, in the face of inelastic demand and technological advances combining to make supply outstrip sales, thereby reducing prices. Many of these aims conflict with one another; reasonable prices for consumers may not guarantee decent standards of living for farmers, while increasing output may not be commensurate with releasing labour from agriculture. In short, it is impossible to avoid conflict between the policies which are necessary to secure output, price and employment goals. The actual resolution of these potential conflicts and the mix of policies varies between states, but in much of western Europe it is made more complex by the existence of a supranational policy, the EC's Common Agricultural Policy (CAP).

The Common Agricultural Policy
The role of the CAP has often been exaggerated, for expenditure on this is less than a half that spent on agriculture by national governments (Cuddy 1981); but it is a key element influencing agriculture in much of Europe (and beyond). While details of policies are given later, some general features of the EC's agricultural programme can be noted here. Development of an agricultural policy was only one of the original principal aims of the EC, but it was ultimately to become the dominant policy. Title II (Articles 34–87) spelt out the aims of such a policy (Hill 1984): to increase agricultural productivity, to ensure a favourable standard of living for the agricultural community, to stabilize markets, to guarantee food supplies, and to secure reasonable prices for the consumer. These aims were contradictory and very general, the real details of CAP being left to detailed negotiations at a later stage.

Negotiation of the CAP was a long and difficult process because of the diverse interests of the member states: the Netherlands wished to expand their modern exporting sector, Belgium wanted defensive subsidies, while Germany, Italy and France sought to protect their marginal, fragmented farms (Blacksell 1981). Matters were further complicated because there were diverse regional interests within individual countries; thus the French Government had to balance the needs of large-scale farmers in the Paris basin against those of a still largely peasant agriculture in the south and west. Agreement on the CAP was eventually secured in January 1962. The main element was a system to guarantee prices, to be implemented nationally at first, but on a Community-wide basis after 1968; this owed much to French and Italian pressure (Blacksell 1981). In practice, prices tended to be set well above current world market levels, and this led to two major problems: higher prices for consumers, and a massive increase in output, yielding 'permanent' and expensive surpluses of some products. This has meant, among other things, that very little EC expenditure has been available for the EC's guidance (or structural reform) agricultural policy and, indeed, for the whole range of non-agricultural policies. Enlargement of the EC has only served to exaggerate these difficulties (Tsoukalis 1981). Details of the CAP and of individual state's policies are investigated in the following sections.

Price support and modernization

Probably the greatest problem for the state has been to reconcile a need to raise or stabilize farm incomes with a relatively inelastic demand for food, both within Europe and on world markets. Historically, farm incomes have been low and, in the 1950s, were less than 80 per cent of non-farm income, in most European countries (Slattery 1966). However, supply tends to outstrip demand since, with rising incomes, families tend to spend proportionately less on food. For example, Pollard (1981) estimates that between 1960 and 1970 the proportion of consumer expenditure on food fell from 35 to 30 per cent. Europe is not alone in this, for the same predicament faces the US Government which also stockpiles costly food surpluses. As a result of global overproduction (relative to demand if not need), and lack of competitiveness in some forms of farming, most countries in Europe have adopted protectionist agricultural policies. These have usually involved some form of intervention in price-setting. In the 1980s this has led to increased global conflict over restrictions on agricultural trade, especially between the EC and the USA.

Price support is not the only form of state response to these problems, for governments have also resorted to demand management (advertising, subsidies to consumers, etc.) and supply management (Bowler 1979). Import restrictions are the main means of supply management, and these were adopted almost universally in the face of the 1930s agricultural crises, even by normally 'free-market traders' such as Denmark and the Netherlands. However, price guarantees – effectively subsidies to producers – have been the main strategy for supporting farm incomes. In the UK the 1947 Agriculture Act established a deficiency payments system whereby annual reviews set guaranteed prices. Consumer prices were fixed in the market but, if these fell below the guarantees, the state paid the difference to the producers. When the UK entered the EC this system was replaced by a very different method of farm price support.

The EC's Common Agricultural Policy is worthy of detailed consideration, not only because it regulates farming in so many countries, but also because of its enormously controversial nature. The CAP has two main branches, guidance and guarantee sectors; the latter is relevant here.

Essentially, the guarantee sector involves central determination of target prices for most agricultural products; these are fixed at national levels plus 'the transport costs to Duisburg' (in West Germany), which is notionally taken as the point of shortest supply. If import prices are less than target prices, then importers pay levies to bring the costs of imports up to EC levels; as the levies are based on the cheapest import prices on any one day, most imports are priced above EC levels. Should the market fail to clear available production, then the EC intervenes to purchase and store what, supposedly, are short-term surpluses. These are bought at intervention prices, which are usually a few percentage points less than target prices. In order to clear surpluses, exporters are paid export refunds for any sales outside the Community at less than EC prices.

While these various prices are supposedly applied uniformly throughout the EC, they are in fact distorted by 'green currencies' and Monetary Compensation Amounts. These are the differences between real exchange rates and the fixed exchange rates by which prices are calculated. As real exchange rates fluctuate considerably over time, this allows states to vary the levels of prices paid to both producers and consumers in accordance with national needs. Some countries, such as the UK, have usually manipulated the system so as to favour consumers, while others, such as West Germany, have favoured producers. In addition to guarantee prices, the EC uses more specialized policies to help support particular groups of farmers. Examples include the compensation (for natural handicaps) paid to hill farmers under the 1975 Less Favoured Areas directive.

How successful are the CAP and other systems of price support? The CAP has been very successful in boosting output, which is not surprising given that target prices have been set above world levels in most years. As a result, self-sufficiency in the EC in many products has increased sharply; for example, between 1956–60 and 1980, self-sufficiency in wheat rose from 90 to 114 per cent, in sugar from 104 to 125 per cent and in beef from 92 to 103 per cent (Hill 1984). This has reduced EC imports relative to inter-member trade, so that well over half of the food exports of Belgium, France, Italy and the Netherlands are now destined for other member states. The guarantee policies have also been relatively successful in stabilizing prices and in providing reasonable standards of living for farmers. However, as prices have been set at the margin necessary to maintain less-efficient small-scale producers, prices in the 1960s and early 1970s were usually well above world market levels. Although the price differential was partly eroded by a rise in world prices in the mid-1970s, the gap has again widened subsequently (see Figure 13); for example, in 1978/9 EC prices compared with world prices were double for maize and for beef, and quadruple for butter. Milk products have been especially protected and accounted for over 40 per cent of total expenditure on CAP at that time.

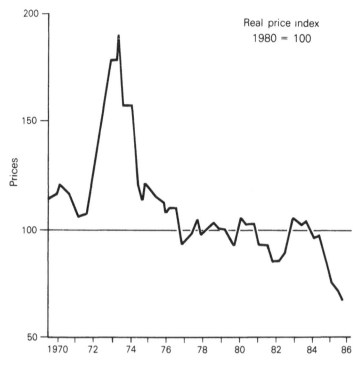

Figure 13   World prices for food commodities, 1970–86

SOURCE: IMF.

On the credit side, then, CAP has fulfilled many of its objectives. On the debit side, however, are the enormous costs of the programme. In common with most domestic protectionist policies, these have 'generated a siege mentality', whereby costs and efficiency are of secondary importance to increasing output (Clout 1981a, p. 88). CAP policies have certainly been fairly insensitive to changing conditions of both production and demand. Huge surpluses have had to be stored, often for years rather than the short periods which were originally expected. This has given rise to the notorious 'mountains' and 'lakes' of food and wine which often cost more to store (or to destroy) than the producer was paid originally. For example, it cost $1.5 billion to buy-in 500,000 tons of surplus butter in 1986. Annual storage costs are about $200 million, yet when eventually sold the butter could realize as little as $50 million.

In addition, the operation of CAP has led to higher prices for the consumer than would have occurred under free-market conditions or, in the case of the UK, if the deficiency price system had been adhered to. Another problem has been continued heavy reliance of some members – especially the UK – on food imports from third-party countries. This has led to high levels of levies on importers in the UK, which has been a major component in its much-publicized budget deficit relative to the EC.

Another criticism is that, despite the fact that aggregate farm incomes have kept pace with non-farm incomes (Klatzmann 1978), the income-support effects of CAP and state policies have been very uneven. Indeed, Hill (1984) considers that income differences between farmers have widened, with the smaller-scale producers and the least-favoured regions performing least well. For example, in Ireland between 1955 and 1977, while the incomes of 5–15 acre farms increased by 2.7 times, those of farms over 200 acres increased by 5.6 times. Nevertheless, although the effects are uneven, the price support system may be just sufficient to keep farmers on the land and, hence, slow down the process of modernization. Certainly an argument can be made that, in comparison with the USA, or the UK before 1972, the CAP has achieved fewer productivity gains than have been secured in other countries.

In response to shortcomings in the price support system, many governments have developed specific policies for modernization. These assume various forms, including the establishment of cooperatives, the provision of advisory services and grants for investment in machinery. Frequently, such modernization policies may be linked to larger regional reform packages, such as the land colonization schemes in Franco's Spain or the consolidation of fragmented plots in West Germany (see p. 146). The EC has developed a number of such structural reforms within the guidance sector of CAP. For example, initially a third of the guidance funds went to assist irrigation and marketing schemes in Italy's Mezzogiorno. Under a later EC farm directive, more widespread grants were made available to modernize low-income farms in 1982 so as to bring them up to the equivalent of providing full-time employment for one person. In 1973, a more selective approach was taken, with funds being channelled into agriculture and into creating non-farming jobs in disadvantaged areas, particularly southern Italy, western France and upland Britain. The effects of such policies, and of the integrated development approach introduced for the Mediterranean region in 1978, seem to have been relatively modest (Podbielski 1981)

Land reform: latifundia and colonization

Land tenure in many regions of western Europe has been inherited from the medieval period and, not surprisingly, is poorly adapted to the needs of modern agriculture. Italy represents a classic example of this, and King and Took (1983, p. 187) write that 'land tenure is the historical foundation of Italian rural social structure . . . . The medieval structure of lord, vassel and serf survives today as the hierarchy of landowner, tenant and labourer.' The actual form of land tenure varies considerably between countries, and in some cases, such as the UK and Denmark, had been subject to thorough reform well before the present century. In southern Europe, the classic divide is between latifundia and minifundia, while fragmentation and pulverization is characteristic of much of northern Europe. Not surprisingly, land tenure has figured prominently in the field of land reform and, in this section, we concentrate on one particular element, the latifundia.

Latifundia have been an important element in the economic, social and political structure of southern Europe. They are the dominant form of landholding in Portugal's Alentejo, Spain's Andalucia and Extremadura, southern Italy and parts of Greece. Their precise form varies but usually involves large estates (over 500 ha being not uncommon), sometimes held under absentee ownership, and either worked by a casual labour force employed on a weekly or seasonal basis, or rented out in small parcels on very short leases. Such a system usually makes for a relatively backward agriculture, with low investment, either because there is abundant cheap labour or because tenants have little interest in improving land held on short leases. Their output per hectare is relatively low, there is chronic un- or under-employment, and they are often characterized by traditional dry Mediterranean farming, involving cereals, olives and sheep. This type of economic system maintains a very polarized social structure with a small, wealthy (sometimes aristocratic) landowning class being opposed to a large rural proletariat. Politically, the landowning classes have tended to be closely associated with, and to have openly supported, dictatorial governments, such as Franco's in Spain or Salazar's in Portugal, while the proletariat has shewn affiliation for left-wing parties.

In the twentieth century, for a number of reasons, governments of different persuasion have attempted to reform the latifundia system, in order to break support for the political left and/or to increase the efficiency of agriculture. Frequently this has been linked to the needs to colonize land so as to settle or resettle large numbers of landless persons. Usually, this involves governments operating under duress, such as in Greece in 1923 when one million people displaced after a disastrous war with Turkey had to be absorbed (King 1977), or as in Finland where 250,000 refugees from Karelia were settled on 50,000 new holdings in 1945 (Mead 1951). The processes involved are illustrated by the contrasting experiences of Italy (an 'effective' top-down reform), Portugal (revolutionary land seizures), and Spain (a failure to reform).

Southern European case studies
In Italy, the full feudal social structure broke down in the late eighteenth and nineteenth centuries but, as the peasantry lacked sufficient capital to buy their own farms, large estates became dominant (King and Took 1983). This produced a highly polarized landholding system: in 1933 there were 3.79 million holdings smaller than 0.5 ha, while only 47,000 farms had over 50 ha although owning 41 per cent of all land. The latifundia system was especially important in the south where it took two distinctive forms (King 1973): *latifondo capitalistico*, mainly in the coastal plains, reliant on

wage labour to produce cash crops, and *latifondo contadino*, mainly in the mountains, leased under share-cropping agreements. Agricultural incomes were generally low and wage-labourers often worked no more than about 100 days a year.

Land reform laws enacted in 1950 provided for land redistribution in selected land reform areas, mainly located in the south. Estates of over 300 ha which were under- or un-cultivated could be broken up into smaller parcels and distributed to individual farmers. In all, some 700–800 thousand hectares were redistributed (Bethemont and Pelletier 1983) to some 113,000 farmers, including about 45,000 casual labourers. In addition, a further 69,000 parcels of land were redistributed or transferred from renting to ownership by small-scale farmers (Clout 1972; King 1971). As part of the reform 'package', each farmer was supposed to receive a new farmhouse and assistance with farm management; these were to be paid for over 30 years.

The success of land reform has to be evaluated in terms of a number of criteria. In political terms it was a resounding success for the Christian Democrats, destroying both the influence of the Communist Party in the south and the power base of the rural aristocracy, hence creating a vacuum in which they could establish a system of clientelism (Allum 1981). However, the economic role of the reform was more ambiguous. Its effects were limited, even in quantitative terms, and rarely affected more than 10 per cent of the land or rural population, even in the reform areas. Indeed, much of the land reform programme was really land reclamation work on previously unfarmed marshland. In national terms, it directly affected only 3 per cent of the territory and 1 per cent of the agricultural population, so that King (1973, p. 224) concludes that it was 'a small step and one that did not basically affect the socio-economic structure of much of rural Italy.'

A major problem was the smallness of the farms created. Most were less than 8 ha and too small to suport a decent family income and reduce the farm's indebtedness, let alone produce a healthy surplus. Not surprisingly, then, some 15 per cent of farmers had already abandoned their new farms by 1962, while others could only continue farming by becoming worker-peasants. In a sense, land reform was too little too late; it created a peasantry at the very time when growing penetration by capitalism would make this non-viable in the face of scale economies in production and the requirements of commercialization. Many part-time farmers were actually supported by welfare benefits or by jobs created by the state in, for example, forestry. This was part of a broader process whereby the Mezziogorno became a welfare region (Pugliese 1985; see pp. 258–9).

In Portugal land reform came later and under very different conditions. The Salazar regime drew political support from both the peasant farmers of the north and the latifundist estate owners of the south. Consequently, it was careful not to alienate either group by implementing the type of radical land reforms that were necessary to permit modernization of the country's grossly inefficient agriculture (see p. 117). Soon after the military coup in 1974, however, agriculture was drawn into a rapid (at times revolutionary) series of political changes. The Alentejo had long been a focus of left-wing opposition to the Salazar regime, and it was here that, in December 1974, the first latifundia estate was occupied by local workers. Between January and March 1985 the occupations became widespread throughout the south.

Central government became alarmed and passed the Agrarian Reform Law in 1975 to regulate land seizures. This allowed some land on larger estates to be 'collectivized', especially if it was under-cultivated. By 1977 some 900,000 hectares had been legally collectivized, together with some illegal seizures, and over 500 collective or cooperative farms had been established (Abreu 1980; Barros 1980). However, by 1977 the era of revolutionary politics was over and the socialist government moved to contain the land reform movement and – it hoped – Communist Party influence in the Alentejo. The Lei da Reforma Agraria in 1977 restricted collectivization to the south and limited the amount of land which could be taken over from existing owners. This marked the peak of the land reform movement and, subsequently, both centre-right and socialist governments have used legal procedures, backed by force where necessary, to decollectivize agricultural land. Thus, by 1984, there were only 330 collectives, with some 400,000 hectares of land.

It is difficult to assess the impact of the land reform movement, given the poverty of official statistics and the highly emotive, political nature of the issues. Land and labour productivity did generally improve and some collectives played an innovatory role in the adoption of new crops such as tobacco. But there was also evidence of mounting costs and of credit being diverted to pay wages rather than being used for essential investments, while many farms were poorly managed because their leaders lacked adequate training. Perhaps the most appropriate assessment is that land reform was far more important politically than economically. The major reason for this was that collectivization occurred independently of, and increasingly against, the wishes of central government. Consequently, essential complementary investment in management and training, agro-processing industries and/or irrigation were never undertaken. In short, land reform never became a wider agrarian reform as had occurred in Italy, no matter how belatedly.

Latifundia were also characteristic of much of central, southern and western Spain; in 1930, over 30 per cent of the land in this region was in estates of over 250 ha. There was considerable agitation for land reform in the late nineteenth and early twentieth centuries, but this was resolutely opposed after 1940 by the Franco regime, which drew political support from the landowning classes. Instead, the regime's policies were directed towards irrigation and colonization schemes. Landless peasants were to be settled in dry farming areas which would be transformed by irrigation. By 1980, 1.35 million hectares of dry farming land had been irrigated, including very large-scale schemes such as those for the Ebro, Jaen and Badajoz basins. However, only 230,000 hectares were redistributed among only 27,000 families, so that 'the state's hydrological policies served mainly to increase the capital assets of the larger holdings located in the newly irrigated areas' (Garcia-Ramon 1985, p. 130).

The 1952 Badajoz plan aimed to irrigate 130,000 hectares and establish between nine and ten thousand new farms, each of 4–5 ha in an area previously characterized by dry monoculture of cereals, low yields and casual labour (Naylon 1966). The scheme did bring benefits, such as increased local employment and the introduction of new and diversified products. But these advantages were not spread as widely as anticipated, with only 32 per cent of the newly irrigated land being redistributed. Programmes were also applied insensitively, for vines and olives were ripped out in some areas even though poor soils meant that these were the most appropriate crops. Irrigation, therefore, does not necessarily hold the key to agricultural modernization, and Cañizo (quoted in Naylon 1973, p. 17) was moved to write that 'water, which is gold in Levante and silver in Aragon, is no more than copper in Castile'. Indeed, irrigation and colonization were treated as substitutes for, rather than complements to, land reform. Eventually the latifundia were transformed, but this was to be through commercialization in the 1970s and 80s rather than through land reforms.

The case studies all display, in their own ways, the intractable economic and political problems involved in, and the potential advantages stemming from, land reform. In the mid-1980s it seems that the need for land reform as a precondition for modernization has been overtaken by the introduction of capitalist relationships in other ways. The case for land reform for social and redistributional reasons is, however, a very different matter and remains an important issue.

Land reform: enlargement and consolidation

The reverse of the problems of the latifundia is fragmentation (very small farms) and pulverization (a farm's land being divided into many small and spatially scattered plots): for a discussion of the semantics surrounding the use of these terms, see King and Burton (1982). These are features which are widespread throughout Europe, both north and south. It is a problem in so far as many such farms are economically non-viable, at least in terms of agricultural production, although they may be sustained by part-time off-the-farm employment (see Table 18). The extent of the problem was highlighted by a survey in 1960 (OECD 1964) which showed that over 50 per cent of farms were economically non-viable in France, West Germany, Greece, Italy, the Netherlands, Portugal, Norway and Italy.

Fragmentation

The two features of fragmentation and pulverization will be considered separately here, although they are often interlinked. The roots of fragmentation are varied but lie mainly in demographic pressures acting upon a fixed resource (land) and in inheritance laws requiring division of farms between more than one descendant of a deceased owner. In terms of size (see Figure 14), the most extreme cases of fragmentation in 1960 were

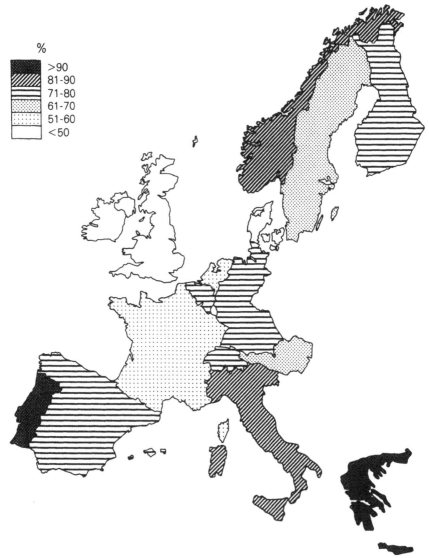

Figure 14   Percentage of farms with less than 10 ha in 1960

SOURCE: OECD (1964).

Greece, Norway and Portugal, where over 90 per cent of the farms were smaller than 10 ha; but West Germany, Finland, Italy and Spain also experienced the same tendency. A more up-to-date picture of landholdings in the European Community in 1982 is given in Table 19; although collected on a slightly different basis, these figures reveal very few major alterations in the basic pattern, even though farm sizes generally have increased since 1960. Italy and Greece still display the highest levels of fragmentation, while Denmark, the UK, Luxembourg and Ireland have the largest farms. France shows evidence of considerable farm-enlargement tendencies, but more than 50 per cent of all farms in West Germany are still smaller than 10 ha. There are, of course, considerable regional variations within this broad cross-national picture (Hudson *et al.* 1984).

Table 19 *Farm size distribution in the European Community in 1982*

| | Percentage distribution by size (ha) | | | | |
	1–5	5–10	10–20	20–50	> 50
Belgium	28.2	19.3	26.3	21.8	4.6
Denmark	2.6	18.3	27.8	38.6	12.6
F.R. Germany	31.9	18.4	22.5	22.8	4.4
France	20.2	13.6	20.6	31.4	14.2
Greece	70.9	20.6	6.5	1.7	0.2
Ireland	15.2	15.9	30.3	29.8	8.8
Italy	68.5	17.2	8.4	4.2	1.7
Luxembourg	17.3	10.4	14.1	37.7	20.5
Netherlands	23.8	19.8	28.3	25.0	3.2
UK	11.8	12.3	15.8	26.9	33.2

SOURCE: Eurostat (1985b).

Smallness is not necessarily a disadvantageous economic feature of farms. There is no such thing as the optimum or ideal sized farm, for this varies according to regional conditions of production as well as according to product. As an extreme example, in the UK the average net income per hectare from horticulture is twenty times that from sheep rearing (Grigg 1984), so that a specialized horticultural farm of about 10 ha may be economically viable. Indeed, this is the case in the Netherlands and Belgium where smallness belies highly efficient and intensive market gardening. However, smallness in many countries does pose some problems for economic development and for household incomes. Small farms often lack access to the credit necessary to finance modernization. Furthermore, the economies of scale of modern farm machinery may make it non-viable to change farm practices on very small holdings. The problems are compounded in that any policy successes in terms of farm-enlargement may be overtaken by economic trends requiring even larger minimum holdings.

Policies for farm enlargement have usually involved two different general measures, either singly or in combination. These are incentives for some farmers to quit agriculture so as to release land for enlargement of other farms, and controls on the sale of land. For example, farmers may be prohibited from selling land to anyone other than neighbouring farmers, or to the state which assembles a land-bank for enlarging other local farms. Given the strong traditional peasant attachment to the land, it is not surprising that such policies have proven difficult to implement. This can be illustrated by policies adopted in a number of countries and by the EC.

In West Germany fragmentation is a long-established problem in many areas, but especially in Saarland and Baden-Württemberg. The first major attempt to ameliorate the problems came with the 1919 Land Settlement Law which gave the state priority in purchasing any farm land offered for sale. This was used to set up younger sons in farming, so as to reduce the pressure for sub-division of their parents' farms (Hirsch and Maunder 1978). There was a general emphasis on maintaining the family farm, and this only changed after 1968 when, in the face of economic development, more attention was devoted to developing larger holdings which provided adequate household incomes. This was to be achieved through farm amalgamation and the creation of non-agricultural jobs so as to stimulate land transfers (Mellor 1978). Farmers who retired early and released their land were also paid state annuities.

Policies in France have followed broadly similar lines. The state established the *Sociétés d'Aménagement Foncier et d'Etablissement Rural* to buy land which could be used for farm enlargement. Greater fluidity in land markets was also encouraged after 1962 by the *Indemnité Viagène de Départ*, a termination payment for those quitting farming. Between 1963 and 1968 alone some 123,000 heads of farms accepted such payments (Franklin 1971), but this may only have financed retirements which would have occurred anyway (Bowler 1979). This is substantiated by Clout's (1975a) study in the Massif Central, which showed that the mean age of farmers receiving these payments was 69.

Denmark, too, has, since 1973, had policies of supporting smaller farms, especially in the face of demand for rural housing and land from non-farmers. In addition to measures encouraging amalgamation, strict controls have limited sales of farms to individuals (hence, excluding institutions) practising full-time agriculture (Hirsch and Maunder 1978). In Sweden, twenty-four County Agricultural Boards have power under the 1955 Land Acquisition Act to encourage farm enlargement, but in practice their influence only accounts for one-third of amalgamations; the remainder have come about through market forces, especially because of farmers leaving for other jobs.

While most farm enlargement policies have been the concern of individual countries, the EC has also developed similar measures under the guidance section of the CAP. The 1968 Mansholt Plan proposed taking 5 million hectares out of agricultural production, mainly in southern France, the Massif Central, Corsica and southern Italy (Hill 1984). It wanted to encourage retirement of elderly farmers with small plots who were being cushioned and kept in farming only by guaranteed prices. In the face of rising unemployment at that time (limiting non-farm job opportunities), and domestic political pressure in Germany and France, the proposal was withdrawn. When the EC did finally agree a policy on this, in 1972, it was a relatively weak compromise. Grants were made available for modernization and for training, but the only really influential measure was payment of lump-sums and pensions to farmers aged over 55 who quit agriculture and sold or leased their land. The emphasis on leasing was important, recognizing the peasants' traditional reluctance to relinquish ownership, even after retirement.

The overall effects of these policies are difficult to ascertain. Farm enlargement had been in progress in many countries from the nineteenth century. Enlargement policies, at best, only assisted a shift which was occurring anyway. The same is true of the incentives for retirement, and, in most countries, there is evidence that these only marginally speeded up the departure of those who had already decided to leave farming. Furthermore, the way in which the land released is utilized is also important. By and large it seems to have been purchased by, or allocated to, the more progressive and innovative farms rather than to those in greatest need of more land (Bowler 1983); this may have served the requirements of modernization but hardly those of equity.

Pulverization
Pulverization is the sub-division of farms into large numbers of dispersed, small plots. This is not a problem in Denmark, Sweden or the UK, where the medieval structure of small, separate field plots was swept away at a much earlier stage. However, in western Europe as a whole it was estimated that more than a half of all farmland needed consolidation in 1950. The average number of plots per farm was four in Italy, six in Portugal, seven in Greece, ten in West Germany and fourteen in Spain (OECD 1969), but these figures conceal enormous regional differences. In Spain, for example, Galicia and the Basque country exhibit high levels of both fragmentation and pulverization (Guedes 1981). In La Estrada in Galicia, O'Flanagan (1982) found that there were 7500 holdings with an average of 30 parcels each. An extreme case is described by Naylon (1959; 1961), that of a 66 acre farm in Zamora sub-divided into 394 parcels, while O'Flanagan (1982) found truly remarkable sub-divisions in the village of Orense in Galicia, for land under chestnut trees was worked by different individuals according to the seasons.

Pulverization is the product of many diverse processes. First, it represents fossilized survivals of a medieval open-field system which is very poorly suited to the needs of modern, mechanized (perhaps monoculture) farming. In addition, partible inheritance laws requiring sub-division of the land between all heirs may lead to excessive pulverization with, in extreme cases, even orchards and rows of vines being divided. Elsewhere, the process of piecemeal reclamation from wasteland may lead to farms acquiring dispersed plots.

Sub-division can be a logical response to environmental conditions – perhaps the needs for transhumance in Alpine regions – or to economic conditions. For example, King and Burton (1982) argue that renting out smaller parcels of land is one way for landowners to maximize their returns in the face of increased demand for land stemming from, say, demographic pressure. Additionally, dispersed holdings offer a way of increasing the range of crops produced and, hence, of spreading risks associated with climate and spreading workloads over the year (Galt 1979). Such conditions often occur in Alpine regions where variations between fields in terms of height and aspect may allow very different types of crops to be grown.

Nevertheless, in most countries pulverization is regarded as a hindrance to development. It can be inefficient in the use of labour time, with long journeys necessary between scattered fields. Moreover, it is inefficient in the use of land itself and, in extreme cases, up to 20 per cent of this can be lost in boundaries (King and Burton 1981). Furthermore, drainage and irrigation projects are more difficult to implement because they require collective action by a number of owners of adjacent fields rather than independent activity by a single farmer. Such an arrangement of property rights will, inevitably, increase the potential for social conflict among owners; this is highlighted in parts of Greece, where up to half the plots are accessible only by trespassing on other people's land (King and Burton 1981).

A number of European governments have developed consolidation schemes to overcome some of the economic difficulties of pulverization. The earliest reforms were in Finland and Denmark, followed by the UK, where enclosure of most of the remaining open fields took place in the eighteenth and nineteenth centuries. However, most consolidation schemes post-date 1945 and they involve varying degrees of state involvement. In Italy, the role of the state is largely that of encouraging individually organized exchanges within the institutional framework provided by cooperatives (Cesarini 1979). In France, Greece, Spain and Switzerland the state may actually organize exchanges, but only if invited to do so by at least half the farmers in a locality. This contrasts with West Germany and the Netherlands where the state has greater powers of compulsion (Mayhew 1970). Two examples are considered here in more detail.

In France and official policy of *remembrement* was declared in 1941 (Clout 1972; Perry 1969), at which time it was estimated that there were 14 million hectares (40 per cent of all farmland) in need of reorganization. By 1969, over 6 million hectares had been consolidated. Under this scheme, more than three-quarters of local landowners had to request consolidation before the state intervened to organize the exchanges. In practice most exchanges have been in the Paris basin, with little reorganization in the south or west where the need was, perhaps, greater, but the operations more costly.

Spanish farms have a very high level of pulverization, with the average number of plots varying from 1.8 in Cadiz (an area of latifundia) to 56.9 in Soria (Guedes 1981). In 1952 the Servicio Nacional de Concentracion Parcelaria y Ordenacion Rural was set up to help consolidate plots, reduce travelling from farm to fields, and provide technical information and infrastructual investment to modernize farming. O'Flanagan (1982) gives an example of the reforms undertaken in La Estrada in Galicia. Between 1963 and 1973, the average number of parcels was reduced from 30 to 6 and the number of farms fell from 98,000 to 22,000; even so, the average farm size remained less than 4 ha. In sheer quantitative terms, this seems to have been the most successful national scheme, with approximately 5.2 million hectares being consolidated by 1982 (Garcia-Ramon 1985). However, the plot exchanges were rarely accompanied by improved marketing or the introduction of new crops, so the overall advances in farming were less than expected. Similar schemes exist in most European countries.

Consolidation schemes do contribute to improving agricultural efficiency (OECD 1972b) and they are generally important in permitting a more rational approach to farming (King and Burton 1982). However, consolidation is a slow process and difficult to implement in the face of often complex legal and traditional landowning and land-use rights. Moreover, state intervention has often encountered fierce opposition because of the peasants' attachments to particular plots, or their fears about the quality of the plots being exchanged (Mayhew 1970). Furthermore, the underlying tendencies to pulverization may be little affected and sub-division may occur again unless – as in Germany after the 1961 Land Transactions Act – powers exist to control this. In addition, and this also applies to farm enlargement schemes, there is the constant problem that technological changes require ever larger and more consolidated fields for efficient operation. For example, in eastern England a single field on a modern cereal farm can be larger than the average farm in, say, southern Germany or northern Portugal. Given that in Spain, for example, the average size of consolidated farms was only some 5 ha, it is evident that reform is still struggling to keep abreast with the requirements of capitalist agriculture.

Integrated development
Land reform measures on their own are usually necessary but not sufficient conditions for agrarian reform, that is, a change in agricultural practices as a whole. Agrarian reform also requires investment in infrastructure, irrigation, marketing, training, and investment in non-farm jobs. It is little

wonder that, in the absence of these, the return on simple land reform schemes has sometimes been disappointing. There has been growing awareness of this failure in recent years and a number of countries have set up agencies for comprehensive rural development planning. Among these are the Cassa per il Mezzogiorno in Italy, the Fund for northern Norway, and the Highlands and Islands Development Board in Scotland. The policies are usually a mixture of agricultural improvements, and investments in manufacturing and tourism. The EC has also experimented with integrated rural development projects in the Western Isles of Scotland, Lozère in France and Luxembourg in southern Belgium, where CAP policy instruments were to be used in harmony with those of the Regional and Social Funds.

Fisheries: national interests and supply limitations

Fish is one of the few food products the consumption of which is increasing in developed countries. Yet, the industry is in a position of nearly permanent over-supply, except for a few species such as mackerel. Aquaculture may change this in the future but, in the mid-1980s it accounted for only about 10 per cent of all fish sales. Instead, traditional 'wild' catches are still dominant, but in this Europe's fishing industry has been faced with increasingly difficult conditions of production. Protective action by non-European states (such as Iceland and Canada) has increasingly restricted European fishing vessels to the north-east Atlantic, which accounts for about 80 per cent of the EC catch. In turn this exerts pressure on that area's stocks and the need to restrict catches. Mediterranean fishing grounds are important for Greece and Italy, but the potential of this zone has been severely reduced through pollution and over-fishing.

There have been increasingly severe restrictions on fishing in the North Sea. The reduction in the fishing fleets of Europe since 1976, the last year of relatively 'open' access, is given in Table 20 (also see Figure 15). France, the UK and Spain have all experienced large declines, numbering between 100 and 200 boats, and several other countries have experienced reductions. Only in Ireland has there been a major absolute increase in the number of vessels. The largest relative loss was in the UK, followed by France and West Germany. These reductions are partly a function of increasing scale and mechanization, most spectacularly exemplified by factory fishing boats, for catches have remained constant or even increased in most countries. However, employment has decreased sharply; between 1958 and 1981, for example, more than 80,000 jobs were lost in Italy and more than 40,000 in Norway (Wise 1984).

The actual effects have varied between countries, as have the forms of adjustment to the new conditions of production. This can be illustrated by some case studies (after Wise 1984). The Netherlands traditionally had a

Table 20   *Numbers of fishing vessels, and fishermen, 1976–82*

	Absolute number in 1976	Absolute change, 1976–82	Percentage change, 1976–82	Number of fishermen in 1982
Belgium	89	+ 7	+ 7.9	865
Denmark	358	− 19	− 5.3	14,500
Finland	10	+ 7	+ 70.0	N/A
France	607	− 125	− 20.6	20,177
F.R. Germany	151	− 33	− 21.8	5,229
Greece	104	− 13	− 12.5	46,500*
Ireland	27	+ 39	+ 144.4	8,975
Italy	247	+ 14	+ 5.7	34,000*
Netherlands	389	− 23	− 5.9	4,206
Norway	633	+ 14	+ 2.2	N/A
Portugal	202	− 23	− 11.4	37,251
Spain	1,844	− 196	− 10.6	106,584
Sweden	65	+ 15	+ 23.1	N/A
UK	630	− 187	− 29.7	23,358

* 1981 data.
SOURCES: OECD (1983b) and Commission of the European Community (1985h).

small but efficient fishing fleet and, although the number of boats fell by one-fifth in the 1970s, the tonnage of the fleet as well as its catch continued to grow. This has been facilitated by the dominance of large companies with sufficient resources to invest in modernization. Spain has the largest fleet in western Europe and even in 1980 had more than 100,000 fishermen. Excluded from traditional fisheries off North America and the coast of Africa, its dependence on the north-east Atlantic has increased; indeed, this was one of the most sensitive issues in negotiations over accession to the EC. France had a relatively poorly equipped fishing fleet in the 1970s and has paid for this with catches stagnating and employment being halved. Although there are differences in the experiences of these countries, they do share increasing dependence on state intervention and on international maritime negotiations.

State intervention in fishing

Although fishing accounts for less than 1 per cent of GDP in most European countries, it is still an important element in their economies. Nationally, the balance of payments contribution is important, and in 1980, for example, only Norway, Denmark and Ireland had a surplus of exports over imports of fish; all other countries – including those with traditionally

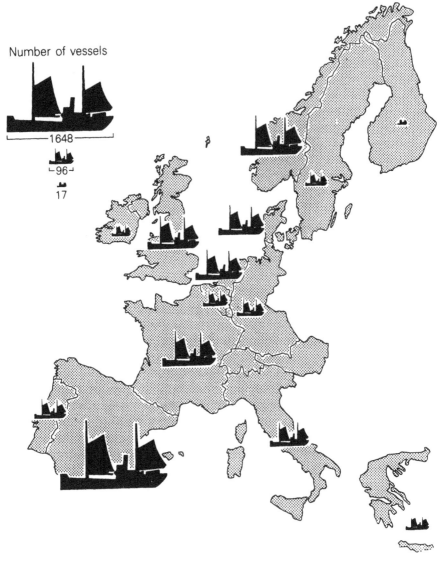

Figure 15 Number of fishing vessels in 1982

SOURCE: OECD (1983b).

important fishing fleets, such as the UK or Spain – were net importers. Employment in fishing is also important in a number of countries, especially Spain (Table 20). While direct employment in the industry is dwindling, indirect employment is still significant; in 1977, some 46,000 were employed in fish processing in the EC (Commission of the European Community 1985h). In addition, fishing remains vitally important in some local economies; and, as these are often in politically sensitive areas such as northern Scotland, the Basque country or northern Norway, the industry has been subject to state intervention.

There are three major elements in state intervention in European fisheries: definitions of territoriality, conservation controls and subsidies paid either for modernization or income-support purposes. These can be examined in the light of their historical evolution. The first international attempt to define territoriality was an 1882 agreement setting a three-mile limit off a baseline of low-water levels. Thereafter, further extensions were declared by those states which had particularly strong interests in fishing. In 1935 Norway declared an exclusive four-mile zone, which was enlarged to twelve miles in 1960. Iceland was next to pressurize for change, extending its boundaries to twelve miles in 1958, thereby provoking the first 'Cod War' with the UK. In the face of mounting pressure from individual governments, in 1964 the European Fisheries Court laid down a new legal framework for fishing; six miles was to be the limit of exclusive rights, while some 'historic rights' had to be recognized between six and twelve miles. This soon proved to be inadequate in the face of continuing pressure on North Sea fisheries and, by 1975, Iceland had unilaterally declared a 200 mile exclusion zone, thereby provoking a second 'Cod War'; Norway took similar action in 1977.

It was inevitable that, with time, the EC would become involved in fisheries negotiation; but initially its principal concern was trade harmonization and liberalization, to create a common market as laid down in the Treaty of Rome. This proved very difficult in the face of very different national systems of protection. However, by 1968 a Common Fisheries Policy (CFP) had been proposed with the aims (necessarily conflicting) of ensuring that supply matched demand at reasonable prices within the EC, while offering reasonable incomes to fishermen. Proposed measures included common organization of markets, a structural (modernization) policy, and equal access for members to all EC fishing grounds. During the ensuing negotiations it proved difficult to secure the necessary delicate balance of members' interests; but with discussions on accession of the UK, Ireland, Denmark and Norway about to begin, a compromise was hastily agreed in 1970. The CFP therefore became part of the *acquis communautaire* (accepted general policy framework) before enlargement. The applicants were faced with a *fait accompli* and had to negotiate from a less favourable base.

Fishing became one of the most difficult and contentious items in the negotiations over enlargement. It was only possible to agree a transitional policy for ten years, with a new policy to be negotiated by 1982. Under the temporary arrangements, the applicants were granted twelve-mile exclusion zones for between one-third (the UK) and most (Norway) of their coastlines. Even this – along with other conditions of membership – proved to be too much for Norway which, after a referendum, decided not to join the EC.

The next important phase of negotiations began in 1977 when the EC countries decided to extend their national fishing zones to 200 miles, following the examples of Norway and Iceland. However, a very difficult problem had to be resolved: were these zones to be reserved for national fleets, as the UK and Eire argued, or for the EC fleets? An extra dimension to the problem was the UK's demand that any system of EC quotas should take into account its earlier loss of fisheries off Norway and Iceland by the application of the 200-mile rule. It was only the approaching end of the ten-year transition phase of the CFP which finally forced a compromise in January 1983. All states had exclusive rights within six miles and they usually had exclusive rights in the 6–12 mile band, except where historical rights required selective access for other members. Within the 12–200 mile zone, members were allocated quotas for specific catches, and these, as would be expected, have proven particularly sensitive to negotiate, especially in attempting to reconcile the interests of Denmark and the UK. Nevertheless, the 1983 agreement has established a Community system for the conservation and management of fisheries resources (Farnell and Elles 1984).

The 1983 agreement also sowed the seeds of future controversy in respect of Spain and Portugal. Spain had a large modern fishing fleet and, before the 200-mile extension in 1977, had operated some 600 boats in what were to become EC waters. After 1983 they were restricted to only 130 vessels, at the same time as they were being excluded from more distant fisheries off the coasts of Africa and North America (Cunningham and Young 1983). As a result the 1985 discussions of Iberian accession to the EC found fishing a difficult area to resolve. In the event, a ten-year transitional period was agreed, during which Spanish and Portuguese fishermen were to obtain some access to EC waters, via a system of licensing.

Given the controversy surrounding access to territorial waters, the other elements of CAFP have received little attention. Nevertheless, a prices policy has been agreed, with the EC paying between 70 and 90 per cent of guideline prices which are negotiated at the beginning of each season. There has also been investment in modernization; in 1981–2, for example, 18.2 million ECUs were spent by the Regional Fund and the European Investment Bank for these purposes. Such expenditure seems likely to increase in importance with the Iberian enlargement of the Community; assistance for modernizing fishing fleets and for onshore fish processing was important in pre-accession aid, and in the early rounds of Regional Fund grants allocated to Spain and Portugal. At the same time, national aid schemes have remained important; for example, France pays fuel subsidies and the UK pays 'temporary' subsidies (valued at some £40 million in 1981–2; Farnell and Elles 1984). As in many other respects, then, EC policies seem to have enhanced existing protectionist measures against non-members, while only securing limited increases in competition within the Community. Fishing seems to be too politically sensitive to be left to the dictates of market forces.

Further reading

1 General reviews of agriculture: H. D. Clout (1981), 'Rural space', in H. D. Clout (ed.), *Regional development in Western Europe*, Chichester: Wiley; A. R. Jones (1984), 'Agriculture: organization, reform and the EEC', in A. Williams (ed.), *Southern Europe transformed*, London: Harper and Row.
2 Land ownership and land reform: R. King (1977), *Land reform: a world survey*, London: Bell; G. P. Hirsch and A. H. Maunder (1978), *Farm amalgamation in Western Europe*, Farnborough: Saxon House.
3 The peasantry: T. Shanin (ed.) (1979), *Peasants and peasant societies*, Harmondsworth: Penguin.
4 The Common Agricultural Policy: B. E. Hill (1984). *The Common Agricultural Policy: past, present and future*, London: Methuen.
5 Fishing: M. Wise (1984), *The Fisheries Policy of the European Community*, London: Methuen.

5 The secondary sector

European industry since 1945

Growth and crisis

In some senses, the 1960s were a watershed in European industrial development. Some economies experienced relative declines in this decade and, after the 1973–4 oil crises, there were absolute declines in industrial employment (although not necessarily in output) in Belgium, Denmark, France, West Germany and the UK. Even southern Europe, which had experienced very high growth in the 1960s, stagnated in the 1970s. By the 1980s it was possible to talk about the deindustrialization (see Blackaby 1979) of Europe; in contrast to the Depression of the 1930s, decline was widespread among most sectors of manufacturing. Even the growth industries of the post-war period, such as vehicle production, petrochemicals and parts of the electronics industry, have undergone large-scale restructuring and job losses.

The economic crisis of the early 1970s, stemming from growing overproduction, collapse of the international monetary system and oil price increases, has already been discussed (see pp. 40–9). The implications were clearly visible in the reduced expansion of trade in manufactured goods: the annual growth rate for these declined from 10.8 per cent in 1963–73 to 5.0 per cent in 1973–80 (OECD 1983c). Between 1973 and 1981 over one million jobs were lost in the secondary sector in the UK, and over 600,000 in both France and West Germany. Only in Ireland did manufacturing jobs increase during this period.

Intensified competition among the developed countries and the emergence of the newly industrialized countries contributed to the crisis of overproduction. At first, competition from the NICs came via import substitution which reduced exports from Europe, but, later, the NICs became a force in exports. For example, between 1960 and 1975 manufactured goods as a proportion of merchandise exports rose from 14 to 82 per cent in South Korea and from 13 to 27 per cent in Brazil. The shift to more global patterns of competition was also evident in multinational investments in Europe (see pp. 72–5). European companies such as Philips and Siemens have international production strategies involving investments in different parts of Europe and in other continents. At the same time, many non-European MNCs have set up production in Europe, drawn by the need to locate inside the protective tariffs operated by individual governments and by the EC. While this is long-established practice for

American MNCs, such as General Motors and IBM, since the 1970s it has become more widespread among Japanese companies; for example, the Nippon Electric Company started to produce silicon chips in Dublin in the 1970s and, in 1985, Nissan decided to produce cars in the UK. Even some of the NICs have decided to invest within Europe, primarily to avoid import tariffs; one notable example was the decision in 1986 by Hong Kong's YGM company to establish a clothing factory on Merseyside.

There have also been changes in the form of European MNCs' overseas investments. Despite the importance of British and French MNCs' activities in their ex-colonies, European MNCs have tended to be more nationally-based or, at best, continent-based than their Japanese and North American rivals. This is clearly illustrated by the European horizons of most European car producers compared with the global horizons of their competitors. However, since the 1960s this has changed (Linge and Hamilton 1981) with growing European investment in North America. For example, between the late 1960s and early 1970s the proportion of West German overseas investment going to North America increased from 17 to 26 per cent. While this was partly due to relaxation of currency controls and realignment of exchange rates in the 1970s, it also reflected the new requirements of global production and the need to secure access to the world's largest single market, the USA.

Some aspects of the internationalization of manufacturing can also be seen in the trade statistics. Manufactured products account for at least 50 per cent of merchandise exports in all European countries (except Norway) and in Switzerland are equivalent to 93 per cent. In absolute terms, a clear lead is held by West Germany with exports valued at $147,003 million in 1983, a figure which was not even exceeded by the USA or Japan. Within Europe, France holds second place, followed by Italy and the UK. Greece brings up the rear with manufactured exports valued at only $2194 million. Therefore, despite the relative decline in the importance of manufacturing, it remains of considerable importance in the economy of western Europe. The international variations within Europe are considered in more detail later.

Technology, R & D and capital mobility

One perspective on the long-term evolution of the international economy is provided by Kondratieff (1935) who identified 50-year cycles of expansion and recession in production. These cycles were linked with technological changes, especially the occurrence of exceptional clusters of innovations (Schumpeter 1939). There was an important cluster of innovations in the 1930s – linked to electrical products and aerospace – which provided a platform for post-war expansion in electronics, synthetic materials, petrochemicals and other industries. Wealth generated by these fuelled demand for, and increased production of, consumer durables. However, in the 1970s there was market saturation and productivity grew faster than demand. Furthermore, the development of microelectronics has contributed to automation and to substitution of capital for labour. For example, in the UK in the 1950s it only required, on average, less than 2 per cent growth in output per annum to maintain employment levels, but by 1980 – owing to technological advances – 4 per cent expansion in output was required (Rothwell 1982). Together, growing automation, production process rationalization and market stagnation have acted to choke off European employment growth. Whether recent clusters of innovations in microelectronics (Freeman 1984) will lead to restoration of global conditions of profitability is arguable.

Maintaining competitivity in technologically advanced manufacturing activities is critical for the European economy. The Commission of the European Community (1984g, p. 5) expressed this in strong terms:

> Despite outstanding performances in certain favoured areas, European industry is falling behind its principal competitors. Of every 10 personal computers sold in Europe, eight were manufactured in the United States. Of every 10 video-recorders, nine were made in Japan. European manufacturers of integrated circuits control only 30% of the Community market and 13% of world sales. In the industry as a whole, the Community supplies only 10% of the world market and 40% of its own market, which accounts for about one-third of world sales. European firms complain that their sales volume and profits are too small to permit the investments which are vital to guarantee their future. Their relative position is therefore declining. All European main-frame computer manufacturers have been forced to enter agreements with American and Japanese rivals to take advantage of their technological expertise. In 1975 the Community's balance of payments in this area was positive. By 1982 it was in deficit to the tune of 10 billion dollars. The long-term implications for employment in the Community are all too clear. The collapse of Community firms will mean that very few manufacturing and service jobs will be created in Europe. At the same time large numbers of jobs will be destroyed in other sectors through the spread of imported new technologies, necessary to maintain competitiveness.

Research and development costs for advanced industrial products have escalated rapidly. For example, a new aeroengine may cost $1 billion to develop, and a new generation of digital telephone switches may cost $600 million (*Sunday Times*, 12 January 1986). As a result, international collaboration has become more commonplace among European producers faced with limited resources and limited domestic markets. In the case of civilian aircraft, individual European companies can no longer compete with the American giants, Boeing and McDonnell Douglas. Instead the main European producers such as British Aerospace, Messerschmidt–Boelkow–Blohm and France's Aerospatiale have combined in projects like the European Airbus. Development costs in the electronics industry are particularly high and, in this field, even European 'giants' such as Siemens and Philips have been forced into collaboration (for example, in their Munich-based 'megaproject') so as to compete with the Americans and Japanese. The scale of research and development has also required considerable state intervention to support the efforts of private companies. As can be seen from Table 21, Europe has been catching-up in this respect, but it still lags behind the USA. West Germany holds the lead in Europe, closely followed by the UK, although an important part of expenditure in the latter is directed at military, rather than civilian, research . Research and development expenditures are particularly low in southern Europe.

Table 21 *Expenditure (as percentage of GDP) on research and development, 1975–84*

	Public-sector expenditure 1975	1984	Total R & D expenditure 1981
Austria	—	0.53	1.17
Belgium	0.73	0.57	—
Denmark	0.58	0.51	1.07
Finland	—	0.62	1.19
F.R. Germany	1.23	1.13	2.48
France	1.17	1.47	2.01
Greece	—	0.13	0.21
Ireland	0.44	0.39	0.75
Italy	0.36	0.72	1.01
Netherlands	0.96	0.97	1.88
Norway	—	0.77	1.29
Spain	—	0.26	0.39
Sweden	—	1.31	2.22
UK	1.27	1.35	2.42
Japan	0.60	0.63	2.37
USA	1.24	1.22	2.51

SOURCE: OECD/*Financial Times*, 30 June 1986.

Research and development is important not only in firms' abilities to maintain competitiveness in national and international marketing; it also has implications for the spatial distribution of production and employment. Basic research is usually carried out in central laboratories, applied research in company divisions, and development work is attached to production units (Howells 1984). In practice this has meant that basic and applied research has tended to be located close to company headquarters in the major European cities such as Brussels, Paris, London or Frankfurt, or in amenity-rich locations on the margins of the metropolitan areas, that is in such cities as Reading or Cambridge in the UK, or Rouen and Amiens in France (Daniels 1982).

Another feature of the post-war industrial economy, linked to technology, has been the rapidity of change, affecting both particular industries and locations. The process of capital arriving in and abandoning regions takes place over increasingly short time periods; this has been termed 'the hypermobility of capital' (Damette 1980). More rapid technological change has increased the pace at which firm's requirements in terms of conditions of production are altered. Hence, in seeking to maximize profit, firms will seek out new locations while abandoning redundant ones at ever-more-frequent intervals. This is particularly well illustrated by the chemicals and petrochemicals industries (Hudson 1983b). Vast industrial complexes at Teesside, Rotterdam and Fos-sur-Mer were developed very rapidly in the 1960s and early 1970s but are now, equally rapidly, becoming technologically (and profitably) obsolete.

EC and state policies

Most countries have well-developed industrial policies reflecting the importance attached to manufacturing development. This stems partly from the belief that manufacturing growth provides the vital motor for economic development. The policies can assume many forms, including both short-term transitional aids and long-term restructuring programmes, and they can be supranational, national or regional. They include:

- investment incentives
- research and development assistance
- labour-market policies (training, trade union laws, migration, etc.)
- stimulation of demand (directly by state purchasing, or indirectly)
- environmental controls
- controls on national and/or foreign capital
- trade promotion and/or production.

The precise form of such policies varies considerably between countries, reflecting differences in both the political framework and in industrial structure in each case. For example, in France state policies traditionally concentrated on protectionism, or on selective financing of particular sectors such as aerospace and telecommunications via, for example, the Fonds de Développement Economique et Social (Tuppen 1983). There has also been an active policy of encouraging mergers so as to create large French industrial combines which are capable of competing on a global scale; examples include Pechiney–Ugine–Kuhlman (metals and chemicals) and Thomson–Brandt (electrical goods and electronics). This is only one example of state industrial policy and, in reality, most industrial sectors in most countries are influenced, to some degree, by state intervention.

Further examples will be considered later in this chapter in relation to particular sectors.

In addition, the EC adds an international level to industrial policy. The Treaty of Rome was not very explicit about how industrial policy should be formulated, but the general aim was to encourage free competition which, in turn, would encourage modernization. This was qualified by a realization that some key areas – such as struggling industries or developing new technologies – require special attention. The Commission monitors trading practices and seeks to fine companies which, for example, operate price-fixing agreements (as with the dyestuffs cartel brought to court in 1969) or exclusive distribution agreements (as in the 1982 case against AEG Telefunken which was considered to have discriminated against retailers). State aids to particular industries are also monitored, especially as these have mushroomed in the face of economic crisis in the 1980s. The EC insists that such aids should be exceptional and limited in duration, although this clearly is open to various interpretations. In addition, detailed guidelines have been laid down for four severely affected sectors: shipbuilding (since 1981); textiles (since 1971); steel (since 1981); and synthetic fibres (Commission of the European Community 1985b).

Finally, in recent years the EC has tried to develop a more positive industrial policy, encouraging growth in new sectors (especially information technology). Measures taken include provision of loans and grants for industrial development, attempting to open up public procurement contracts so as to give equal access for all EC companies, limited protectionist measures against non-EC producers, and attempts to coordinate research and development. The latter is considered to be particularly important, and projects such as EUREKA are designed to encourage joint research between EC countries.

The construction industry

The construction industry is important in all western European countries, yet it assumes a variety of forms. This reflects the largely national character of construction markets, but it is also conditioned by state interventionism, stemming especially from the sensitivity of the industry to seasonal and cyclical fluctuations (Table 22). Demand management has been particularly important, involving both direct measures as in the awarding of public-sector building contracts, or indirect ones such as manipulation of interest rates and tax relief on house buying. The latter has been particularly important in the UK, where the construction industry has been severely affected by the government's stop–go economic policies.

Some indicators of variations in the industry within western Europe are given in Table 22. Italy is by far the largest producer of cement, reflecting both its active construction industry and significant exports, notably to the

Table 22   *Characteristics of the construction industry in 1982*

	Million tons of cement produced, 1982	*Dwellings completed per 1000 inhabitants, 1982*	*Employees in building and civil engineering (000s), 1982*	*Percentage change in activity levels, 1970–80*
Austria	5.0	5.8	N/A	+ 33.3
Belgium	6.3	3.5	202.5	+ 28.4
Denmark	1.8	4.0	132.4	− 28.8*
F.R. Germany	30.1	5.9	1671.1	+ 58.6
Finland	1.8	9.8	N/A	+ 17.3
France	25.2	7.2	1422.8	+ 58.6
Greece	13.2	20.0	N/A	+ 14.1
Ireland	1.6	7.7	78.0	N/A
Italy	40.2	N/A	1481.1	0
Luxembourg	0.3	5.6	14.7	N/A
Netherlands	3.1	8.8	320.9	N/A
Norway	1.8	8.3	N/A	− 10.6
Portugal	5.9	4.1	N/A	+ 77.3†
Spain	30.2	6.2	N/A	0
Sweden	2.3	5.4	N/A	+ 7.0
Switzerland	4.2	7.2	N/A	+ 55.6
UK	13.0	3.3	1051.0	− 13.5

* 1972–80.
† 1970–78.
SOURCE: Eurostat (1984a) and United Nations Statistical Yearbooks.

Middle East and to Africa. Greece, too, has a relatively large cement industry, and in 1986 sought to increase exports to the UK. Other variations in cement production are the outcome of both differences in the scale of the economy and in types of construction materials used; thus cement is a far more important building material in, say, Spain than in Scandinavia where timber frame construction is traditional. Rates of dwellings completed are also indicative of levels of economic development. At one extreme, Greece has a level of 20.0, indicative of high rates of urbanization (Gaspar 1984), the impact of return migration, and the need to upgrade the existing housing stock. In contrast, numbers of dwellings completed in the UK and Belgium were 3.3 and 3.5 respectively, reflecting the economic depression in both these countries. Within particular countries, employment in construction is relatively widespread as is necessitated by the dispersed nature of the industry.

Ownership is relatively dispersed, although precise data on this are difficult to acquire. However, not untypical are Denmark, where 31 per cent of firms have no more than one employee and 99 per cent have fewer than 50 employees, and Sweden where 87 per cent have fewer than 50 employees (Sundin 1984). In contrast, 39 per cent of all new houses in the UK in 1985 were built by only ten companies, with Barratt and Wimpey each accounting for about 6–7 per cent (see Table 23). However, at the other extreme, there is still a very large number of companies producing fewer than twenty houses a year.

Table 23 *Major house builders in the UK in 1985*

House builders	Number of units
Barratt	10,000
George Wimpey	9,600
Tarmac	8,000
C.H. Beazer	4,500
Trafalgar House	3,800
Broseley	3,000
Bovis	2,800
Y.J. Lovell	2,600
John Laing	2,100
Christian Salvesen	2,100
'Top ten' total	48,500
All completions	150,000

SOURCE: *Financial Times*, 4 February 1986.

Within Europe the construction industry has largely remained dominated by indigenous companies operating within national markets. Nevertheless, even the construction industry has shown some signs of increasing internationalization. Northern European companies such as Wimpey are involved in building tourist complexes in the Mediterranean countries. However, internationalization has probably proceeded furthest in respect of European cement producers' involvement in the USA where, by 1986, they controlled almost half the output. This dates from the late 1970s when companies such as Blue Circle of the UK, Lafarge Coppee of France, and Heidelberger Zement of West Germany became forces in the USA. Building firms are less internationalized than cement producers, but there are important examples such as Bovis International of the UK which, in 1986, bought into the Lehrer–McGovern group (partly so as to acquire new technology), and France's Bouygues which has joint ventures in the UK and India.

In general terms, the construction industry has been relatively little affected by technological change. Industrial systems methods of construction have gained a foothold in some markets, but the vast majority of buildings are still constructed using largely traditional methods. It is a labour-intensive industry, which partly accounts for the heavy reliance on immigrant workers. There are British workers on sites in West Germany, although much more typical are the tens of thousands of Italians, Portuguese and Spanish working in France and West Germany. Given current technology, the industry is unlikely to experience any significant automation in the near future. Nevertheless, deskilling is certainly occurring with greater reliance on factory-produced components. At one extreme this may involve window frames or moulded arches, but entire bathrooms and toilets can be assembled off-site. However, on-site assembly is likely to remain labour-intensive.

Industrial systems building offers two advantages: reduction of labour costs and reduction of construction time. This is particularly important given that capital can often be tied up in large-scale construction projects for months, if not years, before any returns are obtained on investment. It is for this reason that European companies in the 1980s began to adopt American techniques of 'fast-tracking'. These involve beginning construction when only the general outlines of the design are known, and using very advanced industrial systems methods. In the Broadgate office complex in London, this approach resulted in savings of one-third of the normal construction time (*Financial Times*, 2 October 1986).

Housing and economic development

Housing development has considerable significance for the economy as a whole. It is an essential element in the assembly of labour power at the locations required by other industries – especially in the creation of new industrial zones, such as growth poles. The cost of housing is also an important component of the overall cost of living which, in turn, influences wages levels (see Bassett and Short 1980). Not surprisingly, then, the state has sought to play a mediating role in housing markets, ensuring the provision of adequate numbers of suitably priced houses in the right locations. This often involves substantial subsidies to housing purchasers, controls on private renting, or programmes of state housing construction.

Cyclical fluctuations in construction also influence other sectors through competition for investment funds. Sometimes funds may flow into construction in consequence of depression in manufacturing, as occurred in the mid-1970s office boom in the UK, following the 1974 oil crisis. However, the possibilities for realizing profits in speculative or office developments can divert investment funds from other industries, as occurred in Portugal in the late 1960s (Lewis and Williams 1982). This has also occurred in Italy where, as Ginatempo (1979, p. 474) reports, 'in terms of local resources, the building sector has absorbed the greater part of investments as a substitute for industrial development.'

International dimensions of western European industrialization

International variations in growth

The earliest centres of non-water-powered industry were the UK coalfields and, by the 1820s, South Wales, the North East and Scotland had become established in steel production and heavy engineering. Later, coal and steel production expanded in the Sambre–Meuse region although, by the second half of the nineteenth century, it had been overtaken by the Ruhr and Saar–Lorraine. At the close of the nineteenth century modern industry was also established in southern Sweden, the Basque and Catalan regions in Spain, the Jura (around Zurich and Basel) and Piedmont–Lombardy in northern Italy. Not only did each of these regions tend to specialize in certain types of production, but there was also specialization within them. In Piedmont–Lombardy, in the twentieth century, Turin specialized in car production, Milan in chemicals, textiles and vehicles, and Genoa in shipbuilding. Within the Jura district there were concentrations of chemicals and pharmaceuticals in Basel and of instrument manufacturing in Geneva. Since 1945, industrialization has spread to virtually every region in western Europe, bar a few exceptions such as north-east Portugal, Estremadura in Spain, and the far north of Scandinavia (see pp. 302–13).

International variations in performance since the first major oil crisis are summarized in Figure 16. Industrial production actually fell in Luxembourg and Sweden between 1975 and 1983, and showed only modest gains in Switzerland, the Netherlands and the UK. As 1983 represents one of the lowest points in the 1980s global recession, these figures do exaggerate the general decline. However, they are indicative of increasingly varied industrial growth rates in western Europe; despite the onset of recession, Finland, Norway (linked to the oil industry), Ireland and Portugal all experienced growth rates in excess of 50 per cent during this period. Therefore, there seems to be a relative, if not an absolute, shift in the location of manufacturing activities towards the presently less-industrialized 'periphery' of Europe.

These changes are relative and, despite the deindustrialization evident in many advanced economies, countries such as West Germany, the Netherlands and Belgium still record the highest proportions of industrial employment and GDP (see Table 24). Nevertheless, the geographical spread of manufacturing since 1945 is also evident in the very limited differences between these older economies, and the more recently industrialized ones such as Portugal and Spain in the 1980s. Therefore, on the basis of aggregate statistics, there seems to be some convergence between core and periphery in Europe following recent industrialization of southern Europe, Ireland and parts of Scandinavia, while some 'mature'

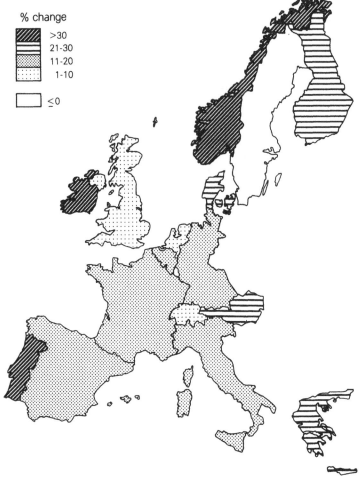

Figure 16   Percentage change in industrial production, 1975–83

SOURCE: Eurostat (1984).

industrial economies such as the UK and Belgium have stagnated. However, to place this in perspective note that just six countries – Belgium, West Germany, France, Italy, Netherlands and the UK – still account for over 70 per cent of western Europe's industrial production. Moreover, the aggregate figures say nothing about the control of industry and different forms of specialization, whether by sector or by stage of production.

Sectoral specialization

The broad sectoral distribution of manufacturing revealed in Table 25 is the outcome of diverse industrial histories (based on local constraints,

Table 24 *Proportion of GDP and of total employment in industry**
1981–4 (percentage figures)

	Proportion of GDP in 1984	Proportion of total employment in 1981
Austria	38	37
Belgium	34	41
Denmark	25	35
F.R. Germany	46	46
Finland	34	35
France	34	39
Greece	29	28
Ireland	25†	37
Italy	40	45
Netherlands	32†	45
Norway	43	37
Portugal	35†	35
Spain	34‡	40
Sweden	31†	34
Switzerland	N/A	46
UK	36	42

* Includes mining and construction.
† 1983 data.
‡ 1982 data
SOURCE: World Bank (1986).

opportunities and state policies) within a changing international division of labour. Nevertheless, some broad generalizations are possible. Processing of food and agricultural products is relatively most important in Denmark, Greece, Ireland and Switzerland, all countries with substantial farming sectors. In detail the types of products are varied; for example, wine, olives and other Mediterranean produce in Greece, compared with meat and dairy products in Denmark and Ireland. Textiles and clothing are most significant in Greece, Portugal and Spain and, indeed, this is a specialization in recently industrialized countries. In contrast, specialization in machinery and transport equipment is more typical of mature economies. This is the sector in which West Germany has European – and perhaps global – dominance, with established industrial traditions and important exports; but France, Sweden and the UK also have a significant presence in this sector. Finally, chemicals have a relatively broad (and maritime) distribution; this includes both North Sea locations in Belgium, West Germany, the Netherlands, Norway and the UK (some of which have indigenous oil reserves), and more recent Mediterranean production centres such as Portugal and Spain. The industrial mix in any particular country is conditioned by many considerations, as is illustrated later by reference to specific industries.

Table 25 *Sectoral distribution of manufacturing value-added in 1983 (percentage figures)*

	Food and agriculture	Textiles and clothing	Machinery and transport equipment	Chemicals	Other manufacturing
Austria	15	9	24	7	45
Belgium	19	9	25	12	35
Denmark	23	6	24	9	39
F.R. Germany	10	5	41	9	34
Finland	11	7	22	6	53
France	16	7	34	8	34
Greece	21	22	12	8	38
Ireland	36	11	15	14	24
Italy	12	18	26	7	38
Netherlands*	19	4	28	13	36
Norway	12	3	28	8	49
Portugal	17	27	12	17	37
Spain	13	15	21	7	44
Sweden	19	3	32	7	50
Switzerland	15	8	25	12	40
UK	14	6	33	10	36

* 1982 data.

SOURCE: World Bank (1986).

The sectoral mix of manufacturing activities in each country is also important in that it helps to explain the differential impact of the global recession of the 1970s. Iron and steel, metal products and many engineering firms were already showing signs of recession before 1973, although their difficulties were considerably increased thereafter. In contrast, textiles, chemicals, and wood and paper products were mainly affected after 1973 (OECD 1983c). Between 1973 and 1977, some two million manufacturing jobs were lost in the EC alone; 30 per cent were in textiles, 55 per cent in consumer industries and 20 per cent in metal products and industrial machinery (Hudson *et al.* 1984). By the 1980s decline had become generalized to most industries, including several sub-sectors of electrical goods, in the face of a general downturn in demand and increasing competition from outside Europe.

The labour process and the organization of industrial production

The history of industrialization can be viewed in terms of distinct phases in the development of the labour process – that is, how labour is deployed in the production process (see Dunford *et al.* 1981; Dunford and Perrons 1983; Perrons 1981). Each phase has involved an increasing division of labour, both socially and spatially, as new ways have been found of extracting profits. The four major phases are usually considered to have been manufacture, machinofacture, scientific management/Fordism and neo-Fordism. These and their relevance for understanding industrialization in Europe are outlined below.

Manufacture was the nearly universal mode of production in the seventeenth and eighteenth centuries. It involved the craft worker being responsible for his or her own final product. In the UK in the early eighteenth century, and in Belgium, Germany and France rather later, the second form of production emerged, machinofacture. Factory production and new technologies led to a partial deskilling of work, but this remained physically united within individual establishments. In order to secure labour supplies, there was a shift to urban locations. These conditions remained dominant throughout the nineteenth century, as new urban-industrial regions emerged in Europe.

Only in the early twentieth century did the third phase emerge, involving scientific management and Fordist assembly methods. Scientific management (Taylorism) involved greater specialization among the labour force while removing decision making from individual production workers to management teams. As a result there was greater job fragmentation, with workers being allocated predefined tasks, while management had control over the production process as a whole. This was enhanced by Fordist assembly-line methods which allowed rapid and efficient recombination of the individual components to produce the final product. As the name suggests, Fordism originated in the motor vehicle industry, but it subsequently dominated many industries, especially during the rise of consumer-durable industries in the inter- and post-war periods.

Fordism resulted in enormous gains in productivity, but this served only to aggravate the problems of overproduction and of saturated markets, as was most clearly revealed in the Depression in the 1930s (Lipietz 1984). In the early post-war period, western European governments seemed to have come to terms with these new conditions of production, using macro-economic policies to regulate the economy (see pp. 78–82). However, the Fordist/Taylorist phase of production was ultimately subject to two weaknesses. One was labour resistance to the pressured working conditions, resulting in go-slows and high levels of absenteeism. To some extent this could be resolved by shifting production to areas of lesser industrial militancy (see pp. 171–2). The other problem, which Aglietta

(1979) terms 'balance delay time', was the less than complete integration of different parts of the production process, so that the overall speed of the assembly line was determined by its slowest parts. As a result, many industries began to encounter falling rates of profit after the late 1950s.

One response to this was neo-Fordism, being the use of electronic information and computer systems to automate production; this allowed for more efficient reintegration of the fragmented elements of production. As a result, the control of production shifted from the shop-floor to a central information centre, while the need for skilled labour was generally reduced (although not for highly skilled technicians). Industry no longer remained tied to centres of skilled industrial labour. Instead, it became rational to divide different parts of production between different locations according to their operating requirements.

This has given rise to a new spatial division of labour. Lipietz (1980a) considers that, typically, there is a three-fold division: between headquarters and research functions (located in 'core regions'), skilled labour processes (in traditional industrial centres), and assembly work (in low-cost-labour locations). Such a division of labour may operate at either the interregional or the international scale, depending on the scale of the company, its product and its markets. While such a division of labour is characteristic of some industries (such as vehicle production), it has been argued that this is not necessarily the predetermined spatial division of labour of neo-Fordism. Instead, automation and deskilling of the labour force can lead to a polarization of production between a metropolitan headquarters and the remotely controlled factory assembly line, completely bypassing the need for skilled labour. This seems to have occurred in many sections of the electronics industry (Massey and Meegan 1979).

The emergence of a new division of labour has been accompanied by continuing internationalization of capital, with different stages of production in some industries being located in different countries. While non-European MNC investment within Europe is not a new phenomenon, this is taking on a different form. Increasingly, Europe is the recipient only of component manufacture and assembly work, with R & D and headquarters functions remaining centralized in Japan or the USA (see the example of the electronics industry, pp. 196–201).

The organization of production

The scale of production

International comparisons of the scale of production are difficult because a variety of size criteria are used in different countries. Turnover, capital valuation and employment have all been used, either singly or in combination, in such definitions, while more sophisticated analyses, such as the 1971 Bolton Report in the UK, emphasized the importance of market shares and personal management. In this section some of the disparate evidence available on the scale of production in western Europe is considered, along with the debate concerning the relative merits of small firms.

Historically, there is a well-established trend to concentration (Mandel 1975) reflecting the advantages of large-scale production for both national and international capital. This is illustrated by the German experience, where the top 50 companies' share of industrial turnover increased from 34.6 per cent in 1960 to 49.5 per cent in 1980. In many cases, state policies have encouraged mergers to create companies capable of competing in national or international markets. For example, in the UK the Industrial Reorganization Corporation was active in this role and, among other initiatives, secured the creation of British Leyland Motor Company in 1968. In France, the Institut du Développement Industriel played a similar role, and helped in the mergers which established Pechiney–Ugine–Kuhlmann in 1971.

Despite these tendencies and the variety of measures used to define scale, most European manufacturing firms are relatively small (Table 26). Over 90 per cent of firms in Denmark, the Netherlands, Sweden, Greece, Italy, Spain and Portugal have fewer than 50 employees. The data for France and West Germany are more difficult to assess, as they include firms without employees but these do indicate the coexistence of large numbers of small enterprises alongside a few large-scale companies.

State policies in many countries shifted to favouring small-scale firms in the 1970s as the growth performance of many industrial giants – whether in steel, chemicals or textiles – faltered. The 'small is beautiful' school of thought stressed that small businesses have a number of functions: as a source of competition for large firms; creating new jobs; acting as seedcorn from which large companies will grow; providing a harmonious working environment and a non-militant labour force; and acting as a seedbed for innovation (reviewed in Rainnie 1985; and Storey 1982). Such a view of small firms is based more on myth than reality. For example, most studies of employment creation stress that their role in this is minor compared with the loss of jobs in larger firms (see Gould and Keeble 1984). This is an oversimplification because small firms are far from heterogeneous; for example, Schutt and Whittington (quoted in Rainnie 1985) identify three main types:

- dependent, which are complementary to or service the activities of large firms
- independent 'type one', which compete with large firms, often on the basis of intense exploitation of labour;
- independent 'type two', which operate in local or specialist markets ignored by large firms.

Table 26 *Scale of industrial production in selected western European countries (percentage figures)*

Denmark, 1978 (% of firms by number of full-time employees)*

1–2	3–5	6–20	21–50	> 50	Total
1	56	28	8	7	100

F.R. Germany, 1970 (% of employees by size of firm)†

1–9	10–49	50–99	100–499	> 500	Total
8	22	30	5	35	100

France, 1980 (% of firms by number of employees)‡

1–10	10–99	100–499	500–1999	> 2000	Total
N/A	18	23	19	40	100

Sweden, 1981 (% of firms by number of employees)§

1–49	50–200	> 200	Total
91	6	3	100

UK, 1976 (% of firms by number of employees)‖

1–99	100–199	> 199	Total
17.1	5.5	77.4	100

Southern Europe (% of firms by number of employees)¶

	1–9	10–40	50–99	> 99	Total
Greece, 1978	93	5	1	1	100
Italy, 1981	85	13	1	1	100
Portugal, 1971	79	16	3	2	100
Spain, 1978	77	18	2	3	100

Netherlands, 1983 (% of firms by number of employees)**

1–4	5–9	10–49	50–99	> 99	Total
63	14	19	3	1	100

Other northern European countries (% of firms by number of employees)‖

	1–99	100–199	> 199	Total
Ireland, 1968	33	17	50	100
Belgium, 1970	33	10	57	100
Luxembourg, 1973	19	28	53	100

---

* Monsted (1984)
† Hull (1983)
‡ Tuppen (1983)
§ Sundin (1984)
‖ Storey (1982)
¶ Hudson and Lewis (1984)
** Kok (1986)

These have different potentials for expansion and different production needs, so care must be taken not to over-generalize about their performance.

Spatial forms of industrialization

One of the most striking aspects of post-war industrialization in Europe has been its changing spatial organization. Many traditional centres of heavy industrial production have been in decline since the late 1950s, while in the 1960s several major metropolitan areas experienced net industrial losses. Instead, there has been a shift from large-scale, spatially concentrated industry to smaller-scale and more dispersed forms. Whereas the 1960s was a decade of large-scale industrial complexes, often in the form of planned growth centres, the 1980s have experienced a shift to smaller-scale production units. At the same time there has been a spatial shift in industry from larger urban centres, to smaller cities and towns and, increasingly, to rural areas.

In the 1970s the most highly urbanized regions in the EC had a declining share of industrial output compared with increases in all other regions, especially the more rural ones (Figure 17). As this analysis is based on large-scale, EC Level-2 regions, it tends to underestimate the shifts which have occurred. Evidence from individual countries, at a finer scale of

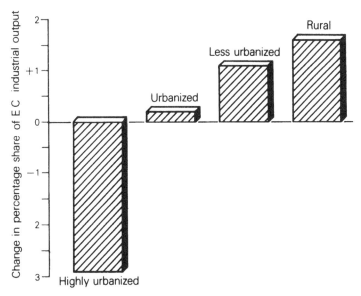

Figure 17 Regional changes in the share of industrial output in the EC, 1970–77, by level of urbanization

SOURCE: based on Keeble *et al.* (1983).

analysis, reveals far sharper urban–rural differences. For example, in Denmark between 1973 and 1978, there was a 21 per cent decrease in manufacturing employment in the Copenhagen region, but rural areas had gains of 13 per cent (Keeble *et al.* 1983). In the UK, over the longer time period 1960–78, conurbations had losses of 26.5 per cent, small towns had gains of 15.7 per cent, and rural areas had increases of 38 per cent (Hamnett 1985). Similar trends are evident in most other western European economies.

Urban–rural shifts in manufacturing are linked to the process of productive decentralization. The pressures for decentralization arise from firms seeking to reduce production costs by cutting real wages or by substituting capital for labour, both of which are easier to achieve outside of existing, large factories. Given the potential to create new spatial divisions of labour, some stages of production may be decentralized. Decentralization can take various forms (Mingione 1983):

- direct decentralization of some stages of production, usually the labour-intensive ones, thereby creating branch plants
- direct decentralization of capital-intensive plants
- indirect decentralization to the informal economy in the metropolitan area
- indirect decentralization to large numbers of small and medium-sized industries in non-metropolitan regions.

Direct decentralization: branch plants

Technological changes, including deskilling and improved communications, have allowed a spatial separation of stages of production. At the same time, the differential production requirements of each stage can make such a spatial separation a logical necessity. Branch plants are the most characteristic form of this, especially those involving assembly work in areas of cheap and flexible labour. The process is most evident in those sectors where technology has either not yet significantly reduced the need for labour (for example, clothing and shoe manufacture), or where technology has deskilled labour without displacing it from production (for example, television-set assembly). At the same time, a number of other conditions have encouraged decentralization, including a lack of low-cost land for expansion in metropolitan areas; this seems to have been particularly important in the urban–rural shift in the UK, according to Fothergill and Gudgin (1982).

The precise form of such decentralization, and the locational strategy pursued, depends on the requirements of the particular industry. For example, when Opel set up a branch plant at Bochum in 1960 they were aiming to create a large-scale enterprise employing 15,000, mainly male, workers. Bochum was a suitable location for it had relatively large labour reserves drawn from a declining coal mining industry and, indeed, 80 per cent of those recruited were ex-miners (Hudson 1983a). However, many branch plants have sought reserves of female rather than male labour, as women have tended to be less unionized, less militant and more amenable to changes in work practices. It has been easier for employers to classify their jobs as lower skilled (irrespective of real skill content) and to pay lower wages. This has typified many of the branch plants established in the UK, whether textile factories in Cornwall (Massey 1984) or electrical goods factories in South Wales. For example, 88 per cent of workers in TV equipment factories and 80 per cent in electrical component factories in South Wales were women (Cooke and Rosa Pires 1985).

Considerable controversy has surrounded the role of branch plants (see Watts 1981 for a summary). The regions in which they are located can benefit from employment creation, the introduction of new technology, and the possibility of some manufacture of components or services being sub-contracted locally. However, they have also been criticized for the quality of jobs provided (deskilled and relatively low paid), for being subject to external control, for having only limited local suppliers, and for offering few opportunities for indigenous entrepreneurship. They are also subject to sudden large-scale redundancies, as large corporations adjust their overall company strategies to changing market conditions. These arguments apply to national and international networks of branch plants.

Direct decentralization: capital-intensive industries
In the 1960s and early 1970s there was decentralization of production in capital-intensive industries – such as iron and steel, chemicals and petro-chemicals – both within European countries, and at the European scale. This often involved the creation of large-scale industrial complexes. Examples include the oil and steel complex at Bari/Brindisi and the oil and chemicals complex at Siracusa–Augusta in southern Italy, the Tarragona oil and chemicals complex in Spain, the Dunkirk steel complex and the Fos steel and oil complex in France.

Decentralization of heavy industries has been encouraged by a number of factors, including a need for deep-water harbours, as a result of increased economies of scale in transport. In addition, increased economies of scale in production led to a need for ever-larger sites, and these were both more readily available and cheaper in peripheral than in metropolitan regions. Another attraction of decentralization, especially from northern to southern Europe, was the possibility of reduced production costs in countries with less strict environmental pollution laws. This is especially important in the petrochemical industries which produce large quantities of waste that are expensive to dispose of cleanly.

The process of decentralization has been greatly assisted in many countries by the role of the state, which has often subsidized the costs of developing green-field sites in peripheral regions. This may involve direct grants to companies, as in the case of, say, the oil and chemicals industry at

Teesside or Milford Haven in the UK. Alternatively, it may involve indirect subsidies via state expenditure on roads, housing or water treatment plants for these new complexes, as at Fos in France or Sines in Portugal (see p. 310). Damette (1980) refers to this as 'the devalorization of capital', effectively a writing-off by the state of part of the value of the capital invested.

Creation of new industrial complexes is very different from the other forms of productive decentralization discussed here, for it involves large-scale capital-intensive plants, which offer relatively little employment. After the initial construction phase, they provide only a small number of jobs, and many of these are for highly skilled technicians who will tend to be immigrants rather than locals. Such complexes also tend to have relatively weak local multipliers, for their links – in terms of inputs and outputs – are mostly outside their region. The scale of the enterprises, and their dependence on external corporate management, also offers relatively few opportunities for development of indigenous entrepreneurship.

Indirect decentralization: within metropolitan areas
Large companies can reduce their operating costs by sub-contracting parts of production to a network of formally independent (but economically dependent) small firms, within the metropolitan area. Such firms may have considerable cost advantages, arising either from their small scale or from their position in the informal economy. They may operate from cheap but poorly built and poorly equipped premises which contravene the legal norms governing health and safety standards, such as would be applied in large, formal factories. Their scale also means that labour relations and salaries are determined in a paternalistic context, leading to a generally more flexible, less militant and cheaper workforce. This is formalized, in some countries, by small firms being exempt from some legal requirements; for example, in Italy firms with fewer than 30 employees have been exempted from some of the national guidelines on minimum wages.

Examples of such sub-contracting are to be found in a number of industries, but are especially important in textiles and metal products. In vehicle production, British car producers have traditionally depended on vast networks of small-scale component manufacturers, many of which were located within the west Midlands. In Italy, the Morini motorbike factory in Bologna sub-contracts almost all of its component manufacturing to small local firms. In textiles, many of the leading fashion houses in London, Paris and Milan are dependent on networks of small-scale clothing workshops or outworkers. This process is particularly facilitated by the small-scale, specialized batch production of many clothing products; that is, the need to produce small quantities of a particular clothing item quickly. This form of outworking has become more important in the European clothing industry in recent years, in response to competition from low-cost producers in the less-developed countries.

Indirect decentralization: diffuse industrialization

Diffuse industrialization typically involves the growth of small-scale, independently owned firms in dispersed locations in rural areas. In common with branch plants, they are dependent on cheap labour; but they are also different from these, and not only in terms of ownership. As Hudson (1983a) emphasizes, diffuse industrialization is dependent on reproducing existing labour conditions (of a dual labour force), whereas branch plants tend to create a proletariat and hence, ultimately, transform local labour markets. Much of the literature on this theme originated in Italy as an attempt to explain rapid industrial expansion in the north-east and central regions (Bagnasco 1977). It has subsequently been extended to other southern European countries, including Spain (Grandados 1984), Greece (Kafkalas 1984) and Portugal (Lewis and Williams 1987).

In the variety of attempts to account for the employment trends in the north-east and centre of Italy, there is general agreement that small-scale establishments and diffuse industrialization have played an important part throughout. The stress on small firms as a source of dynamism went against the grain of much contemporary thinking about industrial development, which emphasized an evolution of industrial economies towards the domination of large organizations. However, the fall in average plant sizes throughout the industrial world and proliferation of research on small manufacturing firms (see pp. 164–7) has prompted a revision of such views. There is also greater appreciation that there are a variety of types of small enterprises. Brusco and Sabel (1981) distinguished between artisanal firms (using traditional methods to supply a local market), dependent small firms (carrying out standardized production under sub-contract to large firms), and independent small firms (innovating in products and selling in a variety of markets). Disagreement over the relative weight of these three types of firms lies at the heart of a debate between two different interpretations of the diffuse industrial growth of rural northern and central Italy.

One view emphasizes the dependent nature of much of the growth of small enterprises in these areas (e.g. Arcangeli *et al.* 1980). The expansion of such firms is explained as a result of a decision to decentralize production (especially from Turin and Milan) by larger enterprises. Small enterprises in rural areas in the north-east and centre could operate with lower initial investment costs than in urban areas, by converting existing agricultural buildings for industrial use. They also pay lower wages by using a combination of family labour, homeworking and 'double labour' (in which industrial workers, or their households, continued to work their agricultural holdings and so have a second source of income). Hence the diffuse industrialization that characterizes the north-east and centre of Italy is analysed much as if the formally independent small establishments were branch plants of firms like Fiat or Benetton scattered among the villages of Marché.

In contrast, there are commentators, such as Fuá (1983), who treat the phenomenon of rural industrialization as the basis for a new model of development, in which the benefits of modern growth are introduced without communal upheaval. While not denying that some of the small firms in the region are dependent on large national and international corporations, their accounts stress the modernization of artisanal production and subsequent cooperation among small producers which give rise to independent growth. Fuá's (1983) explanation of 'the NEC model' blends the initial character of these rural areas, in terms of good infrastructure and access to urban services, with the high proportion of self-employed workers in agriculture to account for the rapid emergence of industrial entrepreneurs during the 1960s and 70s. He points to improvements in communications and the introduction of quite sophisticated production technologies as factors contributing to the development of international exports from such regions. Garofoli (1984) also stresses the development of 'area systems' in which specialization allows small firms to benefit from economies of scale by sub-contracting among themselves. Despite reliance on external (even export) markets the dynamism of such area systems is regarded as essentially endogenous.

However, there is little disagreement between the two views about the labour market conditions that have characterized these regions. Fuá (1983) stressed the importance of a dominance of family-run agricultural enterprises and other forms of self-employment as an initial condition for industrialization. This is significant as a source of 'management experience' which can lead to the creation of small industrial initiatives. It also plays a role in keeping industrial wage claims low, since the new labour force can still live at home, with access to supplementary farming activities, while the lack of social distinctions between new entrepreneurs and workers leads to cooperation rather than conflict (thus trade union membership is low). Their origin in the families of agricultural workers gives these workers a 'safety net' which means that 'they are not "proletarians"' (Fuá 1983, p. 10). Bagnasco (1982) also mentions these features and draws attention to the flexibility of working practices that is found in both agricultural self-employment and small-scale manufacturing work.

Industrial sectors: selected case studies

While the previous discussion has outlined the broad development of manufacturing in post-war Europe, industries have to be considered individually. This is because each European industrial sector has a different global role and has very different conditions of production. Furthermore, these conditions are subject to increasingly rapid technological and economic changes. The positions of particular firms or even entire industries can change dramatically in only a few years, so that today's sunrise industry can soon become a sunset industry. Even some of the largest and best-known manufacturing firms may be rapidly eclipsed – as happened to Alfred Herbert, the UK machine-tool firm in the 1970s, or Kockums, the Swedish shipbuilding company in the 1980s.

Kockums is a particularly illuminating case study. In the early 1970s it was Europe's largest shipbuilding company and the world's leading builder of bulk carriers, turning out a supertanker once every 40 days. Yet, by February 1986, the company had been liquidated (apart from a small remnant, constructing specialized naval vessels). The turning point for Kockums was the oil crisis of 1973–4 which eventually led to a 46 per cent fall in seaborne oil trade over the next decade. As a result, many tankers were 'mothballed' and orders for new ones were rare. In an increasingly competitive market, Kockums were unable to compete with lower-cost producers in Japan and South Korea (see Figure 18) and with highly subsidized competitors in Europe. The company rapidly moved from a position of healthy profits to one of mounting losses and, in 1977, it was nationalized. Thorough restructuring was attempted with some less profitable yards being closed, and two-thirds of the 30,000 labour force being made redundant. However, the yards still lost £976 million between 1977 and 1985, which led to liquidation in 1986.

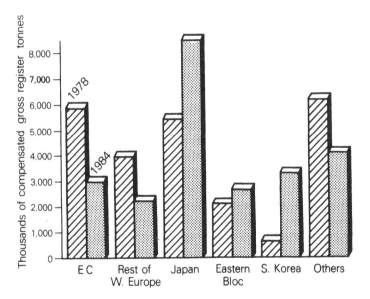

Figure 18   Shipbuilding in selected countries/regions, 1978 and 1984

SOURCE: Lloyds Register of Shipbuilding.

Against this rather depressing example can be set contemporary success stories. Leading companies such as Italy's Benetton in textiles or Sweden's Pharmarcia in biotechnology have expanded from local to European-wide operations in little more than a decade. Both types of examples underline the rapidity of change and the need for detailed examination of particular industries. Lack of space precludes consideration of all industries, but four sectors are discussed here which illustrate different conditions of production and structural adjustment: these are textiles, iron and steel, vehicles and electronics.

Textiles and clothing

In the European Community in the mid-1980s, textiles and clothing accounted for 1.8 million jobs, representing some 9 per cent of the manufacturing labour force (Commission of the European Community 1985c). Even these high levels represent a considerable decline in the relative importance of the industry in Europe for, in the 1970s alone, the production of textiles fell by 15 per cent and employment by an even larger proportion. These trends reflect the emergence of a new international division of labour which has resulted in shifts in production to eastern Europe and to Asia.

Given the relative stagnation of demand since 1973 (expanding at only half the average growth of demand), there has been a radical reorganization of the industry in the face of such competition. Within western Europe there have been shifts to lower-cost centres of production in the south. Another response has been restructuring within countries such as France and the UK, with greater vertical integration and concentration proceeding alongside continued reliance on small-scale clothing factories and outworking. Not surprisingly, the extent of job losses in western Europe's textile and clothing industry have been very uneven. Whereas 954,000 jobs were lost in Belgium, Denmark, France, West Germany, the Netherlands and the UK between 1965 and 1977, 285,000 new jobs were added in southern Europe, in Greece, Italy, Portugal and Spain (Hudson *et al.* 1984).

The branches of the textile and clothing industry have been affected in different ways because it is very amenable to development of new spatial divisions of labour. It involves at least three major stages: fibre production, cloth production and final product (whether home furnishings, clothing or industrial textiles). It also involves the use of natural fibres and synthetic fibres, as well as mixtures of these. As a result, the industry has a complex and varied organization. This is considered in terms of ownership/scale of production and the spatial division of labour.

Scale/ownership
There are major organizational differences between fibre/cloth production and clothing manufacture. The first of these is subject to considerable economies of scale in production, given the possibilities for automation and mass production. In contrast, clothing is a small-scale industry with

tens of thousands of small companies producing limited batches and narrow product lines. In the EC alone there were 7000 clothing companies in 1985, with an average of 100 employees – and this figure excludes firms with fewer than twenty workers (Commission of the European Community 1985c). There has been partial automation even of clothing production, and cutting and sorting of cloth can be guided by computers, although it is only profitable for large runs and, therefore, for only a handful of companies. Apparel production remains labour-intensive: 'fabric is laid out, cut and sewn much as it was 40 years ago' (Toyne *et al.* 1984, p. 16). There is even evidence that homeworking has increased in importance in some countries in recent years (Mitter 1986). The differences between textile and clothing production are reflected in investment levels per worker being almost three times higher in the former than in the latter (Commission of the European Community 1985c).

Despite the tendency to fragmentation, there has been a contrary trend towards greater concentration in the textile and clothing industries. This has been brought about by increased economies of scale in fibre production and for some standard clothing products such as jeans, and greater vertical integration of the industry through from high-street retailing to fibre manufacture. Within Europe, concentration in textiles has proceeded furthest in the UK, with France and the Netherlands revealing medium levels, while Italy actually experienced relative deconcentration owing to extension of sub-contracting and outworking (Figure 19).

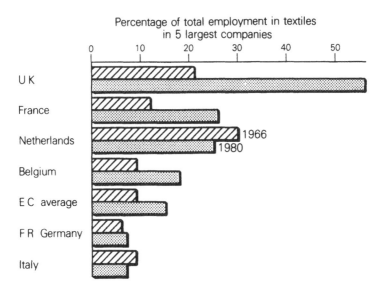

Figure 19 Percentage of employment in texiles in the 5 largest companies in selected countries, 1966 and 1980

SOURCE: based on Toyne *et al.* (1984).

The details of scale and ownership are far more complex, revealing, inevitably, a strong measure of polarization. For example, in France there are a small number of large companies, such as Prouvast, which have vertically integrated fibre and clothing production, but there are also enormous numbers of small-scale clothing firms, especially in Paris. In West Germany there are some very large-scale fibre producers, such as Hoeschst and Anic Fibre, operating alongside very large numbers of small-scale and mostly family-owned clothing factories.

The UK is the exception within Europe. Not only is there a high level of concentration, but it also has two of the world's largest textile and clothing companies (Table 27). Courtaulds is the world's third largest company, although a large proportion of its income comes from its diversified activities, while the Coats Viyella group is a new conglomerate formed by the merger of Coats Patons and Vantona Viyella in 1986. The reasons behind this merger reflect much of the underlying logic of the tendency to concentration: that is, the requirements of internationalization, diversification and integration:

> With operations in 30 countries, most of them complementary to Vantona Viyella, Coats was the perfect partner. Vantona is already the UK market leader in household textiles (Dorma sheets), shirts (Peter England, Rocada, Viyella), uniforms and knitwear. It shares with Courtaulds the role of main supplier to Marks and Spencer and is strong in carpets and hosiery. All these sectors are now capable of being expanded through Coats's overseas connections. In addition, it has gained Coats's high class Jaeger retailing chain

Table 27   *The world's leading textile and clothing firms in 1985*

Company	Nationality	Turnover, 1984–5	Textile business
C. Itoh	Japan	£5.82 billion	Apparel fabric, natural and synthetic fibres
Marubeni	Japan	£3.92 billion	Staple fibres, yarns, fabrics, industrial materials
Courtaulds	UK	£2.2 billion	Apparel fabric, contract and branded clothing, fibres, furnishing fabrics
Burlington Inds.	US	£1.99 billion	Largest and most diversified textile manufacturer in USA
Coats Viyella	UK	£1.69 billion	Fashionwear, home and industrial sewing products, yarns and fabrics
Levi Strauss	US	£1.38 billion	World's largest manufacturer of jeans
J.P. Stevens	US	£1.18 billion	Apparel fabrics, natural and man-made fibres and yarns

SOURCE: *Financial Times*, 12 February 1986.

and Jean Muir name into which it would be possible to put some of its Van Heusen and Viyella clothes. (*Financial Times*, 12 February 1986)

This type of rationalization and the substitution of capital for labour is one strategic response by large companies in western Europe to increased competition from the less-developed countries (other responses are discussed on pp. 178–9). Its importance should not be underestimated for, in terms of job losses, labour productivity gains have been more important than import penetration in many countries. For example, a study of the Manchester clothing industry showed that 61 per cent of all job losses stemmed from labour productivity changes (Gibbs 1983). However, it also has to be recognized that rapid gains in labour productivity (through greater investment and reorganization of the labour process) have been occasioned by increased international competition.

International divisions of labour
The evolution of the international division of labour in textiles production is complex in detail, but Toyne *et al.* (1984) provided the following summary of the stages involved:

Stage	Characteristics	Examples
Embryonic	Simple fabric and garment production from natural fibres for domestic markets; some imports	Least-developed countries
Exporting	Exports of 'mature', mass-produced cheap imports	India
Improved exports	More sophisticated products; development of fibre manufacturing	South Korea
Golden Age	Extensive synthetic fibre production; diversified product mix; emergence of MNCs	Taiwan
Full maturity	Output rises but employment falls as capital investment increases	Japan, Italy
Decline	Employment falls and greater concentration occurs; greater specialization and outward processing are two possible survival strategies	UK, West Germany, France, Belgium, Netherlands

Most western European countries are in the 'decline' phase, although Italy is still at an earlier stage and, indeed, will not necessarily progress beyond this. Other southern European countries such as Portugal and Greece do not fit neatly into this pattern but, being mainly tied into exporting high-quality products under contract, and lacking their own MNCs, probably fall roughly into the same groups as South Korea or Taiwan.

The decline of the mature European producers is fundamentally a consequence of their inability to compete with lower-cost clothing producers, either in southern Europe or in the less-developed countries (although the EC still has a surplus in textiles). In essence this stems from the differential price of labour, for, excluding materials, labour accounts for about four-fifths of the production costs of clothing (Steed 1981). The range of hourly labour costs tells its own story: over $8 an hour in most of northern Europe, $4.48 in Spain, $1.88 in Portugal and only $0.75 in Morocco (Toyne *et al.* 1984). Given the difficulties of automating clothing production, western European countries have found it very difficult to compete in the markets for mass-produced, staple clothing. Not even transport costs are sufficient to outweigh lower wages; shipping cotton shirts to the UK from Hong Kong, for example, adds only 14 per cent to production costs.

The response of northern European companies to this dilemma has been two-fold. One strategy has been to utilize cheaper, foreign workforces. This can be via direct investment in MNC branch plants or via sub-contracting; both approaches can secure markets for a company's basic yarn and cloth. The finished item may then be sold in third-country markets, or reimported. In the latter case – known as outward processing – overseas firms are provided with the fabric and the design, and undertake the labour-intensive task of making up the clothes before they are sent back to the country of origin. This has been particularly important in West Germany where some 70 per cent of firms in the garment industry have some overseas production (Fröbel *et al.* 1980), especially in eastern and southern Europe. The Dutch and Belgian clothing industries are also partly dependent upon outward processing, as are individual companies in other countries.

A second strategic alternative is adjustment of domestic production conditions. This can take different forms, including limited automation or relocation of production to lower-cost regions, as has occurred, for example, in the UK in the 1970s (Massey 1984). There have been shifts in textile and clothing production away from London, West Yorkshire and Lancashire to smaller towns, rural areas and to other Development Areas; this is in response to the need for flexible, low-cost female labour as the competition from other industries, especially offices, has increased metropolitan labour costs. Yet another strategy has been to develop sub-contracting to small suppliers or to outworkers who, frequently, are immigrant women. In the UK, Mitter (1986) found that such workers might be paid no more than £1 an hour, which is on a par with wages in

Hong Kong or South Korea. Outworking and sub-contracting offer the larger firms other advantages, including reduced overheads and flexibility to take on or lay off labour in the face of a notoriously unpredictable demand. A similar strategy is also adopted by the large high-street retailers; Marks and Spencer, for example, who have 29 per cent of the UK retail clothing market, use more than 600 different suppliers.

Partly as a result of the labour strategies outlined above, northern European companies have been able to reverse, or at least halt, losses in their domestic markets in recent years. They have also been aided in this by greater production differentiation, which allows companies such as Laura Ashley or Next to charge relatively high prices for their goods. Above all, they have been assisted by a shift in demand to more rapidly changing fashion goods. As a result, retailers want to place small orders and obtain quick deliveries. In this respect, UK, Italian or French producers can move goods from factories to shop shelves in a matter of days, while Asian competitors take weeks.

The need for small batches produced in short time periods reinforces the advantages of sub-contracting small orders. Developments in new technology have facilitated this, for electronic monitoring systems allow retailers to keep close checks on sales and to reorder quickly. Given the costs of such technology, it tends to increase the relative advantages of large-scale retailers over the independents. Italy's Benetton provides a classic example of the effective operation of a vertically integrated clothing firm. Whereas in 1978 Benetton had fewer than 500 outlets, all of which were located in Italy, by 1986 it had 4000 retail outlets in 57 countries. Its growth has been based on its strong marketing ability, and flexibility in production arising from use of a very large number of small-scale suppliers, mostly in northern Italy.

Trade protectionism
Although the successes of companies such as Benetton have been impressive, since the 1960s most western European countries' textile and clothing industries have been in retreat in the face of cheaper imports. Given the importance of the industry, most governments have developed policies to support domestic producers (Toyne *et al.* 1984). The Belgian Government, concerned with job losses, has provided subsidies for firms, on condition that at least 90 per cent of their workforces are retained after restructuring. In France, government response was the classic strategy of encouraging mergers so as to create large, modern competitive firms. This has not always proven successful, as was highlighted by the collapse of the Agache–Willot group, so that trade protectionism is also favoured. The British Government has also indulged in a mixture of temporary employment subsidies and trade protectionism in order to support the industry.

In addition to these measures by individual countries, the advanced economies have also taken collective action via the Multi-Fibre Agreement (MFA). This was initially signed in 1973, with the aim of securing voluntary export restraint agreements with low-cost producers. The first phase was not especially successful, but the new round of the MFA agreement in 1977 did keep import growth to 5 per cent. In Europe, the MFA was reinforced by EC moves to negotiate voluntary agreements with individual Third World and southern European producers. For example, in 1979 the EC signed an agreement with Portugal for three years of voluntary restraint, during which imports expanded by only 3 per cent. With the enlargements of the EC in 1981 and 1986, it has become more difficult to impose such agreements on southern European producers.

Iron and steel

The UK was the world's dominant steel producer until the early twentieth century, when it was overtaken by the USA. With the later emergence of the USSR and Japan – taking 21 and 16 per cent respectively of world output by 1980 – Europe has never recovered its dominant position. Even so, up to the first oil crisis in 1974 world demand increased by about 6 per cent a year and, in this climate, output in western Europe increased from 63 to 185 million tonnes (Ilbery 1981).

The iron and steel industry has faced a number of difficulties in recent years. Between 1974 and 1984 output fell by 30 per cent (Commission of the European Community 1985d). Changes in long-term demand have moved against steel consumption, owing to the shift to services, the fact that expanding branches of manufacturing (such as aerospace and computers) use little steel, and the substitution of new materials (such as plastics) for steel. For example, in West Germany, steel input per unit of finished product has fallen by 23 per cent in shipbuilding and by 10 per cent in electrical machinery in recent years (Ballance and Sinclair 1983). In addition, western Europe has suffered from intense competition, especially from eastern Europe (Warren 1985) and from a few NICs, such as South Korea, Brazil and Mexico, and has had a declining share of a shrinking world market (see Figure 20). The competition has been felt in different ways, including direct and indirect imports to western Europe. Indirect imports have been particularly important, and it has been estimated that, in the late 1970s, imports of cars alone were equivalent to a loss of markets for steel amounting to one million tonnes (Turner 1982). In addition, western European producers have lost markets in Africa, Asia and Latin America. An example of this was the 1986 agreement between the South Korean companies Pusan and Samsung and the Tunisian SIAPE company to manufacture steel pipes in Tunisia, hence reducing the opportunities for European companies in this market.

Western Europe has found it very difficult to compete against the new

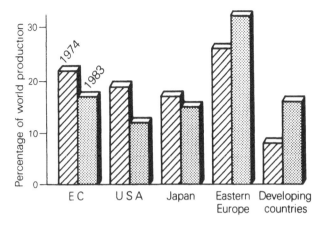

Figure 20 Steel production in selected countries/regions, 1974 and 1983

SOURCE: Commission of the European Community (1985d).

centres of production which tend to have more modern plant, lower wages and, in some cases (as in Brazil), easier access to raw materials. Despite attempts to restructure the European industry – reducing manpower and introducing greater automation – production costs are still 15–20 per cent higher than in south-east Asia and South America. Furthermore, while labour costs are roughly equivalent to those in Japan, the latter is far more cost-effective; for example, in the early 1980s it required 4.5 man-hours in Japan compared with 7.4 man-hours in the EC to make one tonne of steel (Commission of the European Community 1985d). Japan established its leading role in the steel industry in the 1960s when large (8–10 million tonnes capacity) integrated, modern plants were established to serve domestic and export markets. In contrast, the older-established iron and steel industry in Europe tended to be on a small scale, fragmented and characterized by dispersed ownership. Subsequently, much post-war investment has been necessary to concentrate and integrate production in fewer but larger plants, while at the same time closing older and smaller plants, with consequent large-scale redundancies.

The spatial organization of steel production in western Europe
There have been two major forms of spatial reorganization of steel production in Europe since 1945: increased concentration and locational shifts resulting from new conditions of production.

There is not an international spatial division of labour for iron and steel production as there is for textiles. Instead, the industry is characterized by vertical integration and considerable economies of scale. Whereas plants of 600,000 tons were typical in the 1920s, 3–10 million tonnes had become the norm in the 1970s. In part this was dictated by technological developments which increased the optimum scale for joint operation of blast furnaces and rolling mills. While Europe shared in this trend to larger plants, it nevertheless lagged behind some competitors; for example, the ten largest plants accounted for 69 per cent of output in Japan compared with only 35 per cent in western Europe in 1976 (Turner and McMullen 1982).

The history of steel production in the UK is not untypical of post-war developments in Europe. In 1945, an output of 12 million tonnes came from companies with a median size of 0.2 million tonnes but, by 1965, 27 million tonnes came from companies with a median size of 0.4 million tonnes. However, the industry was still not competitive and profits fell sharply in the late 1960s, leading to the 1972 Plan for Steel which aimed to concentrate future production in just five major integrated plants (Warren 1978; Hudson and Williams 1986).

There have also been changes in the geographical distribution of the steel industry within Europe (see Figure 21). Between 1952 and 1972 output increased in all European countries but, even at this stage, differential growth rates emerged. While UK production increased by about 50 per cent, that in France more than doubled. Output also expanded significantly in West Germany, although foreign companies (such as ARBED of Luxembourg) accounted for 40 per cent of this. Even more dramatic was the emergence of southern Europe, with output increasing five-fold in Italy and ten-fold in Spain, although from low bases. In the 1970s, global recession affected most western European countries, with reductions of 20–30 per cent being common by 1982. The UK suffered most for, faced with a poorly competitive industry, the state-owned British Steel Company reduced capacity and output by about a half in a decade. However, the steel industry in Belgium, France and the Netherlands also experienced difficulties. The only countries which continued to expand output through to 1982 were Spain, Italy and, to a lesser extent, Austria and Finland. However, even in these cases there have subsequently been formidable difficulties. Voest Alpine, the nationalized Austrian steel company, lost £550 million in 1985, partly because the state had made it take over an unviable steel mill at Donauretz in Styria. The company's response to this was to try to diversify into commodity chips, which led to further losses. As a result, after the 1985 financial crisis, moves were taken to shed 10,000 workers as part of a four-year restructuring plan.

The iron and steel industry has also undergone interregional shifts following changes in the conditions of production. Unlike the textile industry, which is characterized by a quite complex spatial division of labour, the iron and steel industry has been subject to increased spatial integration of the stages of production since 1945. Therefore, the regional shifts have tended to involve wholesale relocations of production through selective closures and development of, usually, green-field locations. This has led to shifts from the coalfields (as in South Wales, Wallonia, the Ruhr or the Saar) and the ore fields (as in Lorraine or at Corby), and expansion at or near deep-water harbours for ore imports and steel exports. Whereas

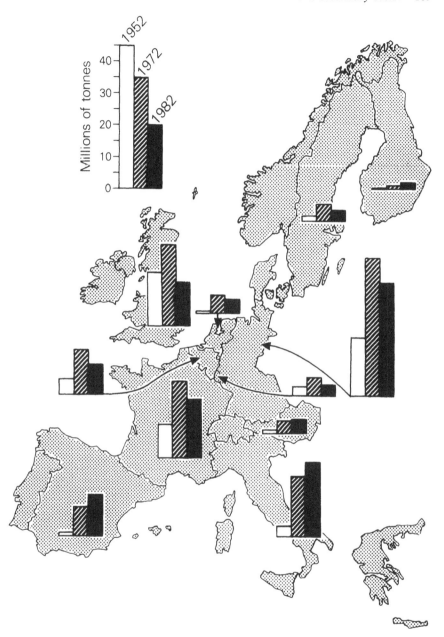

Output of less than 1 million tonnes is not shown

Figure 21 Steel production in western Europe, 1952, 1972 and 1982
SOURCE: UN Statistical Yearbooks and Eurostat (1984a).

in 1938 there was not a single coastal integrated iron and steel works in western Europe, by 1977 there were seventeen (Minshull 1980) located in two major clusters around the North Sea (e.g. Dunkirk, Zelzate and Bremen) and on the Mediterranean shoreline (e.g. Taranto and Fos). The rate at which these shifts occurred was quickened by a need to seek out more profitable locations in an increasingly competitive environment.

Restructuring in the face of global crisis
Despite the first signs of falling demand in 1974–5, several countries already had long-term investment plans in progress so that production capacity continued to increase during the 1970s. As a result, many plants came to operate at below their optimum levels; in the EC, for example, the industry was operating at only 57 per cent of capacity by 1983 (Commission of the European Community 1985d). Prices inevitably fell in the face of intensified international competition, to be followed by accusations of dumping and a drift to greater protectionism. Most companies faced diminishing profits at best, or bankruptcy at worst; for example, the two French state-owned steel companies, Usinor and Sacilor, had losses of £390 million in 1985. Only a few companies managed to stay in profit, such as Germany's Thyssen and the Netherlands Hoogvens with net returns of £113 million and £67 million, respectively, in 1985.

Responses to the crisis were conditioned by the high level of state ownership in the industry, and involved a mixture of government policies and company strategies. The most important responses were trade protection/subsidization, restructuring so as to increase competitiveness, or diversification. Diversification has assisted the German industry where both the largest (privately owned) companies, Thyssen and Krupp, have considerable metal manufacturing interests (implying guaranteed markets) as well as iron and steel plants. The French Government, which nationalized the steel industry in 1981, has also encouraged Sacilor and Usinor to diversify by, for example, taking over the specialized steels and plant construction interests of the Pechiney–Ugine and Creasot–Loire groups (Tuppen 1983). In contrast, the British Steel Company (BSC) in the UK has suffered from lack of diversification and heavy reliance on basic steel production. It has not even had the partial cushion available to the Swedish industry – that is, of concentration on special steels.

Diversification only offers limited support to steel producers, and most countries have experienced considerable restructuring; nowhere is this more evident than in the UK and France. In the UK, the BSC in the 1980s has been forced into large-scale reorganization, involving closure of plants such as Consett and Corby, by the government's insistence that it became self-financing by the mid-1980s (Sadler 1984). Restructuring was achieved by concerted government action to weaken trade union resistance, and a willingness to pay relatively high redundancy payments. The labour force was reduced from 220,000 in 1974 to 54,000 in 1986, by which date productivity equalled the best in Europe. As a result, the BSC made a profit (of £38 million) in 1985–6, for the first time in ten years. In contrast, restructuring in France has encountered stronger resistance. Here the roots of the problem lie in the shift of investment to new coastal plants, such as Dunkirk and Fos, in the 1960s. In the face of the mid-1970s crisis, the logical response, from the steel companies' perspective, was to close old plant in Lorraine and Nord so as to concentrate on the new coastal complexes. Implementation of this plan required about 30,000 redundancies by the early 1980s. There was region-wide opposition to this in Nord and Lorraine, culminating in civil unrest and pitched battles with the police, especially at Denain. This resulted in greater government commitment to create new jobs in these regions. The election of a socialist government in 1981 led to a postponement of cuts, so that when these had to be implemented in the mid-1980s they were particularly severe. By 1985 there were only 4200 steel workers in Lorraine, for example, where there had once been 30,000 jobs (Warren 1985). Cutbacks are also common elsewhere in western Europe. For example, in 1985 Italy's Finsider lost £411 million, while Belgium's Cockerill–Sambre group cut back on its older steel plant in Wallonia.

Given the importance of the steel industry, in strategic and (regional) employment terms, most governments have not been content to rely just on diversification or restructuring as responses to the crises. Instead, most governments have subsidized the steel industry: by writing-off or carrying large debts for nationalized companies, as in the UK; providing aid at low rates of interest, as in France (a plan for £1.7 billion was agreed in 1985); or paying subsidies to private producers, as West Germany has done in Saarland. Most countries have also invoked some form of protectionism. Significantly the EC has also been involved in this.

When the ECSC was created in 1952, its primary aims were to reduce tariffs between member states and encourage trade. However, by 1977, in response to the severity of the steel crisis, the EC was moving to greater protectionism and regulation. An important landmark in this was the 1977 Davignon Plan which included provisions to reduce capacity, coordinate price increases and, utimately, abolish subsidies. However, as a short-term expedient, voluntary export agreements were signed with most leading competitors so as to limit steel imports to the EC. Deepening of the recession in 1980 shattered voluntary cooperation, leading to the EC taking the unusual and dramatic step of declaring a 'manifest crisis' in the industry. This gave it power to set mandatory quotas and production targets for individual member states and to order a cessation of state subsidies (initially, by 1985). Although frequently violated, these have had an effect in reducing capacity, notably in the UK and in France (Morgan 1983). Capacity within the EC as a whole fell from 169 million tonnes in 1980 to 142 million tonnes in 1985 (Warren 1985).

Cars

Vehicle production is one of the most important industries in western Europe, accounting directly or indirectly for some 10 per cent of non-agricultural employment. It also accounts for 10 per cent or more of merchandise exports in several countries, including West Germany, France and Italy (Ward 1982). Yet, in common with many other branches of manufacturing, a phase of expansion up to the 1970s has been followed by more difficult market conditions thereafter.

Between 1950 and 1973 annual vehicle sales in western Europe increased ten-fold, from 1,595,000 to 13,280,000 (Bloomfield 1981) as rising real incomes permitted a widening of the social basis of car ownership. Ownership rates increased from 73 to 225 cars per inhabitant in western Europe between 1960 and 1973 alone (Ballance and Sinclair 1983). This enormous increase in demand was met largely by production within Europe by both indigenous and non-European companies. As a result, Europe's share of global car production doubled between 1950 and 1977 (see Table 28), while the USA's dominance crumbled.

After the oil crisis in 1973, market conditions have proven more uncertain and overall demand has tended to fluctuate. The reasons for this are not difficult to discover. Both oil crises in the 1970s had strong negative consequences for car prices, because of the general effects on incomes and consumption and because of the specific increases in petrol prices. Even the USA giants, Ford and General Motors, recorded trading losses in some years in the 1980s. In addition, demand was already approaching saturation levels within northern Europe so that output became much more sensitive to business cycle fluctuations; quite simply, it is easier to defer replacement purchases than first-time purchases of vehicles (Turner and McMullen 1982).

At the same time, western-European-based car producers came under increasing pressure from other countries, especially Japan, and Europe's share of global output peaked as early as 1970. The Japanese success story is astounding; net imports of 100,000 vehicles in 1950 had, by 1980, become net exports of 3.8 million vehicles, of which about three-quarters of a million were destined for western Europe (Commission of the European Community 1982d). Their competitive strength is based on extensive use of small-scale contractors (Toyota relies on 44,000), combined with automation and intensification of the labour process in their own factories (Ballance and Sinclair 1983). This has given Japanese producers production costs of some 20–30 per cent below those in western Europe. As a result they have made steady inroads into European markets, taking almost 10 per cent of all sales in 1985, and have been able to out-compete western European producers in traditional Third World export markets. Other competitors, such as South Korea, began to expand their exports into Europe in the

Table 28 *World car and vehicle production, 1950–85*

a World vehicle production in 1950 and 1977*

	1950		1977	
	Numbers (000s)	*Market share (%)*	*Numbers (000s)*	*Market share (%)*
North America	8,394	79.4	14,478	35.2
Western Europe	1,595	15.1	12,759	31.0
Eastern Europe/USSR	394	3.7	3,133	7.6
Japan and India	46	0.4	8,604	20.9
Latin America	21	0.2	1,436	3.6
Australia/South Africa	127	1.2	709	1.7
Totals	10,577	100.0	41,119	100.0

b World car production, 1985†

	Numbers (000s)
F.R. Germany	4,167
France	2,632
UK	1,048
Italy	1,389
Spain	1,230
Netherlands	108
Belgium	229
Western European totals	11,204
USA	8,182
Japan	7,647

* Bloomfield (1981).
† *Financial Times*, 14 October 1986.

1980s but – unlike with the textile industry – European car production has not collapsed. Demand has held up relatively well (being broadly static), while high transport costs, combined with protectionist policies, continue to favour investment in Europe – even if this is by American and Japanese, as well as by European companies.

Intensified competition: international survival strategies
The internationalization of production began relatively early with Ford establishing its first branch plant in the UK at Manchester in 1911, and in Germany at Berlin in 1925. General Motors also arrived in Europe at this latter date; it relied more on takeovers, acquiring Vauxhall in 1925 and Opel in 1928. Both American companies chose to produce in Europe

because of the high costs of transporting assembled vehicles and high import tariff barriers – some 212 per cent in Italy and 88 per cent in the UK in 1914 (Ballance and Sinclair 1983).

Given the intensified international competition since 1950, European companies have also developed international networks of production. The range of production locations is considerable: examples are Renault in Spain, West Germany, South America and Taiwan; and Fiat in Spain, eastern Europe and Latin America. In some cases this has been encouraged by the state so as to increase foreign-exchange earnings. For example, by 1965 Renault and Citroen, encouraged by the French Government, had 28 and 19 per cent respectively of production located abroad (Fridenson 1986).

Car production companies have responded in a number of ways to intensified international competition. One response was to develop new models so as to create market leadership through styling, such as the introduction of hatchbacks in the 1970s or technical advances such as the 'lean-burn' engines of the 1980s. Companies may also seek to open up new markets and, most obviously, this occurred in the case of MNC investments in Spain (a highly protected market). Other responses have included strategies to reduce production costs, either through greater economies of scale or by cutting labour costs, or diversification into more profitable activities.

Economies of scale can be achieved in many ways (see IWC Motors Group 1978). The number of parts required can be reduced, and for Ford this was one of the advantages of the Fiesta (1394 components) over the Escort (2140 components). Production costs for components can also be reduced by specialization, hence allowing greater economies of scale in each branch plant, and a distinct spatial division of labour. In the case of the Fiesta, carburettors were made in Belfast, gearboxes in Bordeaux and engines in Valencia and Dagenham. Europe-wide networks of branch plants are now common, although truly global networks still do not exist. Companies have also reduced the total number of models in their range, preferring to offer several variants of one car type rather than a number of different basic models. Even so, it is estimated that a minimum of four basic models are needed still to cover the full range in the car market, and this implies minimum annual sales of between one and two million cars, a figure that few European companies had the potential to achieve in the 1980s. This has pushed many companies into mergers so as to secure an optimum scale of operations. However, these are not always successful, as is evident in the unhappy experience of Citroën's takeover of Chrysler Europe. Another possible strategy is joint development of models, as in the co-production agreements between Leyland and Honda.

Reducing the proportion of labour costs in total production costs has also been an important strategy, but one which can be pursued in different ways. One approach is to locate some production stages in cheap-labour countries; for example, Volkswagen manufactures engines in Brazil and Renault makes gearboxes in Portugal. This is part of the creation of an international spatial division of labour. Alternatively, companies may seek to reorganize domestic production. This may involve weakening trade union power (as at Leyland in the 1980s), reliance on small-scale sub-contractors (for example, Volvo has some 1300 sub-contractors, 70 per cent abroad; see Lindmark 1983), or greater automation. Automation has been especially important in the 1980s, and Fiat has even made a virtue of robotics in their advertising campaigns. Yet another strategy is recruitment of immigrant labour which is usually less militant and more flexible. This has been favoured in West Germany, France, Belgium and the Netherlands where, in the 1970s, it was estimated that 70–80 per cent of production work was undertaken by immigrant workers (Counter Information Services 1977).

Diversification represents a rather different strategy. Fiat are a notable example, and are discussed later in this section. Another example is Daimler–Benz which, in 1985–6, bought MTU (diesel and aeroengines), Dornier (aerospace and electricals), and AEG (electrical goods). This has been part of a process whereby it was transformed from a car company to one of the largest industrial groups in Europe (see Table 9 in Chapter 2). This is a strategy which reduces the risks associated with dependence on car production in an uncertain market.

The car companies
National production patterns within Europe are broadly similar, with increases prior to 1973 being followed by smaller gains or even declines after this date. However, there are important variations within this broad picture. Up to 1979, at least, West Germany and France managed to increase production in absolute terms, but the UK and Italy had declining output in the period 1973–9 (Turner and McMullen 1982). While Italy has managed to reverse this in the 1980s, largely through the revival of the Fiat company, the French car industry has experienced problems during these years. The UK has declined in relative, and sometimes absolute, terms since the early 1970s. Spain has been the major new entrant into European car production. This was based on investment by such MNCs as Ford, Renault, General Motors and Nissan. Spain was attractive because it offered cheap labour, political stability (except in the mid-1970s), and access to an expanding domestic market.

Not all countries have volume car makers, so that markets in, say, Ireland, Greece or the Netherlands are dominated by non-indigenous companies. Moreover, even among the leading car producers there are considerable variations. Less than half the UK market is served by domestic sources, and the leading company is American. In contrast, sales in West Germany are dominated by German companies and domestic

Table 29   *Sales in the major car markets in 1985 (percentage figures)*

	Domestic	Imported	Leading companies share of total sales	
USA	77	23	General Motors	43
			Ford	19
Japan	98	2	Toyota	43
			Nissan	25
F.R. Germany	70	30	VW–Audi	29
			GM–Opel	16
France	65	35	Peugeot–Citroën	35
			Renault	30
Italy	60	40	Fiat	52
			Alfa Romeo	6
UK	41	59	Ford	27
			Rover	17

SOURCE: *Financial Times*, 26 August 1986.

sources (Table 29). Therefore, any analysis of car production has to focus on the individual car companies.

Car sales in western Europe in 1985 were dominated by six large companies, each of which had between 11 and 13 per cent of the market. The market-leaders were Italian (Fiat), German (VW–Audi), French (Renault and Peugeot–Citroën–Talbot) and North American (General Motors and Ford) in origin. No other company had more than 4 per cent of total European sales. These coexist with a number of smaller companies specializing in high-quality production. The major companies have different corporate strategies, and represent a mix of state and MNC capital. While General Motors and Ford have genuine European networks, the European-owned companies still have largely nationally orientated branch plant systems; for example, Renault have 84 per cent of their production in France. Nevertheless, taking a longer-term perspective, the companies have all been subject to concentration and internationalization. These two processes tend to proceed together, for larger companies have greater resources to develop international networks, whether through acquisitions or green-field investments.

Probably the most remarkable case of concentration is to be observed in the UK, where there were over 100 car producers before 1914. After Ford introduced assembly-line production methods (which require a minimum scale of operation), the number of companies rapidly declined to 88 in 1923 and to just four major producers by the 1980s. In Europe as a whole concentration, especially through mergers and takeovers, has led to the twenty major companies of 1960 being reduced to only twelve by 1980 (Ward 1982). Particular examples are the 1969 VW–Audi merger, Fiat's 1974 takeover of IVECO, and the Peugeot–Citroën merger in 1976 followed by their acquisition of Chrysler Europe in 1978. The driving forces behind such mergers are to increase market shares and increase economies of scale. In the mid-1980s, it was estimated that between one and two million units a year was the minimum for a viable volume producer, and only the six largest companies have this number of sales. Moreover, the shift to the European and to the global car concept has significantly increased research and development costs. The R & D costs of the Ford Fiesta are estimated to have exceeded $1 billion, and very few companies in Europe have that level of resources.

Some of these companies will now be considered in fuller detail so as to highlight their different strategies. Fiat provides one of the success stories of European vehicle production. It started production at Turin and eventually established six main plants in this city, and branch plants in the surrounding region. Fiat grew rapidly in the 1950s but it had a very rigid Fordist system of production. This was manageable while it was able to utilize unskilled migrant workers from the south, with little trade union experience. The company also sought greater vertical and horizontal integration and engaged in a number of takeovers, including Magnetti Marelli (sparking plugs) and IVECO (heavy vehicles), while setting up Fiatallis to manufacture earth-moving equipment (Bethemont and Pelletier 1983). Production conditions weakened towards the end of the 1960s as increasing labour militancy, culminating in the 'Hot Autumn' of 1969, led to guaranteed higher wages, labour-market rigidities and, ultimately, overmanning. Amin (1985) characterizes this as Fordism with weak management control. Labour productivity in Ford Germany, at this time, was two and a half times higher.

Fiat responded to these changing conditions in two ways. Greater automation was introduced (for example, electronically controlled welding in 1978) to reduce the labour content of production. The company also decentralized production to the south, especially to small and medium-sized plants in rural areas; while they were assisted in this by government regional development grants, these areas also offered more favourable labour conditions. For example, Amin (1985) found that in southern plants strikes and absenteeism were lower (the former by 30 per cent) and profits higher than the average for Fiat. The two processes were linked because Fiat located a restructured and technically more flexible labour process in the rural south, reliant on semi- or unskilled workers. There was also duplication of activities between plants so as to weaken worker power. Nevertheless, the company still faced falling profits and falling market shares in the late 1970s. The turning point came in 1980, for the management in that year, after a long strike, won the right to determine the number of workers employed and re-established the type of control it had exercised prior to 1968. Large redundancies followed and, in the wake of this, absenteeism was halved and labour productivity was doubled in just three years. With further automation, especially of welding and engine assembly, and the introduction of new models (such as the Fiat Uno), the company has returned to relatively healthy profits. Nevertheless, as a guarantee of longer-term profitability the company has diversified from vehicles, so that Fiat Auto accounts for only 50 per cent of turnover. Among other interests are Comau (production equipment), Telettra (telecommunications), Sorin (bio-engineering) and Fiat Aviazione (aero-engines).

General Motors and Ford present somewhat similar strategies. Until the 1950s they had separate manufacturing/assembly plants in a number of European countries. Then, in the 1960s, both companies sought to increase production and develop integrated European-wide component and assembly strategies, a move which was further encouraged by enlargement of the EC in 1973. They pursued this by initially locating within northern Europe in areas of labour reserves without a tradition of car production. Ford established plants at Merseyside in the UK, Genk in Belgium and Saarlouis in West Germany. General Motors set up branch plants at Merseyside in the UK, Strasbourg in France and Antwerp in Belgium. With rising costs in the 1970s, considerable restructuring has occurred in those plants. Both companies have also set up plants in Spain – Ford at Valencia and General Motors at Zaragossa – as further steps in internationalizing production (Sinclair and Walker 1982; Ward 1982), and creating duplicate facilities so as to weaken the bargaining power of labour. The companies have also benefited from their European operations being parts of much larger global corporations. This has meant that, for example, models like the Fiesta could be developed as global cars with all the economies this implied in terms of R & D, and that, in the 1970s, the General Motors parent company was able to invest huge amounts to re-equip its ailing Opel and Vauxhall groups.

Finally, British Leyland provides a case study of poor adjustment to changing conditions. When the company was formed in 1968, by a merger of BMH and Leyland Motors, it owned a large number of branch plants producing fifteen model ranges, many of which were in direct competition with each other (Law 1985a). Restructuring of the labour process, rationalization of the model range, introduction of new models, and modernization of production were required. The company lacked sufficient resources for this programme and it accumulated huge debts, leading to nationalization in 1975. Its share of the domestic market fell to only 18 per cent in 1982, compared with 40 per cent in 1969. Consequently, severe rationalization plans were carried out with more than ten major branch plants being closed between 1970 and the mid-1980s. Furthermore, because of the need to concentrate limited resources, the company has (against the trend) de-internationalized, selling-off or closing works in

Spain, Belgium and Italy. By the mid-1980s, British Leyland had less than 4 per cent of the European market, was partly reliant on Honda for technology, and the British Govenment was actively seeking to privatize it. The company was no longer a serious competitor at the European scale.

Government policies
The car industry, while not as crisis-prone as textiles, has nevertheless been subject to significant levels of state intervention. This is because of its size, its multiplier effects (which account for 20 per cent of steel and 15 per cent of rubber production in the EC), and the balance of payments costs of imports (Commission of the European Community 1982d). The range of policy initiatives is considerable.

Nationalization as a result of a financial crisis has occurred in many countries, notably of Renault in France and of British Leyland in the UK. Alternatively, the state may encourage mergers so as to create a more competitive industry. Again France, with the Citroën–Peugeot merger, and the UK with the BMC–Leyland merger, provide examples of this. The state may also provide subsidies to attract inward foreign investment. Examples are the Austrian Government's subsidies to General Motors to attract its engine manufacturing plant, and the British Government's subsidies to De Lorean to establish a sports-car assembly works in Northern Ireland. Governments may also use regional policy measures to influence the interregional distribution of production. This was important in Ford's decision to locate branch plants at Swansea and Merseyside within the UK. Similarly, in France, the state agency DATAR compelled Renault in the 1960s to locate branch plants in Brittany and Nantes (Anastassopoulos 1981). Finally, trade protectionism is widespread, particularly against Japan. The EC places an 11 per cent duty on Japanese car imports, while individual countries have their own measures; for example, the UK has negotiated a voluntary export agreement with Japan.

Electronics

Electronics is the sector which typifies the new growth industries of the post-1945 period. Annual growth rates, at least until the 1980s, averaged about 10 per cent (Ilbery 1981). This remarkable expansion was fuelled by several factors; but essentially it was driven by a boom in consumer durables and, latterly, by developments in microelectronics. The pace of innovation has been very rapid, particularly in microelectronics where there have been revolutionary advances in miniaturization and microchip technology. To a large extent consumer electronics and microelectronics need to be considered separately, although recent technological changes are tending to integrate these.

Consumer electronics

Consumer electronics are dominated by some very large, vertically integrated companies, mostly American or Japanese in origin. European companies – with the exception of Philips (Netherlands) and Electrolux (Sweden) – are relatively small. Electrolux actually became the world's largest producer of domestic appliances in 1985 through acquiring Italy's Zanussi and the USA's White Consolidated. At the time these were the 'jewels in the crown' of an acquisitions policy which had encompassed more than 100 companies in 40 countries within two decades. However, most European companies are just too small. For example, the minimum viable scale for television production is 400,000 units per annum, but many European companies are far smaller than this (Ballance and Sinclair 1983). Even some of the larger European companies, such as Grundig and AEG of West Germany, have struggled in open competition with Japanese companies. Both have had to restructure, abandoning some product lines and specializing in more upmarket products involving microelectronic components. In the same way, European companies have failed to compete head-on with the Japanese in a number of markets, including microwave ovens, calculators, stereo equipment and videos.

Japanese dominance has been based on a very careful strategy, involving enormous investment in research and development. Whereas UK companies devoted 1 per cent of their sales' receipts to R&D, Japanese companies have spent 12 per cent on this (Morgan and Sayer 1983). This has provided superior designs while costs have been reduced through high-volume production, automation, and reliance on a flexible and productive labour force, both at home and in offshore assembly plants in South Korea, Taiwan, etc. Furthermore, they have benefited from a virtuous circle of growth, for high-volume sales have sustained R & D expenditure and market leadership. For example, dominance of the colour television market (50 per cent share of world sales in the late 1970s) was used as a base to take a lead in video recorder research and, subsequently, sales. In the 1980s Japanese companies have sought to locate some of their production in western Europe, partly to overcome the protectionist measures instituted as a reaction to their dominance. For less-sophisticated products, such as colour televisions, they have often favoured the UK as their European base. It is the second largest market (after West Germany) and has relatively low labour costs.

The example of television production is particularly apposite since the European industry appears fragmented in comparison with Japan's. Philips is still a major manufacturer of tubes, but most European producers have been in difficulty; for example, Germany's AEG has had to merge its tube manufacturing interests with those of France's Thomson–Brandt. Production of television sets is even more depressing, for western Europe has too many (over 30) separate manufacturers given the size of the market

(Turner and McMullen 1982). That many of these companies have been able to survive at all is largely due to partial exclusion of the Japanese from the market for large television sets until 1980. Prior to this date AEG Telefunken controlled the licences for the European PAL transmission system and allowed Japanese companies only very limited production rights. Since the licensing rights ended in 1980, Japanese companies have made heavy inroads into this market to add to their dominance of small-set production.

European companies and governments have reacted in different ways to competition. West Germany has a relatively open consumer electronics market, and companies such as AEG–Telefunken and Grundig have found it difficult to compete. The British Government has also operated a largely open-door policy. This has been beneficial in that the UK has become an export platform for Japanese companies, although the fragmented British industry, with its weak productivity and poor R & D record, has been devastated. Thorn abandoned tube production in 1976, while Decca's television factory was taken over by Tatung of Taiwan. Some UK companies have reacted by forming joint ventures with the Japanese – such as GEC/Hitachi and Rank/Toshiba – but these have not been notably successful. In contrast, the French Government has been more protectionist. To some extent unique reliance on the SECAM (rather than the PAL) transmission system has isolated their markets but, where necessary, import quotas have been used. In addition, Thomson–Brandt has been encouraged by soft loans and state purchasing to expand and take over smaller producers, such as AEG's tube-making division.

Another company response has been to locate production offshore, in lower-cost locations such as southern Europe or southern Asia. For example, AEG–Telefunken has plant in Hong Kong, while Philips has branches in several countries in southern Asia, including Taiwan, Hong Kong and South Korea. Italian companies in the 1960s were also able to build up exports of 'white goods' through the use of low-cost labour and advanced technology, but by the mid-1970s they were faced with keen Japanese competition and saturated markets. European companies have rarely been able to compete with the Japanese in genuinely open markets.

Probably the one major exception to this last generalization is Philips, which is large enough and has sufficient R & D resources to allow it to compete with both American and Japanese companies. It has been a leading innovator throughout the twentieth century in a range of products, extending from radios and monochrome televisions, through radar and compact discs to videos and telecommunications. It became multinational at an early date, and in the 1980s invests more outside than within the Netherlands. The 1970s were a critical period in the company's inter-nationalization, when it was involved in a 'double capital movement' (Teulings 1984): that is, shifts of production to other developed countries, especially the USA with its important market, and to low-cost production sites in less-developed countries such as Mexico and South Korea. This created a highly sophisticated international division of labour (see Table 30): headquarters in the Netherlands, development and production of advanced electronics systems in several EC countries and in North America, and deskilled production of mass consumer goods in the less-developed countries.

Table 30 *Philips Electronics' international division of labour (percentage figures)*

	Central management	Advanced systems	Mass consumer goods	Totals
Netherlands	47	20	14	18
Other EC	25	42	25	30
Rest of western Europe	4	9	14	12
USA and Canada	4	11	4	5
Less-developed countries	20	18	43	35

SOURCE: after Teulings (1984).

Microelectronics
In 1950 the microelectronics industry barely existed, excepting some small-scale production of transistors, yet it accounted for almost half the output of the European electronics industry in 1980. The home base of the industry was Santa Clara ('Silicon Valley') in California, where pioneer companies such as Fairchild began full-scale production of microchips in the 1950s. The development of integrated circuits in the 1950s was a major technological advance because these had considerable potential applications (Malerba 1985). The pace of technological development has been dramatic and, between 1959 and 1980, the number of transistors or logical functions which could be embedded on one chip doubled every year. At the same time, the development of mass-production techniques and mass-markets resulted in transistor prices falling from $10 in 1960 to $0.01 in 1980 (Ballance and Sinclair 1983). Under these conditions companies such as National Semiconductor and Intel were set up with less than $3 million capital in the late 1960s yet had become important MNCs by the 1980s. In the 1960s, aided by government purchases for defence and civil purposes, US companies established global dominance in microelectronics, and – in contrast to consumer electronics – this was only challenged by Japan in the 1980s.

European companies have played only a limited role in these developments (see Table 31), not least because corporations such as Philips and Siemens were slow to realize the importance of integrated circuits. In addition, the industry in Europe was far less structurally adaptable to change than that in the USA. In the latter, new firms were set up to exploit the technological

Table 31 *Semiconductors at a global scale in the 1980s (percentage figures)*

*a   Shares of world production of semiconductors**

	1980	1985
USA	60	47
Japan	25	39
Western Europe	13	11
Rest of the world	2	3

b   Shares of world production and consumption, early 1980s†

	Production	Consumption
USA	63	53
Japan	26	21
Western Europe	10	18

* *Financial Times*, 30 December 1986.
† Malerba (1985).

advances while, in Europe, adaptation had to occur within existing firms which were less flexible. Division of the European markets into national segments also weakened the position of the European companies. This left a market gap in Europe which was exploited by the USA; for example, by 1968 Texas Instruments had already become the largest producer of integrated circuits in the UK (OECD 1985b).

Since the late 1970s the major innovation has been large-scale integration (LSI), whereby there was convergence of microelectronics and electrical goods as, for example, in advanced telecommunications. Europe has tended to lag behind in this, with a few exceptions such as the Ariane space programme.

Probably the single most important feature of the industry has been a rapid escalation of the cost of research and of economies of scale. Consequently companies such as IBM, with two-thirds of the EC market for computers (Commission of the European Community 1982c), are in a very strong position. Indeed, the entry costs for volume production of basic microelectronics was at least $60 million in 1982 compared with only $100,000 in 1954 (OECD 1985b). Therefore, it has become very difficult for new companies to secure a share of such markets. The Japanese challenge has been based on protecting their domestic market, followed by enormous, long-run research programmes which only began to obtain financial rewards in the 1980s (Morgan and Sayer 1983). No European government or company has devoted sufficient planning or resources to a similar programme, so Europe remains largely technologically dependent.

Not even the giant American and Japanese companies have avoided the growing problems of overproduction in the industry. While IBM managed to sustain profit levels between 1980 and 1984, many others experienced sharp reverses; profit rates for Apple fell from 20 to 5 per cent and for DEC from 16 to 7 per cent (*Financial Times*, 4 December 1985). This is partly a consequence of technological advance for, given improvements in technology, fewer chips are needed to meet even rapidly expanding demand. Furthermore, prices have tended to fall sharply over time in such a sharply competitive industry. Hence there is a premium on being a leader of innovation and a penalty for those whose R & D has left them behind their competitors. This is reflected in the global shares of production and consumption of semiconductors held by the major world powers (Table 31).

Microelectronics in Europe
The industry has been characterized by an international division of labour from its early years: the main stages of production are design of circuits, fabrication of silicon wafers and circuit assembly. Although this has involved some decentralization of production to less-developed countries – especially in the 1970s when labour costs were important – most investment has been in other developed countries so as to secure markets or to avoid protective trade tariffs. This has been reinforced in Europe by the existence of a number of rapidly expanding national consumer markets and the availability of government subsidies to 'buy-in' technology. The most favoured European location for both US and Japanese companies has been the UK, and 'Silicon Glen', in particular, has attracted large investments by such companies as Motorola, Mostel, Honeywell and IBM. The advantages of the UK, other than language and avoiding EC import tariffs of 17 per cent, have been relatively cheap labour and substantial government aid (Morgan and Sayer 1983). Consequently the UK has become an export platform to the rest of Europe. France, West Germany and the UK together account for 72 per cent of production in western Europe.

In the face of American dominance of the industry and American MNCs' considerable presence in production within Europe, European companies have followed diverse strategies. Only Philips and Siemens were really large enough to be competitive in this field; even they have been keen to purchase American technology, which was Philips' reason for acquiring Signetics. Thomson of France has tried to buy its way into this market by purchasing an established producer, Mostek of the USA. Others, such as the UK's GEC and Italy's SGS, have withdrawn into production of specialized semiconductors. However, most European countries have withdrawn completely from manufacturing microchips (see Table 32) and have instead concentrated on defence electronics and telecommunications (which have protected domestic markets) or other specialized products. The difference can be seen in the European and global market shares of companies competing to sell personal computers

Table 32 *Market shares for selected microelectronics products in 1985*

a Telecommunications equipment (world sales 1984–5 $billion)*

AT & T	USA	10.2
ITT	USA	4.7
Siemens	F.R. Germany	3.4
Northern Telecom	Canada	3.3
Ericsson	Sweden	3.2
IBM	USA	3.0
Motorola	USA	2.9
NEC	Japan	2.7
Alcatel–Thomson	France	2.6
GTE	USA	2.3
GEC	UK	1.4
Philips	Netherlands	1.2
Fujitsu	Japan	1.0
Plessey	UK	0.9
Harris	USA	0.8

b Personal computers (% share of European market, 1985)†

IBM	USA	33.2
Olivetti	Italy	10.8
Apple	USA	9.5
CBM	USA	4.6
Apricot	UK	9.5
Bull	France	2.5
Compaq	USA	2.1
Ericsson	Sweden	1.9

c Semiconductors (rank order by sales, 1985)‡

1	NEC	Japan
2	Texas Instruments	USA
3	Hitachi	Japan
4	Motorola	USA
5	Toshiba	Japan
6	Philips	Netherlands
7	Fujitsu	Japan
8	Intel	USA
9	National Semiconductor	USA
10	Matsushita	Japan

over/

d   Computer companies (revenues, 1985, $ billion)§

	From all sources	*Computer-related*
IBM (US)	50,056	48,554
Digital Equipment (US)	7,029	7,029
Sperry (US) ⎱ Unisys	5,527	4,755
Burrroughs (US) ⎰	5,037	4,685
Fujitsu (Japan)	6,563	4,310
NCR (US)	4,317	3,886
NEC (Japan)	9,899	3,762
CDC (US)	3,680	3,680
Hewlett Packard (US)	6,505	3,675
Siemens (W. Germany)	18,575	3,265
CDC (US)	3,680	3,680
Hitachi (US)	20,919	2,885
Olivetti (Italy)	3,070	2,518
Wang (US)	2,428	2,428
Xerox (US)	8,732	1,959
Honeywell (US)	6,624	1,952
Groupe Bull (France)	1,795	1,795
Apple (US)	1,754	1,754
AT&T (US)	34,910	1,500
TRW (US)	5,917	1,450
Matsushita (Japan)	21,221	1,448

* *Financial Times*, 4 December 1985.
† *Ibid.*, 9 May 1986.
‡ *Ibid.*, 30 June 1986.
§ *Ibid.*, 25 September 1986.

and telecommunications (Table 32). The latter is more protected and specialized so that European companies have a greater presence in this industry. One symbol of this was ITT's decision in 1986 to sell its telephone equipment business to Compagnie Generale d'Electricite of France. Mainframe computers have also virtually been abandoned by European producers. The British company ICL was, for a long time, a major indigenous European manufacturer, but it only survived given a preferential state purchasing policy and state aid (some £40 million during 1972–6). Neverthelesss, it almost went bankrupt in 1983, was taken over by STC in 1984, and has subsequently concentrated on specialist computers while also leasing technology from Fujitsu. In 1986 the leading independent European producers were Bull (of France) which, in that year, bought Honeywell's European interests, Siemens (of West Germany) and Olivetti (of Italy).

Governments do play an important role in microelectronics for it is considered a growth industry, has strong linkages with other industries, and has a crucial role in modern defence systems. However, this was a case of too little too late. In the 1950s and 60s the industry was largely ignored by governments, at precisely the time when Japan exercised strong protectionism and the USA had a positive policy of state defence expenditure. As a result, the European companies faced difficulties in the 1980s and only then did governments intervene.

Again, precise government responses have varied considerably within Europe. The UK has mostly encouraged foreign investment via open-door trading policies, aiming to encourage joint ventures. This has had some results, notably the technology-sharing deal agreed between GEC and Fairchild, and Plessey's licensing of Rolm office-information-systems technology. In the 1970s the British Government followed a more positive strategy through the National Enterprise Board. This provided capital for small electronics companies, set up NEXOS to manufacture advanced office systems, and invested an initial £50 million in Inmos to produce silicon microchips. Inmos was a late British entry into the field of mass-produced chips; it was privatized in the 1980s, being sold to Thorn/EMI.

In contrast, the French Government has been more supportive of indigenous firms. It has invested considerable sums in research and development, especially telecommunications, has favoured French companies in its purchasing policies, and has encouraged mergers and joint ventures such as that between Saint Gobain and National Semiconductor at Aix-en-Provence. French companies have also been encouraged to buy into foreign technology. In the 1980s southern Europe has attracted investment in microelectronics. For example, AT&T manufactures semiconductors near Madrid, and Siemens makes data-processing equipment in Spain. The attractions are expanding domestic markets and, for American and Japanese investors, export platforms within the EC.

The approach of the EC to this sector has mainly been that of a facilitator. It has encouraged international joint research initiatives, including the Espirit programme for information technology, the Race programme for telecommunications, and the Brit programme for manufacturing technology. The aims of these have been quite explicit and, for example, the Espirit programme was set up so that Europe could 'become and stay competitive with the USA and Japan within the next decade', especially in high-speed and very-large-scale integrated circuits.

Further reading

1 General review and specific sectors: P. Dicken (1986), *Global shift: industrial change in a turbulent world*, London: Harper and Row.
2 Regional adjustment: R. Hudson, D. Rhind and H. Mounsey (1984), *An atlas of EEC affairs*, London: Methuen; E. Mingione (1983), 'Informalization, restructuring and the survival strategies of the working class', *International Journal of Urban and Regional Reseach*, 7, pp. 311–39.
3 Case studies of industrial sectors: B. Toyne, J. S. Arpan, A. H.

Barnett, D. A. Ricks and T. A. Shrimp (1984), *The global textile industry*, London: Allen and Unwin; R. Hudson and D. Sadler (1983), 'Region, class and the politics of steel closures in the European Community', *Environment and Planning D: Society and Space*, **1**, pp. 405–28; G. T. Bloomfield (1981), 'The changing spatial organisation of multinational corporations in the world automotive industry', in F.E.I. Hamilton and G.J.R. Linge (eds.), *International industrial systems*, Chichester: Wiley; F. Malerba (1985), *The semi-conductor business: the economics of rapid growth and decline*, London: Frances Pinter.

6 The tertiary sector

Introduction

The tertiary sector is involved in the production of non-material products, which may be services such as retailing, education, insurance brokerage or providing holiday accommodation. Services also exist within manufacturing firms (e.g. marketing, R & D), but these have been considered already. Here, following Gershuny and Miles (1983) and others, tertiary sector firms are considered to be those where the majority of the final output is in the form of intangible products.

The service industries have often been neglected by researchers and by policy makers. Manufacturing industry has been prioritized, with emphasis placed on its role in exports, its dominance of wealth-creation, strong backward linkages to other sectors, and the considerable potential for economies of scale and technological changes. However, services have recently been re-evaluated, especially in the 1970s and 80s. It is now realized that manufacturing and services are interdependent and that some services – for example, research or accountancy – support industrial production. Services can also be imported and/or exported and they contribute to the balance of payments. Finally, as service firms sell to other service firms, their growth is not simply dependent on manufacturing.

The share of the service industries in global employment has consistently increased throughout the post-war period, so that in 1983 they accounted for 57 per cent of all jobs in the EC10. In absolute terms, service employment in western Europe increased from about 50 million to some 62 million between 1970 and 1983. However, the rates of increase were even greater in the USA and Japan, underlining the lack of dynamism in the European economy. The expansion of the tertiary sector in advanced capitalist states, in absolute and relative terms, stems from three main changes.

Firstly, the elasticity of demand for services – whether private or public – tends to be greater than for manufactured goods. Secondly, in mature capitalism the proportion of a company's labour force employed in production tends to fall while there are increases in advertising, marketing, research and other 'service' functions (Aldcroft 1980). Many of these services may be 'hived off' to specialist, independent, tertiary-sector companies. Finally, there has been a general tendency for labour productivity to rise much faster in manufacturing than in the service sector, so that the proportion of employment in services has increased more rapidly than its share of output. This is because many branches of the service industries – such as restaurants or education – have not been amenable to significant substitution of capital for labour; but there is evidence that developments in microelectronic technology could change this.

Bell (1974) provided a classic analysis of the expansion of the service sector, which was seen as epitomizing 'the post-industrial society'. He classified services into three major groups: tertiary (transport), quaternary (trade, finance, real estate, etc.), and quinary (health, education, research, recreation, etc.). Employment would continue to increase in the services because these were likely to remain more labour-intensive, and more income-elastic than manufacturing. This view has been challenged, particularly by Gershuny (1978) and Gershuny and Miles (1983), who argued that the emergence of the 'self-service economy' would lead to some decline in service employment; for example, washing machines and video recorders are substitutes for reliance on laundries or cinemas. This shift to home-based consumption and do-it-yourself practices has been encouraged by the introduction of new manufactured products (such as tumble-dryers and videos) and by the rising relative costs of services (because capital could not be substituted for labour as wages rose). Developments in telecommunications could also lead to further substantial shifts to the 'self-service economy' permitting, for example, direct retail ordering from home or direct electronic financial transfers.

The tertiary sector also has an international dimension, even though only 11 per cent of services are internationally tradeable. This represents a considerable increase from only 7 per cent in 1970 (Clairmonte and Cavanagh 1984). Even so, it is far less than the 55 per cent of manufacturing and 45 per cent of agricultural output which is exported. Nevertheless, world trade in services has been increasing at a faster rate than world trade in merchandise in recent years – at some 16 per cent a year compared with 7 per cent. Three-quarters of exports of services originate in Europe, Japan and the USA. Moreover, multinational companies have developed international systems of outlets for the presentation of services. For example, in the Netherlands only about one-fifth of all new MNC branches since 1960 have been concerned with production, being approximately the same proportion as in services, with most of the remainder being trading outlets (Smidt 1985a).

The significance of international trade in services is indicated by its contribution to individual countries' balance of payments (Figure 22). Mediterranean and Alpine countries have large positive balances, mainly from tourist receipts, while France, the UK and Switzerland benefit from both international sales of business services (transport, banking, insurance, etc.) and from tourism. A number of other countries have large negative balances, especially West Germany and Scandinavia. These are countries with relatively poorly developed international business services and net outward flows of tourist expenditures. Ireland, too, has a relatively large

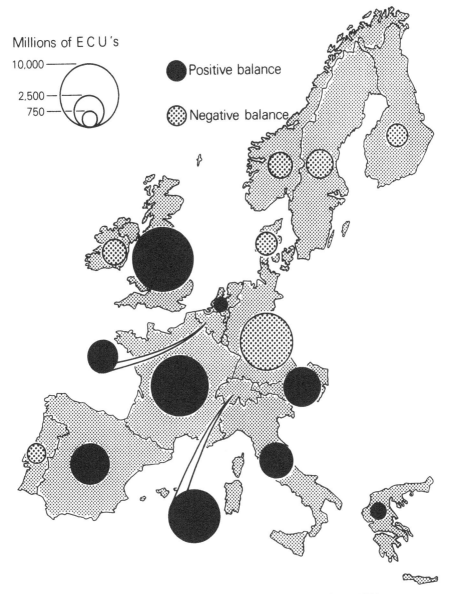

Figure 22 International balance of trade on services, 1983

SOURCE: Eurostat (1984a).

net deficit on international trading of services. This largely arises from negative balances on international freight and passenger receipts, as much of their international transport is provided by non-Irish (especially British) companies (Fitzpatrick 1985). Transport, business services and tourism are the most common international services, while collective services, such as health, are only beginning to be traded internationally.

Services are important to the modern economy in absolute terms, and in the way they support the supply of labour (via education and health) and business expansion (via advertising or research); therefore, it is not surprising that governments have developed policies to promote this sector. Most policies have been nationally orientated, but increasing interest is being shown in the international context. The EC and GATT have both largely neglected the liberalization of services which, consequently, have often remained highly protected in national enclaves; but there is evidence that this is beginning to break down in the 1980s (see p. 25). In future, services are likely to move much nearer to the centre of economic policy debates, both nationally and internationally, as developments in telecommunications increase the flexibility of firm location, and changes in microelectronics technology threaten job losses.

Geographical distribution of services

By 1981, services accounted for 50 per cent of all employment in every country in northern Europe, except in Ireland and Switzerland (Table 33); the latter is deceptive, as there is a high level of double-job holding between sectors in this country. Southern Europe lags behind – especially Portugal and Greece which, respectively, have relatively large manufacturing and agricultural sectors. In contrast, services accounted for over 50 per

Table 33 *Proportion of national employment and GDP in services, 1960–1981/3 (percentage figures)*

| | 1960 | | 1981/3 | |
	Employment	GDP	Employment, 1981	GDP, 1983
Austria	30	42	54	58
Belgium	44	53	56	63
Denmark	45	58	58	72
F.R. Germany	38	41	50	52
Finland	33	48	54	60
France	39	50	53	62*
Greece	24	51	35	53
Ireland	39	52	45	—
Italy	29	47	44	54
Netherlands	46	45	54	63
Norway	44	58	56	55
Portugal	27	39	37	51
Spain	27	—	46*	60
Sweden	41	53	61	66
Switzerland	38	—	49	66
UK	48	54	56	66

* 1982 data.
SOURCES: World Bank (1984, 1985).

cent of GDP in 1983 in every country, although, again, southern Europe lags behind northern Europe – especially Denmark which has a remarkable 72 per cent of GDP in this sector.

Over time the services have come to account for increasing shares of both employment and GDP in most countries. One exception is Norway, where their proportion of GDP has fallen since 1960, as a result of North Sea oil developments. In general, however, there is a tendency to greater uniformity in the broad sectoral composition of employment and GDP in western Europe, although this may stem from contrasting changes; for example, in the UK stagnation of manufacturing is a primary consideration whereas in Spain the flight from agriculture is important. At the regional level there are even greater variations within Europe; for example, the proportion of employment in services in such places as Hamburg is three times larger than in southern Italy (Hudson *et al.* 1984).

The structure of service employment

Sectoral variations

There are a number of ways in which services can be classified (for example, see Illeris 1985), but Gershuny and Miles (1983) provide a useful approach. They sub-divide services into four categories: distribution (transport, wholesaling and retailing), producer (financial, research, marketing, etc.), social (health, education, etc.) and personal (domestic services, laundries, entertainment, etc.). Alternatively, Massey (1984) considers that there is a basic difference between producer services (direct aids to production) and services concerned with the reproduction of labour power (that is, retailing, personal private services, and collective services). The actual categories are broadly similar even if their theoretical connotations differ.

All the broad categories of service employment have increased over time, as is verified by Gershuny and Miles' (1983) analysis of trends in selected countries (Table 34). While there were small gains in distribution and somewhat larger ones in producer services, the most significant increases were in personal and, above all, collective services. Particular sectors are discussed separately later, but some general comments are necessary here. Producer services have been expanding as a result of the hiving-off of some functions from manufacturing firms and, in some countries, through expanding exports of services. Marketed services, such as retailing and personal services, have been subject to increased automation and to greater economies of scale (for example, in hypermarkets) so that employment has tended to stagnate. Finally, collective services have expanded as a result of the growing involvement of the state in directing economic and social affairs and, in particular, through the

Table 34 *The changing structure of service employment in selected countries, 1963–78*

	% Share of total employment, 1978			% Change in share of total employment, 1963–78		
	Distribution	Producer services	Personal and collective services	Distribution	Producer services	Personal and collective services
Belgium	13.7	14.6	30.6	3.6	3.6	7.0
Denmark	12.6	14.3	37.3	– 1.2	2.1	14.5
F.R. Germany	12.7	11.8	24.6	0.1	1.5	6.6
France	14.4	14.7	27.2	2.2	3.1	5.6
Ireland	18.0	11.4	27.6	0.7	1.5	1.9
Italy	11.3	9.8	24.6	4.1	8.3	
Netherlands	17.8	14.7	30.3	4.5	8.3	
UK	16.0	12.4	29.2	0.2	1.0	8.1

SOURCE: after Gershuny and Miles (1983).

expansion of its role in collective consumption and legitimation (see pp. 77–82).

Spatial divisions of labour

There have also been changing divisions of labour within each of the broad sub-sectors discussed above. Between 1960 and 1980, some 80 per cent of the expansion of service-sector jobs in the OECD countries was in white collar jobs, especially professional, technical and administrative grades, while manual and clerical jobs tended to stagnate or grow more slowly (Bannon 1985). This is borne out by studies of individual countries; in Ireland, for example, professional jobs expanded by 55 per cent in the 1970s, compared with only 32 per cent for clerical jobs.

There are also variations in the locational flexibility of services. Some types, such as personal services and distribution, have a greater need for direct access to their markets than is the case with, say, producer services. At the same time, the growing scale and complexity of many service-sector corporations means that a number of distinctive stages exist in the presentation of services, ranging from headquarters control to direct customer access. For example, a large insurance company may have distinctive headquarters, central data processing, and sales divisions. Each of these has different locational requirements, so that a spatial division of labour may develop within particular industries. Therefore, there may be a double-polarization: between types of service industries and between stages of production within particular industries.

Lipietz (1980b) has shown how this type of double-polarization can produce distinctive, if complex, interregional variations in service employment in the case of France. In broad terms there are four types of regions in France, represented here by selected examples. Paris has the largest proportion of service jobs but it also has an 'over-representation' of business and financial services, and of senior staff. This is the region where conditions most favour the expansion of locationally flexible services, and of headquarters functions. Nord is representative of the older industrial areas. Service employment accounts for less than a half of all jobs, and there is 'under-representation' of most sectors of skilled jobs. The Centre represents rural regions which have gained considerably from industrial decentralization from Paris. However, it has failed to benefit significantly from development of related producer services or of more highly skilled posts. Finally, Languedoc–Rousillon represents 'the sunshine regions' where new concentrations of health, educational and tourist complexes are being established, with some provision of higher-skill posts.

This broad regional distribution of service employment is also to be found in other countries, although in less extreme form than in highly centralized France. For example, in the UK the South-East has 43 per cent of national population but 54 per cent of service employment (Daniels 1982), with high levels of professional and scientific jobs. This arises from centralization of higher-order functions in both the public and private sectors. Similarly, the East Region in Ireland (including Dublin) has 61 per cent of all producer services, although it has only 49 per cent of all services (Dineen 1985).

Skill polarization within the service sector is also associated with a division of labour in terms of gender. A large part of increased female participation in the labour force is accounted for by the tertiary sector. However, women tend to be found in lower-order, especially clerical, posts, rather than in higher-order jobs. This applies both between and within sectors. For example, in the UK male employment fell while female employment increased in distribution, between 1961 and 1980, but the growth of male employment surpassed that of female employment in the professional and scientific services (Daniels 1983). Within-sector differentiation is even more marked: in France in 1975, only 23 per cent of senior staff were women, compared with 64 per cent of basic employees (Lipietz 1980b).

There are divisions of labour within the service industries in terms of skills, sex and spatial distributions and, to a degree, these tend to coincide, producing a differentiated regional system of service employment. However, there are also differences between the major branches of service employment, as is revealed in the following sections.

Producer services

Producer services have been a major growth area. Their main market is government or other firms (Damesick 1986), whether in agriculture, manufacturing or the tertiary sector. They also exist within manufacturing companies but only separate producer service firms are considered here. Nevertheless, it is a broad category which can be sub-divided into at least three types: finance (banking, credit, insurance, real estate, etc.), professional services (R & D, marketing, advertising, data processing, legal services and accounting), and 'other business services' (cleaning, personal travel, transport, security, etc.).

These are diverse activities with very different features. Cleaning and security largely involve low-skilled, low-paid manual tasks, whereas financial and professional services involve a number of highly skilled, well-paid jobs but with sharp polarization between these and clerical employment. In the case of banking, insurance and accountancy, automated data processing systems have led to some displacement of lesser skilled jobs, as a means of reducing costs. For example, bank statements can be produced by computers which can also handle most routine recording of financial and other transactions. Automation can reduce the need for counter-staff; an example is the use of electronic 'cash points' for executing cash withdrawals from banks and other financial agencies. Against this, however, many forms of professional services – such as research and development or legal services – have proven resistant to automation and are still highly labour-intensive.

Internationalization and world cities

A feature of financial and professional services has been a tendency to increased internationalization. This has partly been led by the need of manufacturing MNCs for international advertising, marketing and other agencies, which has produced a hierarchy of international financial centres, one classification for which is shown in Table 35. Within Europe, according to this, only London counts as a first-order supranational financial centre, but a number of other cities have important if lesser roles. International banking probably became truly global in the 1970s when OPEC surplus oil revenues needed to be recycled and there was a growth of eurodollar markets leading to an end of US dominance over trading in dollars (Thrift

Table 35 *Hierarchy of international financial centres in 1980*

a   Supranational centres

First order:	LONDON	New York
Second order:	AMSTERDAM	FRANKFURT
	PARIS	Tokyo
	ZURICH	

b   International centres

First order:	Bombay	BRUSSELS
	Chicago	DUSSELDORF
	HAMBURG	Hong Kong
	MADRID	Melbourne
	Mexico City	Rio de Janeiro
	ROME	San Francisco
	Sao Paulo	Singapore
	Sydney	Toronto
	VIENNA	
Second order:	Bahrain	Buenos Aires
	Kobe	Los Angeles
	LUXEMBOURG	MILAN
	Montreal	Osaka
	Panama City	Seoul
	Tapei	

SOURCE: Reed (1983).

1986). Advances in telecommunications and electronic data processing aided this global shift.

International trade in services is still highly protected, however, and liberalization has proceeded in a piecemeal fashion. Nevertheless, in the mid-1980s there was an attempt by some of the more-developed countries to include services in the GATT negotiations. Part of the pressure for this resulted from

> ... the global outreach desired by multinational banks and corporations in the telecommunications sector, and later by companies providing insurance, accounting and other business services. In turn, arguments of national interest have served to buttress these efforts. Increasingly, US export advantage is seen as having shifted to service transactions. It is, therefore, considered unfair to have US markets open to goods while foreign markets are closed to US services. *Financial Times*, 11 February 1986

Similar arguments would seem to apply to some European countries such as France, Switzerland and the UK.

Some of the more radical deregulation and internationalization of financial services has occurred in banking (Coakley 1984). For example, in 1985–6 alone, Sweden, Norway and Portugal all took steps to license foreign banks to operate in previously closed domestic markets. In the UK, deregulation took the form of the 'Big Bang' of 1986 which led to liberalization of international competition, mergers of stockbrokers and stockjobbers, and greater competition (between, say, building societies, banks and insurance companies) in financial markets. In West Germany financial deregulation in 1985 led to foreign banks being allowed to lead-manage deutschmark Eurobond issues.

Liberalization has tended to lead to two linked phenomena: concentration and diversification. Diversification is the obvious outcome of the breakdown of barriers between financial sub-sectors, while concentration follows from the need to create larger companies both to trade across a broader range of activities and to compete in international markets. Examples of both processes abound. For instance, in Sweden there were 450 savings banks in the 1950s but only 150 in 1985 and, in the face of deregulation, further mergers occurred such as that between Sundsvallsbanken and Uplandsbanken in 1986. There have also been a number of significant international acquisitions in the 1980s, such as the Amsterdam Rotterdam Bank's takeover of the European Banking Co., a London merchant bank. In the UK itself, the Big Bang deregulation was preceded by mergers between building societies and by diversification of financial companies' interests with, for example, Lloyds Bank and the Prudential Insurance Company both buying chains of estate agents. Deregulation in Denmark during 1985–6 was also followed by diversification: the Baltica insurance company bought into stockbroking and banking, and the Topskring insurance group set up its own bank, Topbank.

Deregulation is likely to lead to a further concentration of key decision-making powers at the international scale. The dominance of 'the world cities' (Thrift 1985) such as London, Tokyo and New York has continued to strengthen *vis-à-vis* other financial centres, with London acting as the centre not only of the traditional 'sterling' area, but also for much of western Europe. It also has a special time niche, partly overlapping with operations in both New York and Hong Kong. In terms of stock-exchange activity, although Frankfurt, Milan and Paris all have important roles, London has become the major centre. Stock-exchange dealings have tended to be divided by time-zones, with Tokyo, New York and London each being dominant at different times of the day, although Frankfurt is also important within Europe.

The City of London has become the leading world banking centre in the post-war period, recovering a position it had lost to the USA in the 1920s. This was based on the creation of Euromarkets (to provide multi-currency loans to countries and to large corporations) in the 1960s and 1970s, which were based in London. Two potent indicators of this are that the Deutsche Bank moved all its non-DM Eurobond business from Frankfurt to London in the early 1980s, and that the USA's Morgan Stanley shifted its European base from Paris to London in 1977. As a result, London accounted for 27 per cent of international bank lending by 1982 (Coakley 1984). Its nearest global rival, New York, accounted for only 17.5 per cent and its nearest European rival, Paris, only 7.2 per cent. In the 1970s and 80s the growth of London-based banks also outstripped that of the powerful Swiss banks, which were hampered by state regulation. However, City of London banking dominance does not equate with British dominance *per se*, for the City has become a banking enclave, a centre for packaging and booking loans from international sources. For example, in 1983 UK-owned banks accounted for only 21 per cent of all lending from the UK, a figure which was matched by American-owned banks and surpassed by Japanese banks with 27 per cent. Whereas there were only 73 foreign banks in London in 1970, there were 428 by 1983 (Bateman 1985).

Insurance markets, too, are becoming more international, not least because of the needs of international manufacturing companies. With growing deregulation and internationalization, financial strength has become concentrated in the USA and the UK, and eight of the world's twenty largest insurance brokers operate from the City of London (Table 36), with the remainder being North American. In terms of insurance companies and underwriters, it is again London-based corporations, such as the Royal Group and Lloyds, which are dominant in Europe. International acquisitions, such as that by the Dutch Aegan company of Union Levantina de Seguros of Spain, are becoming increasingly common as European and global insurance networks are being created. Lloyds has a particularly firm grip on the market for reinsurance.

For similar reasons accountancy is internationalized. An outstanding

Table 36   *The world's twenty largest insurance brokers in 1984*

			1984 gross revenues ($ millions)
1	Marsh & McLennan	US	1,120.9
2	Alexander & Alexander	US	576.0
3	Johnson & Higgins	US	390.5
4	Frank B. Hall	US	372.8
5	Fred S. James	US	292.8
6	Sedgwick Group	UK	286.8
7	Reed Stenhouse	Canada	255.3
8	Corroon & Black	US	199.1
9	Willis Faber	UK	177.8
10	Rollins Burdick Hunter	US	133.4
11	Hogg Robinson Group	UK	111.6
12	C.E. Heath	UK	104.0
13	Minet Holdings	UK	102.7
14	Jardine Insurance Broker	UK	98.9
15	Steward Wrightson	UK	88.6
16	Bayly, Martin & Fay	US	77.9
17	The Crump Cos.	US	69.6
18	Arthur J. Gallagher	US	64.2
19	Sodarcan	Canada	52.0
20	Bain Dawes	UK	45.4

SOURCE: *Financial Times*, 14 April 1986.

example is the UK's Price Waterhouse, with offices in some 95 countries. It has also diversified into legal services, management consultancy and tax guidance.

The same pattern is repeated in the case of professional services. For instance, there has been a rash of mergers and takeovers among advertising agencies in the 1980s in order to create the 'global reach' which is considered essential to be competitive. This is particularly important as their major clients are MNCs, whose activities extend across a number of countries. Unilever has sales in some 75 countries but the contracts for advertising its products are handled by only four agencies. The resulting concentration has largely benefited the USA which is the home-base of nine out of the ten top advertising agencies (Table 37). The only major European competitor is the UK's Saatchi and Saatchi which, in 1986, became the largest agency in the world through its acquisition of Ted Bates. Another response to competition – in common with financial services – has been diversification, and companies such as Saatchi and Saatchi have set up management consultancy and market-research agencies so that they can offer broad 'packages' of professional services to their major clients.

Table 37 *The world's ten largest advertising agencies in 1985*

			1985 gross revenues ($ millions)
1	Young & Rubicorn	USA	536
2	Ogilvy Group	USA	481
3	Dentsu	USA	473
4	Ted Bates	USA	466
5	J. Walter Thompson	USA	451
6	Saatchi & Saatchi	UK	441
7	BBDO International	USA	377
8	McCann–Erikson	USA	345
9	D'Arcy Masius Benton & Bowles	USA	319
10	Foote, Cane and Belding	USA	284

SOURCE: *Financial Times*, 7 May 1986.

Taken together, the distribution of producer services indicates that London is Europe's only true world city. It is arguably *the* global financial centre, and this is based on its dealings in money (especially eurocurrency), securities, commodities (especially the futures markets for products such as sugar and coffee) and services (especially insurance, reinsurance, air travel, accountancy and legal advice) (see Thrift 1986). This is not to say that UK companies are dominant, for many foreign companies are active in the City, and have taken over established City firms. Nevertheless, despite being challenged by the rise of New York in the 1970s and 80s, London is still the dominant world centre for a number of producer services.

For this reason its property market is also becoming internationalized. Foreign companies now own a large share of London property; a notable example was the purchase by RODAMCO (Netherlands) in 1986 of the Haselmere property group. In turn, however, UK companies have been active in foreign property markets. Brussels was a major target in the early 1970s but, when rents fell in the late 1970s, investment switched to the USA (Bateman 1985).

Offices and office employment

Most producer services are office-based and there is no doubting the importance of offices in the modern economy. In terms of employment, it has been estimated that in the OECD countries in the 1970s, office employment increased by 45 per cent compared with only 6 per cent for total employment. Furthermore, in the late 1970s 'office-type' employment accounted for about 40 per cent of all jobs in more-developed countries, such as France and the UK, and 30 per cent in less-developed ones such as Ireland (Gershuny and Miles 1983).

Much of the interest in offices concerns how the emergence of new spatial divisions of labour (influenced by developments in office systems and telecommunications) and state policies have affected centralization and decentralization. Offices, traditionally, have exhibited a tendency to centralization in central urban locations, although there are differences between higher-order managerialist functions and the more routine clerical jobs. The highest-order functions, especially those involved in international trading of services, tend to be concentrated in capital cities and those with truly 'global' reach tend to cluster in a few select locations, such as London, Paris or Frankfurt. Capital cities offer a number of advantages, including the availability of a pool of skilled labour, access to other producer services (which frequently form important clients) and the advanced transport and communications systems of metropolitan areas.

With the increasing diffusion of telecommunications, locations are becoming more flexible, excepting where personal contacts are important. This relates to the spatial division of labour within the service industries, for personal contacts tend to be more important for higher-level office jobs. Goddard (1973) illustrated this with reference to an eight-tier hierarchy of office jobs. The top level had five times as many contacts as the fifth level, while below the sixth level there were virtually no contacts. The lower tiers of routine clerical operations exhibited the highest degree of locational flexibility. In contrast, higher-order posts requiring frequent contacts are likely to be spatially clustered, and in London some 80 per cent of all business journeys took less than 30 minutes. The degree of concentration in a less-developed economy, such as Ireland, is even more marked. Only 1 per cent of Dublin offices used services from outside the capital, while only 14 per cent of participants in Dublin office meetings came from outside the city (Bannon 1979). The precise location of offices within the city centre varies but, commonly, there is a specialized zone adjacent to the central business district.

A very broad measure of the relative importance of office centres at the European scale is provided by office accommodation costs (see Figure 23), although these are also influenced by exchange rates and local planning restrictions. On this basis, London emerges as Europe's primary office centre, followed by Geneva (with the demand exerted by numerous international agencies), Paris and Frankfurt. (Total office costs, which also incorporate salaries and company cars, present a different picture, with London occupying a much lower position and Brussels a much higher one.) At the other end of the scale are Athens and Lisbon which operate more as national than as international office centres. At the national scale, a number of case studies exist which illustrate the concentration of offices generally and especially of higher-order office jobs in capital cities. For example, 31 per cent of French office jobs are in Paris and 38 per cent of

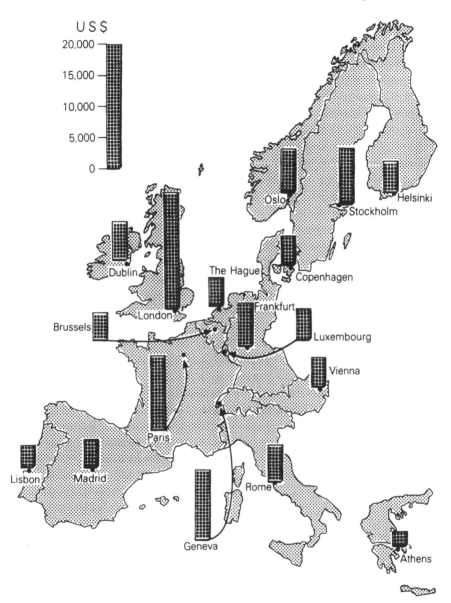

Figure 23 Average office accommodation costs in selected western European cities, 1986

SOURCE: based on *Financial Times*, 15 May 1986.

UK office jobs are in London. Office centralization is particularly marked in less-developed economies, such as Ireland, where the capital city is very dominant in the national economy (Bannon 1979). This is borne out by Gaspar's (1977) study of Portugal, which showed that 57 per cent of the 220 largest companies had their headquarters in Lisbon.

Office decentralization

Although a high degree of centralization of office employment still exists, there has been decentralization in many countries. For example, in the UK total office employment grew by only 0.4 per cent in the 'cores' of the largest metropolitan areas but by 20 per cent in their suburban 'rings' between 1966 and 1971 (Goddard 1979). Decentralization has occurred for a number of reasons. In part it is due to suburbanization of jobs and of population, especially of the more affluent. This is important for services such as legal advice and property/estate agencies which require proximity to customers. In addition, changes in cost structures have made city centres less attractive. Wage rates, especially for female staff, also tend to be comparatively higher in the larger cities relative to suburban locations or provincial towns. The costs of accommodation also tend to be significantly higher in the major cities. For example, in the UK in the early 1980s office rents per square foot were £25 in the City of London, £7–10 in a suburban location such as Croydon, and £3–5 in provincial cities (Daniels 1982). Not surprisingly, then, many companies have sought to decentralize those divisions of their office establishments which do not require the intensity of personal contacts available in city centres. This has been further encouraged in many countries by state policies to encourage or compel office decentralization.

A number of countries have developed a mixture of controls and incentives in order to encourage office decentralization. In France, controls on new office development were introduced in 1958, followed by taxes on new office buildings in 1960, the establishment of tertiary growth poles (the Métropoles d'Équilibre) in 1965, and the establishment of the Association Bureaux-Provinces in 1974 to assist firms wishing to relocate (Daniels 1982). In the UK, the Location of Offices Bureau was set up in 1963 to encourage decentralization, while a system of Office Development Permits was introduced in 1965 to control building in London (and later in Birmingham). The 1972 Industry Act also provided some financial assistance to encourage offices to relocate in Development Areas. Grants for decentralization are available in Ireland and the Netherlands and, in the latter, the government has also promoted the creation of out-of-town office centres, such as Rijswijk or Zoetermeer near The Hague. La Defensé has similarly been developed as a suburban office complex in Paris.

Most assessments of the effectiveness of office decentralization policies stress their limited achievements. The controls are usually quite simple to evade. In London in the period 1965–72, some 41 Office Development Permits were refused but only six firms left the City; the others eventually found other means of securing adequate accommodation (Goddard 1979). Even where firms decentralize, the desire to remain within easy travelling distance of the metropolis usually restricts the distance they are prepared to move (Pye 1977); in the case of London, 85 per cent of firms moved less than 60 miles (Daniels 1982). Similarly, decentralization policies in Paris have had a minimal effect. Few firms took up the investment aids, while the demands to build new office accommodation in the city centre were largely acceded to (Bateman 1985). Ultimately, it seems that market forces rather than state policies have exerted the greatest influence on office locations (Pickvance 1981). This has been particularly true in the late 1970s and the 80s when many governments eased constraints on office developments in the face of economic recession.

Non-marketed services: the public sector

Non-marketed services is a varied sector which includes central and local government (both generalist administration and special functions such as planning or social services), the armed forces, and public corporations such as railway companies and postal services. In the mid-1970s the public sector accounted for some 20 per cent of employees in the EC, and this reflected annual growth rates of 3 per cent or more in most European countries in the 1960s and early 1970s (see Table 38). As a result, the share of public employment in total employment approximately doubled between 1950 and 1980 (Gershuny and Miles 1983).

Table 38 *Public sector employment in selected countries**

	Percentage of total civilian employment in 1975	*Percentage annual change, 1960–70*
Belgium	9.0	3.8
Denmark	19.6	—
France	12.6	3.9
F.R. Germany	10.6	3.3
Ireland	9.3	—
Italy	14.0	4.1
Netherlands	16.1	3.3
UK	19.3	2.8

* Excluding the armed forces.
SOURCE: Gershuny and Miles (1983).

The increase in state employment reflects the expanding role of the state in terms of social investment, collective consumption and legitimation (see pp. 77–82). In Europe about one half of state employment is in collective consumption, especially education and health provision. This has been underpinned by expansion and development of the welfare state. Although neither health nor educational provision are exclusively the prerogative of the state, they do largely fall within its domain. Other important areas of welfare state expenditure are the social services and housing administration. In addition, there is considerable state employment in central and local government administration, in economic planning and in such legitimation functions as policing.

Since the 1970s, however, the seemingly constant expansion of state employment has slowed and, in some cases, been reversed. Government budgets came under pressure as a result of energy price increases and general inflation, while recession reduced their revenue base. As a result, there have been rising budget deficits leading, in many cases, to significant reductions in levels of public sector employment. One particular response to this crisis – epitomized by the policies of the Thatcher government in the UK – has been privatization, or the selling of public agencies (or the more profitable parts of their activities) to private interest groups. In the UK this has involved telecommunications, some public transport and the cleaning and catering functions of the health service. In France, the right-wing government elected in 1986 has also fostered privatization.

Some public services, such as hospitals and schools, are relatively ubiquitous, being located in local or regional catchment areas. However, other functions such as central government administration, or the headquarters of the postal service, have a high degree of locational flexibility. This is a feature shared with private-sector offices. Not surprisingly, therefore, public services exhibit a high degree of regional concentration, and in Ireland, for example, some 55 per cent of public administration jobs are in the East Region (Bannon 1985). The public sector is also highly concentrated in France, Spain and Portugal, all of which have or had centralized administrations and weak local and regional governments. In contrast, countries such as West Germany or Switzerland, which have federal structures and lack an obvious 'primate' city, tend to have a much more even geographical distribution of public service employment.

Given the weight of public services, and the direct control that governments exercise upon them, it is not surprising that state policies have been designed to influence their locational distribution for regional policy purposes. The principal aim has been decentralization so as to relieve pressure on office space and on limited labour reserves in the capital cities, while securing some job dispersal to areas of high unemployment. This was very much the policy of the British Government in the 1960s and 70s. Civil service dispersal was recommended by the Flemming Report in 1965 and some 50,000 jobs were affected in this way (Rhodes and Kan 1971). Later, the Hardman Report recommended dispersal of more higher-grade civil service jobs but, in the late 1970s and 80s, this policy was abandoned in the face of deepening economic recession which increased unemployment in almost all regions. In the Netherlands the 1960 Drees Commission proposed decentralization of government services from the Randstad. Between 1963 and 1973, fifteen government services and some 2000 jobs were decentralized but, in the 1980s, policy implementation has slackened (Smidt 1985b). A number of other European governments, including those of Italy, Sweden and Spain, have attempted some decentralization of public sector employment in the 1960s and/or the 70s (Daniels 1979). The impact of decentralization varies according to whether new functions are created, as in Spain with the advent of elected local and regional government, or whether responsibilities are transferred to lower levels as in France in the early 1980s.

Consumer services I: retailing

Corner shops and international multiples

Retailing is probably the largest single group of marketed services. In the mid-1970s, along with wholesaling, it accounted for some 12 per cent of total employment in the EC, while there were over three million retail outlets (Gershuny and Miles 1983). The industry is subject to increasing economies of scale and, as a result, employment has been relatively stagnant and has even decreased in recent years in some countries. There have also been changes in the pattern of ownership, with shifts from traditional small-scale family units to larger retail outlets, multiple chains and some multinational groups.

The pressures for reorganization of retailing stem both from the demand side and from the supply conditions in which firms operate. A major change in demand has arisen from the growth of 'single-visit' trips which offer savings in time and convenience compared with traditional, multiple-shop visits. This has encouraged concentration of retailing in fewer but larger units (supermarkets, hypermarkets, etc.), often in out-of-town locations. The new pattern of demand has been facilitated by rising standards of living, improvements in personal mobility and changes in household technology. Improvements in mobility have been based on the diffusion of car ownership; whereas, in the early 1950s, only between 2 and 6 per cent of households had cars in even the more-advanced European economies, by the late 1970s levels had reached 50–70 per cent (Dawson 1982). This permits families to travel longer distances to large distribution centres. With respect to household technology, the availability of freezers in over 50 per cent of homes in most northern European countries has been important, enabling fewer (but bulk-buying) trips to be made.

While demand conditions have facilitated a shift to larger-scale retailing, the driving force behind reorganization has been firms' operating conditions. Large groups have had better access to the substantial capital needed to establish modern retailing units. Moreover, vertical integration of retailing and wholesaling offers advantages in terms of securing markets and guaranteeing supplies, but this option is only available for larger companies. Therefore, small-scale units have higher overhead costs, limited economies of scale and fewer opportunities to substitute capital for labour than have larger units. Consequently, they have not been able to compete with larger shops in terms of price. Many have formed contractual chains (such as 'Mace' grocers) to try to secure economies of scale in purchasing. However, only where they offer specialist goods, locational convenience (as in corner shops), or willingness to stay open long, unsociable hours (relying on family labour) can they compete with the large companies.

The emergence of large-scale department stores and of multiple retail chains can be traced back to the late nineteenth and early twentieth centuries, especially in the UK. However, such developments were retarded in some countries, such as in Belgium and West Germany, by state policies which, for political reasons, sought to protect small-scale shopkeepers. Since 1945, however, concentration in retailing has been evident, although it is more marked in some countries than in others (see Figure 24). While independents account for over 50 per cent of units in France, Italy, Ireland and Belgium, they account for less than one-third in countries such as West Germany and the UK where, instead, about half the units are parts of corporate chains. The advance of the multiples in particular countries has been very rapid: for example, in France their share of the food retail market increased from 10 per cent in the 1950s to 50 per cent in 1980 (Tuppen 1983). In most countries, over 10 per cent of total retail sales in the 1970s were accounted for by no more than ten companies; but, in terms of particular products, the domination can be far greater and, in the UK for example, Marks and Spencer account for over 50 per cent of underwear sales. Table 39 shows the largest European retail companies, and underlines the high level of concentration in the UK.

Supermarkets have been particularly successful in supplanting corner shops in the food market; in the UK, for example, their share of the grocery trade increased from 44 per cent in 1971 to 59 per cent by 1979. However, hypermarkets have been the major development of the 1970s and, unlike supermarkets, they sell a wide range of non-food products. Definitions of hypermarkets vary, being over 1000 square metres in West Germany, over 2500 in Belgium and over 5000 in the UK. However, according to EC guidelines (minimum size of 2500 square metres) there were some 1500 in the Community in 1980. Over 400 of these were in France and the UK had 150. By 1985, according to UK definitions, the largest hypermarket chain was owned by Asda with 86 units. In contrast there were only a few hypermarkets in Italy and Ireland. Not surprisingly, these differences are reflected in average employment in retailing units, with the UK figure being about five times larger than that for Italy.

Another trend has been the growth of international multiple chains. Companies such as Benetton (Italy), C & A (Netherlands) and Laura Ashley (UK), have retail outlets throughout Europe. France provides a

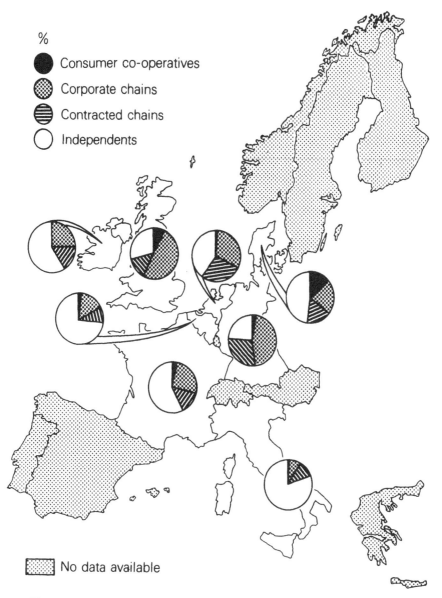

%
● Consumer co-operatives
◉ Corporate chains
⊜ Contracted chains
○ Independents

▒ No data available

Figure 24   Major types of retailing in selected western European countries

SOURCE: based on Dawson (1982).

good example for, in the late 1970s, it had some 2000 foreign-owned retail units; among UK interests were some 45 Burtons, 51 Laskeys, six Habitats and three Marks and Spencers. Given the high levels of disposable incomes in West Germany, that country has been particularly successful in attracting the upper-end-of-the-market retailers such as Burberrys, Louis Vuitton and Cartier. One important feature to emerge

Table 39   *Major European retailing companies in 1986**

			Turnover ($ million)
1	Carrefour	France	6204
2	Migros	Switzerland	5694
3	Marks & Spencer	UK	5635
4	J. Sainsbury	UK	5151
5	Edeka Zentrale	F.R. Germany	5126
6	Tesco Stores	UK	5062
7	Karstadt	F.R. Germany	4843
8	Ahold	Netherlands	4618
9	Dee Corporation	UK	4315
10	ASDA/MFI	UK	3796
11	Great Universal Stores	UK	3573
12	Sears	UK	3436
13	Kaufhof	F.R. Germany	3422
14	Boots	UK	3207

* Companies with principal interests in retailing
SOURCE: *Financial Times*, 'Top 500', 1986.

from this reorganization of retailing is a new spatial division of labour. Management and highly qualified marketing and buying jobs are becoming concentrated in a few locations, mostly in national capitals, although international headquarters do exist. At the same time, most jobs in supermarkets and chain stores are unskilled shelf-filling or semi-skilled cashier posts. Such jobs are often taken by part-time and/or female workers.

State policies and retailing

Retailing is rarely characterized by public ownership in western Europe but it is influenced by state policies. In particular, state policies contribute to the macro-economic framework for retailing, influence the distribution of employment for social and regional purposes, and have a considerable bearing on intra-urban locational patterns.

There are instances of state regulation of entry to retailing, such as in Italy where shopkeepers require a licence to trade; but most countries do not have such negative controls. Instead, the macro-economic context of retailing is shaped by taxation laws at local, national and EC levels, and by rules on competition. Retail price maintenance practices are followed in several countries, including West Germany which permits local association agreements; but the general drift of national, and especially EC, policies has been the removal of restrictive practices and of price-fixing. In addition, retailing may be affected by macro-economic policies designed to squeeze or ease credit, or to increase or decrease real spending power. The UK has had a particularly notorious 'stop–go' approach to economic policy (see p. 29) and this has clearly influenced retail demand.

Retailing does not figure large in regional policy formulation, having been neglected along with most other branches of the service industry. This is surprising given the sheer volume of employment involved although, given the nature of the industry, this is mostly ubiquitous. Only a small part – the central accounts of large companies, etc. – is locationally flexible. However, there are regional policies, as in West Germany, where subsidies are paid to retailers in some depressed areas; social policies, as in France, where pensions are paid to help small-scale shopowners to retire; and restructuring grants, as in Denmark, where loans and grants are paid to help modernize or rationalize small businesses (Dawson 1982).

Public policies have probably been most active in respect of the intra-urban locations of new retailing developments. After 1945 there was a need to rebuild the war-damaged centres of many cities, but some countries have continued to favour regeneration of existing city-centre shopping areas over out-of-town expansion. This has been prevalent in the UK, where out-of-town shopping centres and hypermarkets have been strongly opposed by most local authorities. At the same time, local authorities and private development companies, singly or in partnership, have been active in redevelopment schemes such as the Arndale Centre in Manchester or Eldon Square in Newcastle. By 1977 some 33 of the 37 largest shopping centres in the UK had experienced either major or minor redevelopment. Characteristically, this involves a mixture of pedestrianization, provision of covered precincts, enlargement of car-parks, development of leisure and recreational features, and 'street landscaping'. While the UK has been to the forefront of such redevelopments, with about two-thirds of schemes over 20,000 square metres in the EC (Dawson 1982), they are also to be found in other major European cities such as Brussels and Lyon. The Dutch Government has, since 1974, made subsidies available to renovate run-down shopping centres. Outside of the UK there has been a move to develop hypermarkets in city centres; French examples are Euromarché in Evry and Auchon in La Defénse (Tuppen 1983).

Out-of-town retailing developments are more common than city-centre redevelopment in many European countries. These can take the form of shopping complexes or of single hypermarkets, although they are tending to merge. Some hypermarkets sub-let space to small shops, and some new shopping centres contain hypermarkets. Hypermarkets are most highly developed in France where their growth has been exceptional; the first hypermarket was opened in 1963 but, by 1980, there were 413 (Commission of the European Community 1982f). This approach has been favoured by the central government, although the Loi Royer in 1978 tightened up on the development of new retail centres; this led to increased emphasis on city centres. The majority of local authorities in the UK have tried to limit such developments, instead preferring to protect existing investment in the city centre. Thus, London has sought to fossilize its retail pattern, while Paris has opted for planned decentralization (Metton 1979). In practice most countries have adopted a combination of locational strategies.

Consumer services II: tourism

Tourism can be defined as involving temporary, short-term movement of people to destinations outside the places where they normally work and live; this can be for business or holiday reasons. While tourism can be both national and international, the discussion here concentrates on the latter.

This is one of the high-growth industries of the post-1945 period. Whereas there were some 25 million tourists globally in 1950, by 1980 the number had risen to 285 million and about a half of these were in western Europe. Individual countries have experienced even sharper increases, and in France, for example, the number of holidaymakers increased from fewer than two million in the 1930s to 29 million in 1980 (Tuppen 1983). Expansion has been continuous, being halted only briefly by recession in the mid-1970s. The growth of tourism is based on a conjuncture of increased demand and a reduction in costs.

There is a strong positive elasticity of demand for tourism. In the 1980s, EC citizens spent some 7 per cent of their household budgets on holidays (Commission of the European Community 1985e). This was based on rising standards of living, considerable increases in personal leisure time and early retirement. Of course, there are variations by country and by social class; for example, in France 81 per cent of middle and lower managers took holidays in 1980 compared with only 17 per cent of farm workers (Tuppen 1983). Affluent families in northern Europe are probably accustomed to two or more holidays a year, while the unemployed and those on low incomes take no holidays.

Business incentive and conference travel also adds to tourism demand. At the international scale, London and Paris have benefited most from the estimated 14,000 international conferences which are held each year (Law 1985b). The scale of these can be enormous; for example, the International Rotary meeting at the National Exhibition Centre in Birmingham attracted 23,000 delegates and generated £6.2 million of revenue. Many cities have built special conference facilities to attract this business for the rewards can be considerable. Brighton, for example, is estimated to obtain £30 million in revenue and 40 per cent of hotel bookings from conferences (Law 1985b).

A significant factor in the expansion of tourism has been a reduction in international transport costs. Chartered flights have brought down the real cost of air fares, while rising car-ownership rates have permitted more flexible and more distant holidaymaking. In general, the geographically peripheral European regions are more dependent on air travel and the more accessible ones on road travel; while 65 per cent of visitors to Greece arrive by air, 93 per cent of visitors to Austria arrive by road.

Costs have also been reduced by the emergence of large-scale, vertically integrated holiday companies such as Thomsons or Intersun of the UK, Holland International of the Netherlands, and Scharnow–Reissen of West Germany. These companies sell packages of travel, hotel and insurance cover, the costs of which are reduced through the large scale of their operations. This provides a strong bargaining position *vis-à-vis* hotels and airlines, and the ability to plan ahead and to reduce margins on each individual element of a package. There has also been considerable expansion of transnational hotel groups since 1945. While the USA dominates, France and the UK have major international hotel chains (see Table 40) such as Club Mediterranée (56 hotels abroad) and Trust House Forte (53 hotels abroad). Several of the larger hotel groups are linked with airlines or other service activities.

Origins and destinations

To place international tourism in perspective it is important to emphasize that, in simple numerical terms, domestic tourism is more important in most countries. For example, in West Germany 92 per cent of all tourist overnight stays were by domestic visitors (Boissevain 1979). Even in a mass tourist country such as Spain, 36 per cent of all holidaymakers were nationals, and in Italy only 81 million of a total of 505 million overnight visitors were foreigners (Bethemont and Pelletier 1983). In terms of international tourism, the pattern is one of mass movements from northern to Mediterranean or Alpine Europe (see Figures 25 and 26). Major destinations are Spain, Italy, France and Austria and the main sender nations are West Germany, France, the UK and the Netherlands. The broad distribution changes only relatively slowly but, in the 1970s and 80s, small adjustments are evident such as the growing attraction of Portugal and the UK, and the decreasing relative attraction of Ireland. The rate of growth of tourism in particular countries can be phenomenal; in Spain, for example, it spiralled from some six million in 1960 to 42 million visitors in 1982 (Valenzuela 1985). In terms of the origins of tourists, as expected, it is the larger, urban-industrial nations such as West Germany, France and the UK which are prominent.

There are a number of different types of holidays, including visits to historic cities, walking or touring in attractive rural areas, winter sports, and Mediterranean sun and beach holidays. Of these the most common

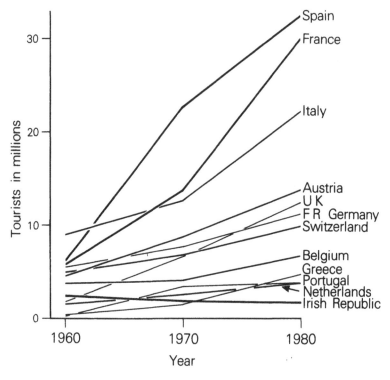

Figure 25   International tourist movements in selected western European countries, 1960, 1970 and 1980

SOURCE: after Ilbery (1981) and OECD (1985a).

form, especially for those taking only one holiday a year, is Mediterranean seaside tourism. This is the typical package holiday based on a cheap chartered aircraft or long-drive coach, and basic accommodation. Accommodation may be serviced, but in recent years, with rising labour costs, self-catering has become more important. The growth of modern Mediterranean tourism is typified by Spain. In 1959/60 as part of its policy of internationalization of a previously strongly autarkic economy (see p. 32), the Spanish government devalued the peseta and this opened the way to development of cheaper package holidays in the 1960s. The attractions initially were sea, sun, cheap food and wine, but recreational facilities such as golf courses and tennis courts have been added. There has also been a growing tendency for northern Europeans to purchase villas as second homes or retirement homes in Spain.

The roots of Alpine tourism lie in the nineteenth century when the Alps were locations for summer relaxation by the wealthy elites of Europe (Barker 1982). By the 1870s the arrival of railways in the Alps and the emergence of the first modern travel company, Thomas Cook, was leading to increased tourism, although it was still restricted to the middle classes

Table 40 *Transnational corporations and tourism, 1978*

a Transnational hotels abroad by country or region of origin

	No. of parent groups	Transnational-associated hotels abroad	
USA	22	508	(49.6%)
France	8	156	(15.2%)
UK	13	147	(14.4%)
Other European	14	87	(8.5%)
Japan	7	23	(2.2%)
Other developed market economies	8	65	(6.3%)
Developing countries	9	37	(3.6%)
Totals	81	1023	(100%)

b Transnational-associated hotels abroad, by main activity of parent group

	No. of transnational corporations	Transnational-associated hotels abroad	
Hotel chains associated with airlines	16	227	(27.1%)
Hotel chains independent of airlines	56	687	(67.0%)
Hotel development and management consultants	3	15	(1.5%)
Tour operators and travel agents	6	46	(4.5%)
Totals	81	1025	(100%)

SOURCE: UNCTC (1982).

and to exclusive resorts such as Montreux. Early in the twentieth century, winter tourism increased in importance, being boosted by the organization of the first Winter Olympics at Chamonix in 1924. Finally, since the late 1950s there has been growth of mass tourism based on package holidays, especially as rising standards of living have allowed more families to enjoy second holidays. Winter tourist resorts can take a number of different forms. Préau (1970) differentiates between resorts based on existing settlements such as Chamonix, and green-field developments such as Les Belleville. These have different economic organizations, especially in what Pearce (1981) terms 'integrated' resorts, where a single promoter (usually a large metropolitan group) undertakes development of an entire green-field complex. In such developments there are likely to be few openings for either local businessmen or local workers.

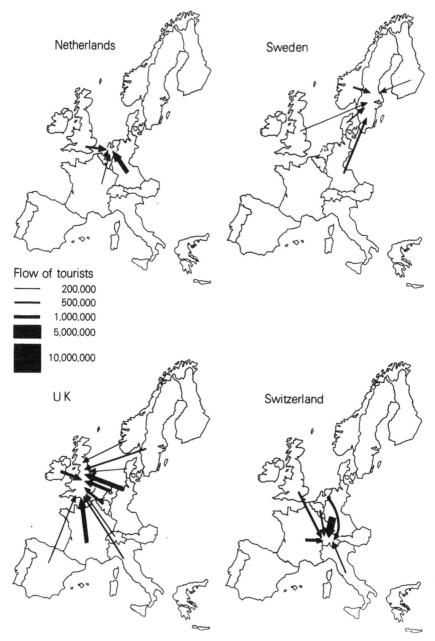

Netherlands

Sweden

Flow of tourists

——— 200,000
——— 500,000
▬▬ 1,000,000
█ 5,000,000

█ 10,000,000

U K

Switzerland

Only annual tourist flows larger than 200,000 are shown

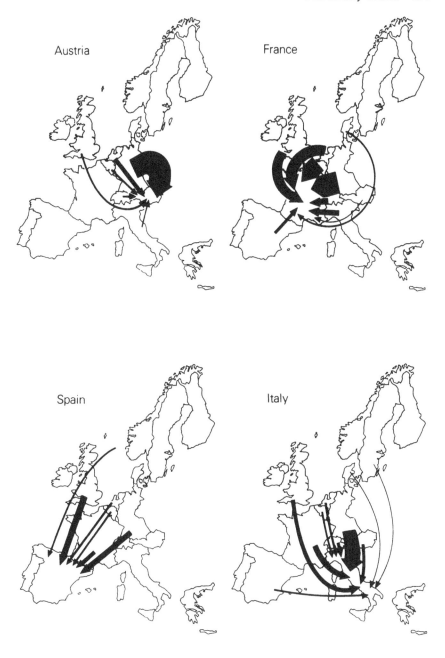

Figure 26 Major international tourist flows to selected western European countries, 1984

SOURCE: OECD (1985a).

Political economy of tourism

Tourism is a diverse industry and, in terms of its structure and capitalist relationships, there is a continuum between the large hotel (owned by external capital) and an apartment in a family home which is let to casual visitors. The former may be owned by a multinational company, while the latter, employing family labour and family capital, has features in common with the peasant family farm (Vincent 1980). However, since 1945 tourism has been subject to increased penetration of capitalist relationships. This is particularly evident in the growth of package-holiday companies with transnational operations. It can also be seen in the presence of a network of international hotel chains (Table 40). Some of these are linked to airlines or other service activities. Indeed, the prevalent trend is towards hotels selling not only accommodation but also packages of services on-site (conference facilities, etc.) and off-site (car hire, theatre tickets, etc.). In a broader sense, tourism is also part of the process of internationalization in modern society, being a means whereby values and consumption habits are spread throughout Europe (Lanfant 1980). Against this, most tourist firms are small-scale. For example, 80 per cent of such firms in France employ fewer than ten people (Tuppen 1985).

Tourism brings both advantages and disadvantages to an economy. Boissevain (1979, p. 133) is among those who consider tourism to be economically beneficial, writing that 'wealthy locals, affluent tourists, and foreign social scientists may lament the coming of mass tourism to the European South. The less well-off, who constitute the overwhelming majority of the region, when they have experienced its impact, have welcomed it.' The three major positive impacts are the balance-of-payments contribution, employment provision and income multipliers.

Western Europe as a whole has a net positive balance of $9.7 billion on tourism (OECD 1985a). Countries such as Spain and Italy have small-scale expenditure but large net receipts from international tourism, while West Germany and Scandinavia have unfavourable net financial flows (Figure 27). In the middle are such countries as the UK and France with large financial inflows and outflows. Over time there has been a sharp deterioration of the balance of payments on tourism in Scandinavia and other northern European countries, and significant gains in southern Europe and the UK. These net earnings/losses are important in two senses. Firstly, they can contribute to the overall balance of payments and, in much of southern and Alpine Europe, are equivalent to over 10 per cent of merchandise exports. This can be sufficient to cover net deficits on merchandise trade and is an important element set against total GDP or total exports in such countries as Austria, Greece, Portugal and Switzerland (Table 41). In this way, tourism has helped to pay for technology and capital goods imported by countries such as Spain and Portugal during

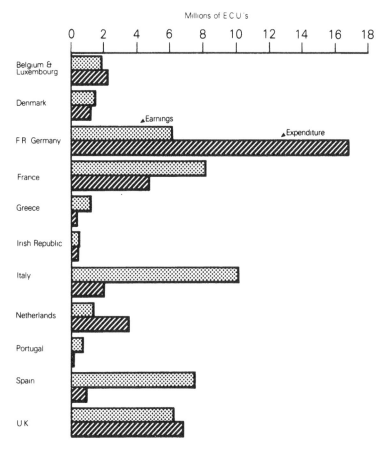

Millions of E C U 's

Figure 27 Earnings from and expenditure on international tourism, 1983 (in European Currency Units)

SOURCE: Commission of the European Community (1985e).

their industrialization in the 1960s. Against this, a country can have a net tourism deficit which is a drain on the national economy. In West Germany this amounts to one-third of the visible trade balance. Secondly, tourism has a high income elasticity of demand, especially compared with most manufactured goods, and it therefore presents opportunities for future expansion of exports.

Tourism is also a major potential source of employment. In the EC in the 1970s, tourism accounted, directly and indirectly, for some 8.5–10 million jobs. In addition to jobs in hotels and restaurants, there are also indirect employment spin-offs in manufacturing industries supplying tourist resorts, and induced effects through the multiplier effects of income earned in the tourist industry. The indirect effects may even outweigh the direct ones. For example, in France there were 300,000 directly employed in tourism in 1980 and over one million indirectly employed (Tuppen 1983). A large proportion of the labour required is semi- or unskilled, which makes tourism particularly suited to the needs of less-developed regions, although this also has disadvantages. In addition to employment generation, tourism provides a boost to income levels, again via direct and indirect effects. These are notoriously difficult to quantify, but Archer (1977) has estimated that the tourism income multiplier in the UK is of the order of 1.68 to 1.78. Even more difficult to quantify are the social benefits of tourism, but these do exist. For example, in southern and northern Europe many small-scale farmers only survive by virtue of income earned from tourism, while tourism contributes to maintaining traditional landscapes in upland Britain.

Tourism also has a number of economic costs, especially in that the balance-of-payments benefit is reduced by a 'leakage' of growth (Airey 1978). In Italy in 1975, for example, tourist receipts were $2578 million but overseas payments of $1050 million reduced the net balance to the economy to $1528 million (Mathieson and Wall 1982). The 'leakage' of growth stems from a number of considerations:

- costs of imported goods, particularly food and drink (This is especially high in regions with only limited agricultural capacity, such as the high

Table 41 *Tourism receipts in relation to GDP and to exports,*
1983

| | Tourism receipts as a percentage of: | |
| | | Exports of goods |
	GDP	and services
Austria	7.8	16.9
Belgium/Luxembourg	2.1	2.2
Denmark	2.3	5.8
Finland	1.0	3.2
France	1.4	5.0
F.R. Germany	0.8	2.6
Greece	3.4	16.4
Ireland	0.3	5.1
Italy	2.6	19.4
Netherlands	0.4	1.8
Norway	1.2	2.5
Portugal	4.1	13.3
Spain	4.3	20.9
Sweden	1.2	3.0
Switzerland	4.2	9.1
UK	1.2	4.0

SOURCE: OECD (1985a).

Alps or, for example, Corsica. In the latter 90 per cent of food has to be imported – see Kofman 1985.)

- foreign-exchange costs of foreign investment in developing tourist facilities
- remittances of profits abroad by foreign companies owning hotels and other facilities
- remittance of wages by expatriate workers, whether these are unskilled cleaners or highly paid managers. (In Corsica, for example, only 20 per cent of salaried workers in tourism are Corsican – see Kofman 1985)
- management fees and royalties for franchised businesses
- payments to overseas airlines and to tour operators
- overseas promotional and publicity costs
- extra expenditure on imports by nationals resulting from earnings from, and the demonstration effects of, tourism.

The extent of the leakages depends on the type of tourist development. Among the more important considerations are the structure of the regional economy and whether food supplies, construction, hotel furniture, etc., can be supplied locally; that is, the propensity to import. In this the siting of the development is important, for more isolated resorts – such as the Greek islands – are more likely to depend on expatriate labour and less likely to be able to sub-contract to local suppliers. Ski resorts provide another example, for high-elevation resorts, sited well away from existing settlements, tend to have very limited multiplier effects (Barker 1982). Finally, the structure of local capital markets is important. Where there are well-developed indigenous capital resources – as in northern Italy or in Catalonia – local investors may take the lead in development. Elsewhere, as in Andalusia or much of southern Italy, there may have to be greater reliance on external investment sources (de Kadt 1979).

Another problem associated with tourism is the dependency of growth upon external (and often fickle) sources. The choice of tourist destinations is susceptible to large fluctuations from year to year, especially in the 'beach and sun' Mediterranean package market where there is little product differentiation. Small price changes, following currency devaluations or revaluations, can have a considerable impact on tourist numbers. Countries can also fall out of fashion (as Monaco has in the post-war period), with the number of hotels being reduced by a half by 1980. Political uncertainties can also reduce tourist flows, as happened in Portugal after the 1974 military *coup*.

Tourism has a number of other disadvantages. It creates jobs but their quality is often poor. Most jobs are relatively unskilled and there is a tendency for better-paid jobs to be taken by expatriates. Employment tends to be seasonal for tourism is seasonal (see Figure 28). This is most marked in the Mediterranean countries but it also applies in northern Europe. The Alpine countries have two peaks to the tourist season but this

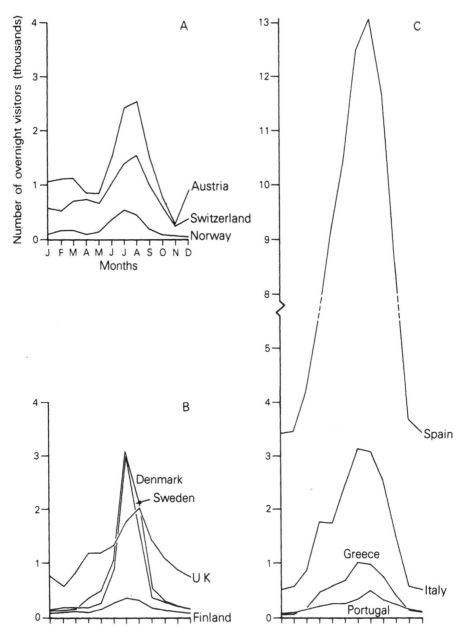

Figure 28   Seasonal variations in international tourist arrivals
SOURCE: OECD (1985a).

can cause special problems: although winter tourism is a boon at a time of slackness in agriculture, summer tourism coincides with the harvest period, leading to greater reliance on migrant labour. Where small businesses are involved in tourism, there is often a sharp gender division of labour. For example, Hadjimichalis and Vaiou (1986), writing about tourism on the Greek island of Naxos, state that:

> Rooms-to-let is a household operation run almost entirely by women. They are rooms within or near the family house which are rented during the summer. Cleaning rooms and serving guests is regarded as an extension of daily housework, 'naturally' women's work. Negotiating prices and making contacts with the authorities is usually left to men.

This is likely to offer only limited opportunity for development of the economic potential of the individual.

Tourism also tends to be spatially polarized – being concentrated in a few historic cities, upper Alpine regions or the Mediterranean coastline. This brings intense pressure to these areas in terms of congestion, the need for new infrastructure and pollution costs. The extent of the latter should not be underestimated; for example, the 4.5 million visitors to the Austrian Alps each year leave behind them an estimated 4500 tonnes of rubbish. Tourist development in such areas may conflict with industrial or agricultural development, competing for the same 'optimum' sites (on the coast, for example), limited water supplies and, possibly, limited labour reserves (A. M. Williams 1986). While tourism can attract labour away from agriculture, leading to abandonment of some marginal farming areas as in parts of Portugal's Algarve (Cavaco 1980), good agricultural land near the tourist resorts may be left uncultivated in expectation of future gains from a sale to a tourist company. In addition, tourism brings about rapid cultural change as commercial values and products replace traditional values and crafted goods. The speed of such change usually allows little time for adjustment by the local populace.

State policies for tourism

Governments have sought to promote tourism for a number of reasons (Pearce 1981): to improve the balance of payments, facilitate regional development, diversify a country's economic base, increase income levels, expand state revenues and create new jobs. The latter has been particularly important in the 1970s and 80s, with rising levels of unemployment.

The precise form of state intervention varies between countries, but it mainly constitutes aids to the private sector rather than outright public ownership or strong central controls. An important aid, which has been adopted by every National Tourist Board, is overseas promotion. This usually involves advertising campaigns, such as Spain's 1980s image-building slogan of 'Everything under the Sun'. This may be backed up by special incentives, such as the provision of coupons to foreign tourists enabling them to purchase petrol or tourist services at reduced costs. Other policies have included measures to attract tourists to remoter, little-known regions. This can be done via advertising campaigns, and the British Tourist Authority has been particularly keen to attract foreign tourists away from London. Alternatively, new attractions can be created in these regions; for example, this was the aim of the Portuguese and Spanish Governments in developing *pousadas* and *paradores* (state-owned, subsidized luxury hotels in remote interior areas). Governments may also seek to aid the industry by promoting out-of-season holidays, encouraging national tourists to go to domestic resorts (via advertising or, indirectly, via controls on taking foreign exchange abroad), and providing special educational and training courses.

The major element of state expenditure has probably been investment in infrastructure and in tourist facilities, sometimes as devalorized capital. Infrastructure costs have fallen directly on the state because it requires enormous expenditure to construct new or improved airports and roads so as to open up regions to tourism. For example, the construction of Malaga airport in Andalusia was crucial to development of the Costa del Sol, for it obviated a reliance on Seville airport which involved over three hours of travelling to the resorts. Several governments have also provided loans or grants to private companies to facilitate construction of accommodation and other facilities for tourists. Examples abound. Italy's Cassa per il Mezzogiorno spent between 7 and 10 per cent of its budget on tourist facilities in the 1960s (White 1976). In 1973 the Swiss Government allocated 500 million francs to open up mountain regions to tourism, while the Société de Credit Hotelier advances loans to develop accommodation (Ilbery 1981). In Spain, since 1942, up to 70 per cent of the costs of hotel building could be advanced by the state (Naylon 1967). Perhaps the most ambitious schemes have been in France, where the government was worried by the post-war relative decline of the Riviera as a tourist area, following rapid advances in other Mediterranean countries. One response was region-wide plans, as in Rousillon–Languedoc, to develop large tourist complexes using a combination of public and private finance.

The EC plays only a minor role in regulating and promoting tourism. It has had some influence in easing customs checks, facilitating social security payments abroad and, via the Social Fund and the Regional Development Fund, in promoting the development of tourist services (Commission of the European Community 1982e). However, even by the mid-1980s, despite agreement to a general EC framework for tourism (see Table 42), relatively little progress had actually been made in liberalization of the international presentation of tourist services.

Table 42 *The European Community policy framework for tourism*

a Freedom of movement and the protection of EC tourists

- easing of customs checks
- reduction of police checks at frontiers
- social security provisions for tourists
- assistance for tourists and regulation of car insurance
- protection of tourists' interests (e.g. in complaints about the shortcomings of tourist services)

b Working conditions for those engaged in tourism

- right of establishment and freedom to provide tourist services
- vocational training grants and mutual recognition of qualifications
- aid from the European Social Fund
- promotion of staggered holidays
- harmonization of taxation
- promotion of energy efficiency

c Common Transport Policy and tourism

d Safeguarding the European heritage and tourism

- environmental protection
- arts heritage

e Regional development and tourism

- ERDF assistance
- EAGGF assistance

SOURCE: based on Commission of the European Community (1985e)

Further reading

1 General: J. I. Gershuny and I. D. Miles (1983), *The new service economy: the transformation of employment in industrial societies*, London: Frances Pinter.
2 Producer services: N. Thrift (1986), 'The geography of international economic disorder', in R. J. Johnston and P. J. Taylor (eds.), *A world in crisis: geographical perspectives*, Oxford: Basil Blackwell; P. W. Daniels (1982), *Service industries: growth and location*, Cambridge: Cambridge University Press.
3 Retailing: J. A. Dawson (1982), *Commercial distribution in Europe*, London: Croom Helm.
4 Tourism: E. de Kadt (1979), *Tourism: passport to development?* Oxford: Oxford University Press.

Part Three

The Regional Dimension

7 Uneven regional development and regional policies

Regional inequalities

International variations within western Europe and the idea that these accord with a core, semi-periphery and periphery classification have already been emphasized (see pp. 82–5). However, many economic processes involving capital accumulation or the organization of labour operate at the sub-national level. These regional formations of capital and labour have their own role in the national and the international economy. They also condition how the national economy is affected by changes in the global economy, and how it responds to these.

Defining regions

Prior to the development of industrial capitalism, regional differences were mainly between urban and rural areas or between types of agricultural regions (Johnston 1982, pp. 133–4). Thereafter, the development of industrial capitalism was characterized by concentration tendencies in the process of capital accumulation. This saw the emergence of core regions (defined in relation to capital accumulation) and of some specialized industrial regions (often based on coalfields or mineral deposits). Later, there was selective decline in these industrial regions and expansion of others in response to the emergence of new divisions of labour and production requirements. There were transfers of value (interpreted crudely as profits and capital) from the peripheral regions to the core; this augmented accumulation in the centre while distorting it in the periphery.

The regions referred to here are not fixed and immutable objects; rather, they must be seen in relationship to processes of economic and social change. Massey (1984) sees this in terms of rounds of investment which, potentially, create new divisions of labour and new bases of social organization. The existing character of these areas is also important for it conditions the emergence of new economic patterns. Regional patterns appear complex because, at any one time, 'more than one new spatial structure may be in the process of establishment, be undergoing change, or be disappearing with decline' (Massey 1984, p. 122). Regions may be embedded into a multitude of spatial structures – relating to agriculture, energy, and different branches of industry and services – and their roles in each of these may be distinctive. In terms of industrial production alone, regions can be distinguished by sectoral specialization, ownership and control of capital, and stages in the division of labour.

Furthermore, 'the layers of history which are sedimented over time are not just economic; they are also cultural, political and ideological strata' (Massey 1984, p. 120). Regions have different class structures stemming from their sectoral specialization and their role in the division of labour. They also have varying degrees of class solidarity, reflecting historical differences in economic and social development. This, in turn, has significance for contemporary economic structures such as the potential for local entrepreneurship; the ability of organized local groups to resist state or company policies, especially relating to closures and redundancies; and the opportunities for part-time working in agriculture and manufacturing. In short, social conditions in the region help shape the social organization of production within local enterprises. This is perhaps most clearly illustrated by diffuse industrialization (see pp. 171–2).

Given the complexity of the economic and social processes which define local economies, is it possible to identify, in some sense, real regions? In a simplistic sense this is logically impossible because different types of economic activities have different spatial divisions of labour and, moreover, these change over time. Sayer (1984, p. 227) argues that any attempt to define a set of regions would 'cut across many structures and causal groups in a "chaotic" fashion'. However, regions are real in other senses for these are the entities to which state and EC policies relate. Regional consciousness (see pp. 249–52) also conditions economic change, especially the way in which local social groups respond to threats of job losses. There are, then, important reasons for considering sets of regions as one possible 'cut' through the economic and social processes that operate in space.

National and regional inequalities

In 1982 the regional variations within countries in terms of incomes were at least as great as those between countries (Table 43). Taking the EC10 average as an index value of 100, national incomes range from 44.4 in Greece, to 125.6 in Denmark (see pp. 82–5 for a fuller discussion of national differences in GDP). In contrast, there were intranational (regional) income ranges such as 42.0 to 101.8 in Italy and 82.8 to 234.3 in the Netherlands. Enlargement of the EC has tended to increase these differences over time. For example, accession of Greece took the gap between the poorest and the most prosperous region from 1:6 to 1:10, while Iberian accession increased the ratio to 1:12.

Regional inequality is not simply a case of gross regional income differences between, say, north-eastern Portugal and northern Germany. Instead, regions such as these have fundamentally different employment structures, class structures, access to consumer goods and to health and

Table 43 *International and intranational GDP per capita levels* in the European Community in 1982*

Country	Average GDP per capita	Lowest GDP per capita region	Value	Highest GDP per capita region	Value
Belgium	98.0	Luxembourg	73.8	Antwerp	122.9
Denmark	125.6	Øst for Storebaelt	107.5	Copenhagen	149.5
F.R. Germany	122.6	Lüneburg	82.6	Darmstadt	153.0
France	114.7	Limousin	90.3	Ile-de-France	162.0
Greece	44.4	Thrace	29.9	Eastern Central and Islands	49.7
Ireland	60.7	N/A		N/A	
Italy	70.5	Calabria	42.0	Valle d'Aosta	101.8
Luxembourg	100.8	N/A		N/A	
Netherlands	100.3	Friesland	82.8	Groningen	234.3
UK	98.2	Northern Ireland	71.4	South East	108.5

* EC average = 100.

SOURCE: Eurostat (1985b).

educational services. If non-EC countries such as Switzerland or Sweden are also considered, then the gap between the poorest and the wealthiest regions of Europe seems even greater.

The availability of Eurostat Level II regional statistics allows for a detailed analysis of regional patterns. GDP per capita statistics provide a broad indicator of levels of economic development within the European Community (see Figure 29). These reveal a centre–periphery spatial pattern in the EC12 with the lowest levels being in the Mezzogiorno, Iberia, Greece and Eire. The highest levels are in parts of West Germany, the Netherlands, Denmark and northern France. In the UK only the South East and Scotland (through oil production) are above average.

Labour market conditions are indicated by unemployment and migration statistics (see Figures 30 and 31). The latter, poignantly, has been labelled 'unemployment of departure' and, although migration is conditioned by more than just economic conditions, it is a strong indicator of labour-market opportunities. Unemployment levels in the EC10 again approximate to a centre–periphery pattern, with rates being highest in northern Britain, the Mezzogiorno and Greece; the exception is the incidence of unemploy-

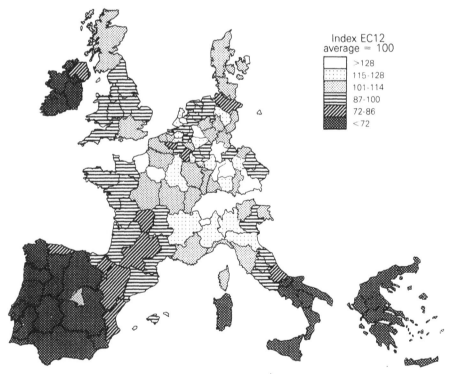

Figure 29 Regional distribution of GDP per capita, 1977–81

SOURCE: Eurostat (1986).

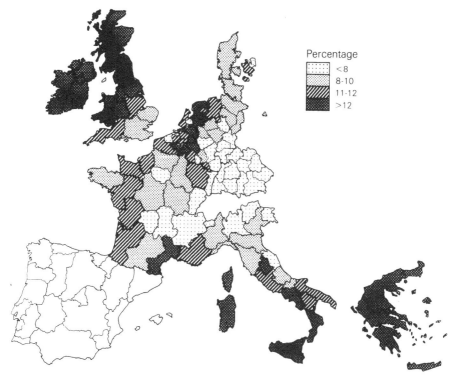

Figure 30   Regional distribution of unemployment rates, April 1985

SOURCE: Eurostat (1986).

ment in parts of the Netherlands and Belgium. A 'core' of low unemployment exists in West Germany, eastern and southern France, and northern Italy. Finally, net migration rates in the early 1970s reveal a complex pattern, although continuing losses in southern Italy, much of the UK, Greece and Iberia do suggest a centre–periphery pattern. A review of the 1960–75 period reveals a much more distinctive centre–periphery pattern, with large net migration losses in most of Iberia, the Mezzogiorno, rural northern and western France, Ireland, most of the UK and much of Greece (Haselen and Molle 1981).

Together these indicators suggest a centre–periphery spatial pattern of regional development. Rokkan (1980) considered that even the EC6 had a distinctive core which he termed the 'Lotharingian axis', composed of the Rhinelands, northern Italy and the Paris region. Enlargement to ten members added the south-east of the UK and parts of Denmark to that core, while considerably enlarging the periphery. The Mediterranean enlargement has almost exclusively (bar Madrid and parts of Catalonia) involved further additions to the periphery. This is verified by Keeble *et al.*'s (1982) analyses of economic potential (population weighted by GDP). Economic potential is at a maximum in Rheinhessen–Pfalz, while other

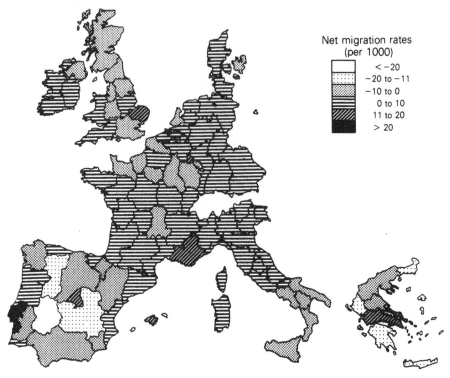

Figure 31   Regional distribution of net migration rates, 1970–75

SOURCE: after Haselen and Molle (1981).

areas with high values are Hamburg, Zuid Holland, Antwerp and Ile de France. At the other extreme, with values less than 20 per cent of the maximum, are Ireland, northern Britain, southern Italy and Greece.

Enlargement of the EC could be expected to have increased the range of interregional differences but, surprisingly, regional differences have also increased *within* the EC9 over time (Camagni and Cappellin 1981). In the 1950s and 1960s, per capita regional disparities within the EC tended to fall because of out-migration from, and rising productivity in, the peripheral regions. However, this was reversed in the 1970s when the relative positions of all regions in Ireland, Italy and the UK worsened, while all regions in West Germany and Denmark and some regions in Belgium (excepting Wallonia) and the Netherlands made relative advances. These movements partly reflected changing national economic fortunes and shifts in international exchange rates, but there is a depressing tendency to regional polarization. Another important feature of the 1970s was the pace at which economic adjustments occurred between regions. The hypermobility of capital (Damette 1980) meant that the relative strengths of regions – or at least parts of them – ould change relatively quickly as ever-more-rapid rounds of new investment or disinvestment altered their roles.

A simple typology of economic regions

The simple centre–periphery classification is a spatial description and it is inadequate as a basis for analysing regions in relation to economic processes. The problem is how to obtain a classification which is detailed enough to have some meaning in relation to underlying economic and social processes, while permitting generalization. Any detailed consideration of economic processes (as in Part Two of this volume) suggests a number of typologies, each highlighting an aspect of agriculture, industry or services, or, alternatively, some combination of these. However, a simpler approach is required for our purposes.

A number of simple regional classifications do exist for western Europe. For example, Minshull (1980) identified four types of regions: regions of dynamic growth, stable rural regions, older industrial regions and remoter regions. This is a useful pragmatic scheme but it lacks any clear relationship to regional economic processes. Clout (1981c) favoured another pragmatic classification: overdeveloped regions, neutral regions, depressed regions, underdeveloped regions, intermediate regions and frontier regions. Intuitively, these regions make sense but, again, there is a lack of clear, explicit reference to the economic processes which have shaped them.

More satisfactory is Lewis' (1984) differentiation between regions of slow accumulation and regions of rapid accumulation. These can be further subdivided according to the type of accumulation process. Slow accumulation regions are those characterized by pre-capitalist production (in agriculture) or by obsolete capital (especially in mining or in specialized industrial sectors). Rapid accumulation regions fall into three types: metropolitan (with high concentrations of ownership and control of capital), planned economic complexes, and areas of dispersed economic growth (see Table 44). This provides a model for the following discussion.

These regions are meaningful in several senses. They are differentiated by their role in capital accumulation – central or peripheral, rapid or slow (or even disinvestment). In turn this influences labour-market conditions, the incidence of unemployment, and national and international migration. Economic change has different requirements and poses different challenges in each of these regional types. This has led to distinctive political reactions to the process of economic adjustment and to different types of economic policy responses. These are exemplified in terms of particular regions in Chapter 8.

Regional issues and regionalism

There is mutual interdependence between changes in economic and social

Table 44   *General classification of economic regions*

---

a   Rapid accumulation

● Metropolitan
  Examples: London, Paris, Frankfurt, Randstadt, Milan-Turin, Athens

● Newly-emergent zones of accumulation
  Examples: north-eastern and central Italy, parts of southern Italy, western
  Portugal, western France, eastern Ireland, southern Norway, East
  Anglia, much of Switzerland and Denmark

---

b   Slow accumulation

● Obsolete capital
  Examples: Ruhr, Saar, southern Belgium, UK coalfields, north-east France,
  Luneburg (Netherlands), Basque country

● Rural regions, less articulated with capitalist development
  Examples: interior of Spain and Portugal, northern Scandinavia, Greece
  (excepting Athens and Thessaloniki), western Ireland

---

structures. This is particularly well-illustrated with respect to regional issues which have been prominent in the political agendas of most western European countries since 1945. In a sense this is a new emphasis to long standing social and economic inequalities. Lewis (1984, p. 140) writes:

> Clearly intranational differences in levels of employment, types of employment, migration rates and living conditions are a long-standing feature of all countries. However, regional disparities have not been regarded as creating a problem – let alone one which the state apparatus by itself could solve – until recently.

One reason for the political highlighting of regional issues has been the growth of popular protest against the economic policies of large companies and the state. These range in intensity from parliamentary lobbying to illegal occupations of productive facilities, and to violent mass demonstrations. The intensity of opposition partly depends on local class structures and solidarity. Hudson and Sadler (1983) have shown this to be important in the more muted reaction in the UK, compared with France, to steel closures in the 1980s. Governments have to be seen to respond to such pressures if their role in social and economic regulation is challenged (see pp. 77–9). The pressures are intensified when some political parties have a distinctive territorial base of electoral support. Examples include the Christian Democrats' careful cultivation of voters in the south of Italy, PASOK's campaigns over rural issues in Greece, and the responses of both the major political parties in the UK to the upsurge of Scottish nationalism in the 1970s.

Regional issues gain even greater significance where they are linked to regionalist or separatist political movements. The 1960s and 1970s witnessed a particularly vigorous growth of ethnic separatism in western Europe (C. H. Williams 1986). By the mid-1970s Krejci (1978) was able to identify 37 distinctive ethnic groups in western Europe, all but four of which had secured some form of self-government. A number of theories have been advanced to try to explain the rise of separatist movements. Among these are the demonstration effect of decolonization of the former possessions of the European imperial powers, realization that discrimination by the core is affecting the socio-economic position of peripheral regions, and the fact that the uneven development which is inherent in capitalism is necessarily nationalism-producing (see C. H. Williams 1986). It is not surprising that uneven development should produce popular reaction, but why should this focus on ethnic identity rather than, say, religion? For Knox (1984) the answer lies partly in non-economic reasons, for he sees ethnic separatism as one response to the internationalization of culture and consumption under global capitalism. In addition, the processes of establishing national bureaucracies and developing national culture in western Europe's nation states has provided a direct challenge to regional cultures.

However, the economic roots of the separatist movements should not be underestimated. Hadjimichalis (1983, p. 137) considers that regional social movements 'arise from both objective conditions of uneven regional development and subjective conditions of rising regional consciousness'. With the burgeoning of the state's role in economic management, it has become increasingly likely that the central state will clash with regional economic interests. Williams (1980, p. 156) expressed this as follows: 'Much of the explanation for nationalist resurgence rests, then, not so much on the internal dynamics of core–periphery relations, but on the capacity of the declining state to meet the needs of territorially-concentrated ethnic groups.' It was precisely these kinds of crises – the failure of the state to deliver on the economic front – which contributed to protests in Lorraine over closures in the steel industry, and to expanding electoral support for the Welsh Nationalist Party in the 1970s. The clash of regional and national interests is not limited to disadvantaged regions. Indeed, more-prosperous economic regions may consider that central government is making unjust demands on them so as to favour poorer regions. Such was the case in Scotland, Catalonia and the Basque country.

Regionalist movements are widespread in Europe and assume a multiplicity of forms. The actual form of regional protest varies considerably and does not permit any simple generalizations as to the causes of non-violent as opposed to violent movements. All that can be done here is to indicate (after C. H. Williams 1986) some examples of these: there have been violent regionalist movements in the Basque country, Corsica and Northern Ireland; non-violent resistance movements in Brittany, Catalonia and Wales; and party-political opposition in Alsace, Galicia and Scotland.

While cultural issues may be pre-eminent in some of these movements, economic issues are never far from the foreground of the agenda. Ethnic separatist groups and non-ethnic regional groups (such as those in the north of England or south of France) have sought to influence central state policy with respect to the level of resources made available to, or retained in, a region. Employment is usually the prime issue and state provision of employment has become highly politicized. Nowhere is this more evident than in respect of regional policies or in the regional dimension of the policies of publicly owned companies.

Regional economic policies

Evolution of regional policy

State intervention via regional policy is not simply required as a response to local economic problems or even to local or regionalist pressures, although the latter can influence the precise form which it takes. Instead, regional intervention is essential for the functioning of the capitalist economy. This may involve measures to reduce inflation pressures (in land and labour markets) in the more dynamic regions, as well as to secure labour reserves in other regions. The role of the state in this is neither independent of class struggle (which it reflects) nor dominantly influenced by any one fraction of capital or labour (see Amin 1983b). Instead, regional policies change over time reflecting both the requirements and political power of capital and labour, and the need to reproduce the system of capitalist relationships as a whole. According to Bleitrach and Chenu (1982, p. 152):

> [The origin of regional policies] goes back to the great crisis of 'overaccumulation of the 1930s'. Since then, it has been possible to observe the development of both industrial and financial concentration to previously unheard-of extents and the role of the state – in the economic sector in particular. State intervention in the economy is no longer a matter of contingency (as in a war economy, for example), but is structural: it has become necessary for the expanded reproduction of the relations of exploitation.

Post-war developments in regional policy accord with the major fluctuations in the western European economy as a whole, and fall into three distinct phases (Nicoll and Yuill 1980). The first phase (1945–50) was one of tentative policies, for this was a period of post-war recovery when national growth priorities held sway. There were some national and international schemes to assist migration to areas of labour shortages (see pp. 259–60), but otherwise there were few attempts to develop regional policies. The main exceptions were the Cassa per il Mezzogiorno in southern Italy (concentrating initially on agricultural development) and the UK's programme of loans and controls to facilitate industrial mobility to the Development Areas. The second period, from the 1950s to the early 1970s, was one of active and innovative regional policies, which intensified with the slide into recession after the mid-1960s. Most governments developed some form of regional policy to encourage new employment creation in less-developed regions and/or limit growth in metropolitan regions. Finally, since the 1973–4 oil crisis, regional policies have entered a more muted phase. In an era of recession, priority has shifted from reducing inequalities to promoting national growth. Expenditure on regional policy has been reduced as part of an overall reduction in public expenditure (see p. 79), while controls on development in more prosperous regions have been relaxed. In the context of reduced state resources and a declining pool of potentially mobile firms, emphasis has also shifted to encouraging indigenous development potential, especially in the form of small and medium-sized enterprises.

The machinery for implementing regional policy reflects general administrative structures in particular countries. For example, it is no surprise that France has one of the most organized and most centralized systems of spatial economic policy. The programme of Amenagement du Territoire assigns every part of the country a specific role in the national economic plan, while most significant decision making is routed through Paris. The UK, too, has a centralized regional policy system, with most decisions being channelled through the Department of Industry, even though the overall programme has evolved in a highly pragmatic and fragmented manner. The contrasts, in terms of decentralized decision making, are provided by Switzerland and West Germany. In the former, communes and cantons have a high degree of autonomy in policy making while, in the latter, the Lander have freedom to develop economic policies within general guidelines laid down by the federal government. In the phase of muted spatial policy since the early 1970s, the level of administrative control has been affected by two opposing forces. On the one hand, the move to control public expenditure has led to greater centralization but, at the same time, the 'gap' left by the eclipse of regional policy has led to greater emphasis being placed on local economic initiatives, funded and controlled by individual communities. This also accords with the demands of many localities for greater control over the management of their economies at a time of economic crisis, and for greater freedom to pursue their own political and economic goals.

Types of regional policy measures

Regional policies are designed to influence the distribution of capital and/or labour between regions usually in the interests of the economy as a whole. This can involve social investment (subsidies to firms, etc.), social consumption (provision of infrastructure, retraining of labour, etc.) or mediation of some of the effects of uneven capitalist accumulation (temporary job subsidies, etc.).

The major policy mechanisms include controls or incentives and vary between sectors, between private and publicly owned bodies, and over time. There are five main policy types: financial incentives for capital; locational constraints; infrastructural provision; direction of nationalized corporations and government departments; and schemes for labour mobility or retraining. Any government's regional policy 'portfolio' is likely to include a combination of some of these policy types (see Table 45). Most policies have concentrated on manufacturing industry but, in response to structural changes in the economy (see p. 203), the tertiary sector has secured greater attention in the 1980s. Agriculture has largely been neglected by regional policies *per se*, although agricultural policies themselves have regionally differentiated effects. Each of the major types of regional policy will now be considered in turn.

Financial incentives
Financial incentives can influence the location of capital, either diverting mobile firms to, or encouraging new starts in, less-favoured regions. These include grants and soft loans (below market rates of interest) which may be for equipment, buildings or land. Most such incentives are related to the size of the capital investment and are frequently criticized for promoting capital-intensive projects (such as petrochemical plants) which provide relatively small numbers of jobs. Only rarely are the incentives directly related to the employment created by the investment; this is common practice only with some of the Italian incentives and, for a short period, with the UK's Regional Employment Premium. Other forms of financial incentives are tax concessions (important in Italy and Ireland) and transport subsidies to reduce some of the costs of peripheral locations. Promotion of the advantages of particular regions is also common, and Alsace, for example, has its own promotion bureaux in Japan and the USA.

There are differences among the policies according to whether they are automatic or discretionary; the former are simpler and are usually preferred by capital, while the latter lend more flexibility to governments' regional economic strategies. Another difference between national schemes is whether they apply to large proportions of national territory (as in Ireland and Italy) or to only a few regions (as in Denmark and the Netherlands). The extent of spatial coverage by such schemes is shown in Figure 32. Yet another difference pertains to whether the schemes cover only manufacturing (as in the UK before the mid-1960s) or also include services, as in Austria, West Germany and the Netherlands (for a comprehensive account see Yuill and Allen 1986).

Table 45 *Major regional incentives in EC countries in 1984*

	Incentive	Eligible activities*	No. of awards in 1984
Belgium	Interest subsidy (Flanders)	E, M, C, S, T	361†
	Interest subsidy (Wallonia)	E, M, C, S, T	129‡
Denmark	Investment grant	M	96
	Municipality soft loan	M	8
France	Regional policy grant	M, S, T	827
	Regional employment grant	M, S, T	N/A
	Local business tax concession	M, S, T	222§
F.R. Germany	Investment allowance	E, M, T	2,580
	Investment grant	M, T	923
	Special depreciation allowance	E, M, C, S, T	15,901
	ERP regional soft loan	C, S, T	10,674
Ireland	IDA new industry grant	M	238§
	IDA international service programme	S	26§
Italy	Cassa grant	M, S	2,710§
	National fund soft loan	M, S	1,536§
	Social security concession	M, S, T	N/A
	Tax concession	E, M, S	N/A
Luxembourg	Capital grant/interest subsidy	M	40§
	Tax concession	M	8§
	Equipment loan	M, T	206§
Netherlands	Investment premium	M, S	151
UK	Regional development grant	M, S	25,747
	Regional selective assistance	E, M, S, C, T	813
Greece	Investment grant	E, M, T	} 1,279§
	Interest rate subsidy	E, M, T	
	Increased depreciation allowance	E, M, T	N/A
	Tax allowance	E, M, T	N/A

* E = extractive; M = manufacturing; C = construction; S = services; T = tourism.
† 1981 data.
‡ 1982 data.
§ 1983 data.
SOURCE: based on Yuill and Allen (1986).

Locational constraints

Locational constraints are less widely used than are financial incentives, which is hardly surprising as they are a one-sided policy instrument. They can be used to constrain building in, or change of use of existing buildings within, one (usually the metropolitan) region, but they exercise no influence on the destinations of 'deflected' businesses. In fact, one of four outcomes is possible. Firstly, the enterprise may find a way of circumventing the legislation; this frequently happened in the UK where companies denied permission to build new premises were able to move to existing buildings. Secondly, the enterprise may be prepared to quit the metropolitan region but there is no guarantee it will go to a 'problem' region, even if financial incentives are available to attract it there. Instead – and again this is UK experience – it may only move a short distance to an 'intermediate' economic region. Thirdly, multinational capital denied permission to locate in the metropolitan region may shift its investment to another country. With the expansion of networks of MNC branch plants throughout Europe, this option has become more widely available. Even the threat of such action is usually sufficient to make governments reconsider their decisions. Finally, companies may decide not to expand

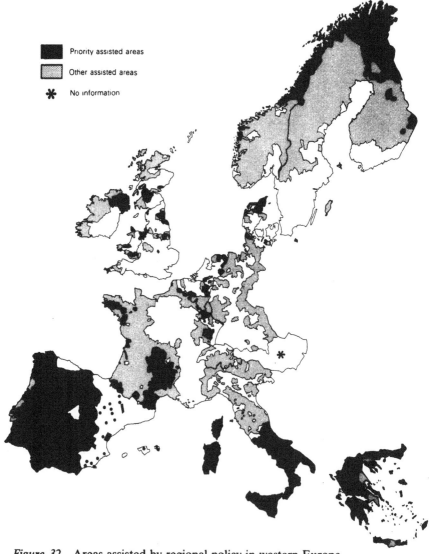

Figure 32   Areas assisted by regional policy in western Europe

SOURCE: based on Yuill and Allen (1986), Wiberg (1984), and other sources.

and instead remain in their existing premises. Both the latter options help explain why many governments have not adopted this policy instrument.

Only four EC countries have had significant policies of locational constraint. The UK was first into this field in 1947 with a system of Industrial Development Certificates for manufacturing firms, followed some two decades later by Office Development Permits. France adopted controls with a system of *agréments* in 1955 to regulate new and existing floor spaces; initially for manufacturing, this was extended to services in 1958. The controls were later supplemented by financial disincentives: since 1960, for example, there have been special taxes on floor space in the Paris region. Italy developed a system of controls for the Milan and Turin areas in 1971. This was prompted by the experiences of the 'Hot Autumn' of 1969 (see p. 42), for congested urban living conditions and a housing crisis were considered to have contributed to industrial unrest. Finally, in the Netherlands in the mid-1970s, a Selective Investment levy was introduced for new office construction in the Randstadt; this was backed up by a system of licensing. In addition, Finland, Sweden and Norway have introduced locational controls on office developments in the 1970s (Ilbery 1981). In the economic context of the post-1974 period, it is not surprising that most of these locational constraints have been loosened.

Infrastructural investments
The first two policy instruments discussed have both involved attempts to influence directly the location of individual companies. An alternative strategy is to create an environment which is generally more favourable to economic development. This involves state investment in infrastructure and is associated with more general attempts to modernize a region. Such an approach is necessarily permissive and, while in some cases it can succeed, it may also result in creation of vastly under-used infrastructure. The range of investment options is considerable and includes water and sewerage treatment, transport (roads, railways, ports, canals, airports, etc.), and provision of subsidized industrial land and/or buildings. Industrial estates have been particularly important, having been pioneered in the UK during the 1930s Depression. Originally conceived as ways to attract manufacturing, the estates have become more diversified over time, reflecting structural changes in the economy. In the mid-1980s attention is increasingly being focused on office parks or on science parks (such as that at Cambridge in the UK or the Tecnopolis Novus Ortis near Bari in Italy) (see Table 46). In all these examples, the aim is to socialize parts of the costs of private capital, thereby increasing the likely profit margins from operating in particular locations.

Growth centres represent an extreme form of this policy approach. These are green-field industrial complexes (mostly based on oil or steel) which offer some external economies between firms as well as general urbanization economies. There is usually an implied belief that their scale

Table 46   *The growth of science parks in selected European countries, 1980–85*

	Number of parks 1980	Number of parks 1985	Number of establishments on parks in 1985
F.R. Germany	0	18	269
UK	3	13	180
France	3	8	320
Belgium	4	5	76
Netherlands	0	3	42

is sufficient to cross the threshold of self-sustained growth (Moseley 1974). A large part of the development costs is socialized. The state may provide a port complex, new roads and cheap land and subsidize construction costs. If the site requires assembling a new labour force, the state may also invest in the necessary means of collective consumption such as hospitals, schools and training centres and, if required, public housing. A number of countries have adopted such strategies, including Spain, Italy, France and Portugal (see Lewis and Williams 1985). France has a particularly well-developed growth-centre strategy, including both industrial poles (as at Fos, Dunkirk and Le Havre) and métropoles d'équilibre for tertiary activities (Bleitrach and Chenu 1982).

Other countries have adopted versions of growth centres but only as a means to concentrate limited resources; examples are Ireland's growth centres at Galway, Ennis, Shannon, Limerick and Cork (Breathnach 1982), the 312 development centres in West Germany (Blacksell 1987), the 18 development centres in the Netherlands, and Denmark's concentration of subsidies and investment in growth centres of about 5000 inhabitants (Toft Jensen 1982). These, however, are very different from the large investments in green-field industrial complexes which have occurred in locations like Taranto, Sines and Huelva.

Growth-centre strategies have been criticized for a number of reasons, including their failure to develop local multipliers, their lack of spread effects to the surrounding region, and low employment generation in capital-intensive complexes. Hence, Damette (1980, p. 85) considers that the major function of growth centres (and of other forms of regional policy) is the 'rapid development of sites corresponding to the optimal location in terms of profit maximization'. Growth-centre complexes were mostly developed in the 1960s and early 1970s, a period of cheap energy and of expansion in the oil and steel industries. In the 1980s these are the very industries which are most severely affected by recession, and so this policy instrument has largely been abandoned.

Nationalized corporations and government employment

Most governments in western Europe are substantial employers in their own right, either via local and central public administration or through nationalized companies (see p. 83). They therefore possess the potential to influence directly the interregional distribution of employment. This can be highly politicized and, in extreme form, public job creation can be used by political parties to cultivate regional electoral support; the most notorious example is probably the clientelism of the Christian Democratic party in southern Italy.

Although most governments pay some attention to the need to redistribute public-sector jobs interregionally, few have developed formal policies in this field. Italy is the outstanding exception, for the state holding corporation, IRI, and nationalized companies such as Italsider (see p. 81) have been directed to locate a significant proportion of investment and/or jobs in the Mezzogiorno. The exact proportion has varied but has been as high as 60 per cent. The UK has pursued such policies rather half-heartedly. Following the Hardman Report some civil service jobs were decentralized from London (an example being the Vehicle Licensing Centre in Swansea), but this policy was abandoned in the 1980s. The National Enterprise Board has also tied its investments in some companies to locational requirements; for example, Inmos (silicon chip producer) was encouraged to locate in South Wales rather than Bristol. However, in the UK the most significant regional political battles have involved the investment policies of nationalized companies, such as British Steel, British Shipbuilders and the National Coal Board. This policy approach has generally been more important in the 1960s than in the 1980s – that is, at a time of expanding rather than of contracting public expenditure.

Labour mobility

Most regional policy measures seek to influence the location of capital, but some are directed at labour mobility. The aim is to encourage migration so as to reduce unemployment in less-developed regions while easing labour shortages elsewhere. This approach has been strongly opposed on social grounds, because it disrupts traditional communities and leads to excessive pressure on infrastructure and housing in the metropolitan region. The question is highly politicized, as is typified in the campaigns in France to establish the right 'to live and work at home'. In the UK the argument has been cast as a choice between 'taking work to the workers' rather than 'taking workers to the work'.

There are few examples of this type of policy instrument. The UK briefly operated a Labour Transfer Scheme in the late 1920s to help out-migration from depressed areas and, in the 1970s, had a small-scale Employment Transfer Scheme to assist financially interregional labour movement. Sweden has also had a scheme to assist interregional migration. Capital has not usually sought to promote greater use of this policy, perhaps because international and regional migration flows were sufficient anyway to meet their needs in the 1950s and 60s. Furthermore, many firms have preferred to decentralize parts of production to the labour reserve regions.

Regional policy: an evaluation

Western Europe has experienced some 40 years of regional policy, but how effective has this been? The variety of objectives and policy instruments, and of local and national contexts, makes it easier to address this question in the case studies in Chapter 8, but some general remarks are in order. Regional disparities have certainly not disappeared and, in the EC9 for example, there has been sharp polarization; whereas the ten poorest regions in 1970 had income levels less than one-third those of the ten richest regions, by 1979 this proportion had fallen to just one-quarter (Lewis 1984). The main contribution of regional policy has been improvement in regional infrastructure and promotion of industrial mobility. However, while regional policies may have speeded up interregional relocations, the critical factor in many moves may have been the need for firms to seek out flexible labour reserves in these regions (Massey 1979). At the same time, the limited stages of production located in the less-developed regions have led to the criticism that regional policy has fostered branch-plant economies, externally controlled and with a preponderance of low-skilled, low-waged and/or female jobs.

Regional policies were designed partly to diminish regional crises, but they have largely failed in this. Hadjimichalis' (1983, p. 130) explanation is that 'accumulation, geographical transfer of value and political/ideological mediation could not operate in harmony, since the system is confronted with a multitude of demands.' The pressures on regional policy have been manifold: more and more regions have experienced economic difficulties and have required state assistance, while new forms of regional crises have appeared in depressed rural areas and inner cities, in addition to the traditional declining industrial regions. Multiple demands on state intervention (in health, housing, etc.) have meant that it is increasingly difficult to earmark funds for regional policy, especially at times of public-expenditure cuts.

Ultimately, however, the limitations of regional policy are more fundamental. The state is seeking to influence the decisions of privately owned capital, but its only instruments are a range of financial inducements and some relatively weak controls on industrial location. Where policy objectives are not in harmony with the dictates of capital accumulation, the state is ultimately powerless to intervene. This is especially so given the internationalization of capital. As an extreme example, it can be argued that regional policy in a less-favoured region in southern Europe can be undermined by the single decision of a large MNC to relocate in another country. In an era of global production strategies, regional policy is likely to play a supporting role to capital, rather than being directive. Furthermore, the occurrence of a global economic crisis and high national unemployment rates also tends to make governments prioritize national over regional aims in their policies for nationalized industries.

'Regional' policies in the European Community

Most regional policy expenditure is at the national level, but there are some European attempts to promote regional development. EFTA has only limited interest in these (for example, grants for industrial estates in Portugal), but the EC has a number of 'regional' policies. These are part of a larger group of EC expenditures on 'structural' purposes, either to ease adjustments to common market conditions or to promote growth and competitiveness. The range of loans and grants, and their growth between 1973 and 1983, is summarized in Table 47. The largest expenditure is undertaken by the European Investment Bank, but the most rapid growth of expenditure in the 1970s was on the Social and Regional funds. All EC expenditure necessarily has a spatial impact, but discussion here is limited to the major instruments which have an explicit spatial dimension: the European Regional Development Fund (ERDF), the Social Fund (SF), the European Investment Bank (EIB) and the European Coal and Steel Community (ECSC).

European Regional Development Fund
The general idea of a European regional fund was set out in Article 2 of the Treaty of Rome, which stressed the need for harmonious development. However, other preoccupations kept this in abeyance for a number of years and serious discussion began only in 1971; even then, it was the prospect of the first enlargement which actually spurred the Community to consider regional policies seriously. Guidelines for the ERDF were agreed eventually and the policy came into force on 1 January 1975. In part, the ERDF was set up for overtly political reasons, being one mechanism for transferring funds to the UK which was expected (correctly) to have a net EC budget deficit. There was also a concern that adjustment to increased competition, and to potentially greater capital and labour mobility, could lead to regional polarization which would foster disenchantment with Community membership.

The aims of the ERDF were to complement, not replace, national policies, not least because West Germany (the largest contributor) refused to agree to the larger fund originally proposed. Two types of investments were financed: in infrastructure (which actually absorbed about two-thirds of all expenditure), and in industry and services. In the case of industrial projects, the ERDF funded only up to 20 per cent of total costs, and could

Table 47 *Community grants and loans for structural purposes, by instrument, 1973–83 (in millions of ECUs)*

	Grants								Loans				
	Agricultural Guidance Fund	Social Fund	Regional Fund	European Monetary System + earthquake interest subsidies	Energy demonstration projects	Research and development*	European Coal and Steel Community budget*	Total budget aid†	European Investment Bank‡	European Coal and Steel Community	Euratom	New Community Instrument‡	Total loans
1973	183.6	167.7	—	—	—	—	55.2	407.0	652.5	292.2	—	—	944.7
1974	254.8	234.7	—	—	—	92.5	54.0	636.0	846.4	410.2	—	—	1256.6
1975	247.2	330.4	249.90	—	38.9	127.3	47.7	993.7	971.6	1098.1	—	—	2015.7
1976	310.4	364.4	387.90	—	38.5	147.0	41.0	1289.2	1086.0	1081.0	—	—	2167.0
1977	312.7	499.0	385.80	—	20.0	209.4	50.8	1477.7	1352.6	989.6	96.90	—	2439.1
1978	269.2	568.3	561.40	—	36.7	242.6	96.9	1775.1	1966.5	797.7	70.30	—	2834.5
1979	507.3	774.5	970.40	200.0	70.1	237.7	120.3	2880.3	2141.8	675.8	152.50	277.0	3347.1
1980	627.7	1014.2	1126.40	197.0	78.1	284.3	131.9	3456.6	2702.3	1020.3	171.50	197.6	4091.7
1981	725.4	1003.8	1708.96	193.2	59.2	351.7	210.0	4252.4	2516.6	387.6	364.29	539.8	3808.3
1982	783.0	1532.0	1895.00	212.7	79.8	389.2	264.2	5155.9	3453.2	668.9	361.60	791.0	5274.7
1983	943.7	1876.1	2121.20	214.4	82.0	462.6	234.0	5934.1	4255.7	776.6	366.50	1211.8	6610.6

* ECSC research aid is included under 'Research and development'.
† Excluding supplementary measures and special measures in favour of the UK and the Federal Republic of Germany:

1980: UK = 194 million ECU   1982: UK = 1804 million ECU.
1981: UK = 1244 million ECU   1983: UK = 1400 million ECU   F.R. Germany = 273 million ECU.
‡ Including special loans following earthquakes: southern Italy (1981–3); Greece (1982).
SOURCE: Commission of the European Community (1985g).

not surpass 50 per cent of the amount of assistance provided by the national government. Projects were submitted through national governments but the Commission had the power to select the schemes to be supported.

Defining the areas which would be eligible to receive funds was the most difficult aspect of establishing the ERDF. The EC set about this in a relatively objective manner and the 1973 Thomson Report used a number of indicators to identify 'problem' regions. The indicators were GDP per capita below the EC average, percentage employed in agriculture above the EC average, more than 20 per cent of the workforce employed in declining industries, high unemployment rates, and out-migration rates in excess of 1 per cent a year. Not surprisingly, the resultant map of eligible areas covered most of the UK and Italy, western France, and substantial areas of other Community member states, Given the small scale of ERDF funds, these clearly were far too extensive. Eventually a compromise was reached whereby individual member states were allowed to designate the areas eligible for assistance, but each country was given a quota of total expenditure (see Table 48). The largest share went to Italy, followed by the UK and France, while no other country had more than 10 per cent. In practice, the eligible areas were most of Ireland and Italy, western France, northern and western UK, and small areas in Belgium, Denmark, Germany and the Netherlands.

By the late 1970s the ERDF was in need of revision. Given the limited resources available, deepening regional crises in Europe, and the accession of Greece, the Commission decided to concentrate expenditure. The revised quotas in operation by 1982 are shown in Table 48; Belgium, Germany, Luxembourg and the Netherlands no longer qualified for quota funds, while the share of France had been considerably reduced. Instead, the funds were concentrated on the poorest regions (see Figure 33), so that the quotas for Ireland, Italy and the UK were increased, while Greece was given an initial share of 16 per cent. In addition, 5 per cent of all funds were held back in a non-quota pool so that the Commission had some flexibility to allocate funds as special needs occurred. Greater selectivity also lay behind comprehensive and integrated EC programmes for selected areas. Initially, special integrated plans were established for Naples and Belfast but, later, integrated Mediterranean programmes were launched in Greece, the Mezzogiorno and southern France.

Two further revisions of the EDRF have occurred since the mid-1970s. Firstly, in 1985 quotas and non-quotas were replaced by a new set of ranges, establishing minimum and maximum shares of expenditure for each country. This brought countries such as Belgium and the Netherlands back into guaranteed funding although still broadly in line with the old system of national quotas. The proportions of each national population living in areas eligible for ERDF assistance in 1985 are shown in Table 48. Secondly, the Iberian enlargement in 1986 added large areas of less-developed and/or depressed regions which, consequently, required further

Table 48 *European Regional Development Fund national allocations, 1975–86*

	Quota allocations				Expenditure 1975–84‡		Eligibility§
	1975	1982	1985 range	1986 range	Millions ECU	Percentage	
Belgium	1.5	0	0.9 –1.20	0.61–0.82	114	1.0	33.1
Denmark	1.3	1.3†	0.51–0.67	0.34–0.46	137	1.1	23.8
F.R. Germany	6.4	0	3.76–4.81	2.55–3.44	545	4.7	37.1
France	15.0	2.4*	11.05–14.74	7.48–9.96	1683	14.6	41.4
Greece	N/A	16.0	12.35–15.74	8.36–10.64	1094	9.5	65.5
Ireland	6.0	7.3	5.64–6.83	3.82–4.61	713	6.2	100.0
Italy	40.0	43.7	31.94–42.59	21.62–28.79	4353	46.4	40.7
Luxembourg	0.1	0	0.06–0.08	0.04–0.06	12	0.1	100.0
Netherlands	1.7	0	1.01–1.34	0.68–0.91	156	1.4	18.7
UK	28.0	29.3	21.42–28.56	14.50–19.31	2736	23.7	37.7
Portugal	N/A	N/A	N/A	10.66–14.20	N/A	N/A	N/A
Spain	N/A	N/A	N/A	17.97–23.93	N/A	N/A	N/A

* Overseas departments only
† Greenland only.
‡ Projects only; excluding programmes.
§ Percentage of national population living in areas eligible for ERDF assistance in 1985.
SOURCES: Commission of the European Communities (1985f; 1986).

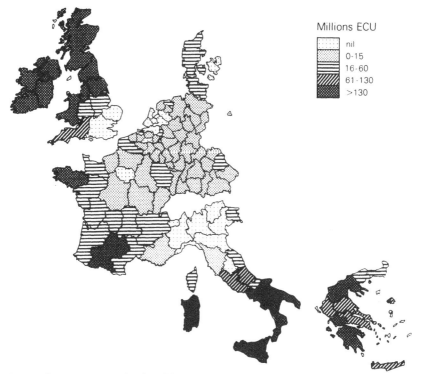

Millions ECU

	nil
	0-15
	16-60
	61-130
	>130

Figure 33 European Regional Development Fund regional allocations, 1981–4

SOURCE: Eurostat (1986).

revision of the national 'shareouts', with notable reductions in the shares of Italy and the UK.

Between 1975 and 1984 total expenditure on the ERDF amounted to 11,700 million ECUs. It is no easier to assess the effectiveness of this expenditure than it is national governments' regional policies. However, the Commission's estimate that some 185,000 'permanent' jobs were created in industrial and service enterprises shows how inadequate this has been, given that unemployment exceeded ten million as early as 1982. Ultimately, the share of total EC expenditure allocated to the ERDF – about 5 per cent in the 1970s – was too limited to allow it a major role, and it has provided useful support for existing policies rather than acting as an independent or an innovative force.

European Investment Bank
The EIB was established under Articles 129 and 130 of the Treaty of Rome with the aim of providing loans for three types of projects: modernizing old industries, stimulating projects which were of interest to a number of member states, and encouraging development in assisted regions. The Bank was to provide loans where it was difficult to secure funding from

commercial banks, subject to a proviso that it could only lend up to 50 per cent of the total investment.

In view of the particularly severe problems faced by Italy and, later Ireland, 'soft' (low-interest) loans could be provided for these countries. In practice the major recipients have been Italy, especially the Mezzogiorno, the UK, especially the north and west, Ireland and France (see Table 49 and Figure 34). Over time there has been a discernible trend to greater concentration of loans in lower-income countries. A wide range of projects has been supported by the EIB, including irrigation schemes, industrial complexes (such as the Taranto steel works), railway improvements (as between Naples and Reggio), and cross-frontier motorways (such as that linking Paris and Brussels). Since 1978 the EIB's role has been supplemented by the New Community Instrument, which is also used to channel funds to poorer regions, especially for infrastructure.

Table 49   *ECSC, Social Fund and EIB expenditure on regional development in the EC9 (percentage figures)*

	Portion of total ECSC expenditure on restructuring, 1978–80	Portion of total EIB expenditure on regional projects, 1978–80	Portion of total Social Fund expenditure 1981–83
Belgium	1.9	0.3	1.9
Denmark	0	2.0	2.2
F.R. Germany	13.8	0.1	6.7
France	18.6	10.6	16.3
Ireland	0.7	14.1	10.3
Italy	2.0	45.4	31.4
Luxembourg	2.3	0	0.04
Netherlands	1.1	0	1.3
UK	59.6	27.5	29.9

SOURCE: Commission of the European Community (1985g).

While the EIB serves a useful role in filling gaps in the commercial money markets, its role is relatively limited. Its funds are constrained and only some 20,000 million ECUs, were earmarked for specifically regional development projects between 1958 and 1985 (Commission of the European Community 1985f). Therefore, it supports the investment strategies of private capital or local authorities, rather than directing overall capital investment in the Community.

European Coal and Steel Community
The ECSC was established to facilitate free trade in coal and steel and to permit rational industrial planning (see p. 34). It was also involved

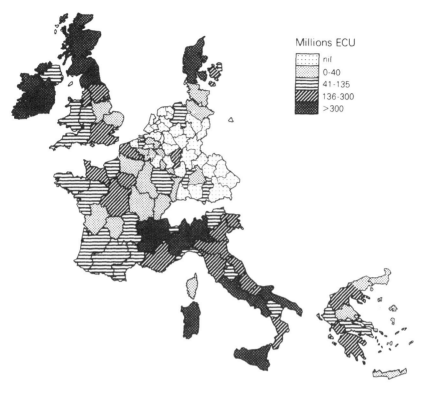

Figure 34 European Investment Bank and New Community Instruments regional allocations, 1981–4

SOURCE: Eurostat (1986).

initially in assisting labour mobility so as to ease labour shortages in parts of the industry. As the economic climate changed in the 1960s and 70s, and national governments moved towards reducing production, the ECSC has also adopted a social role, especially in ameliorating some of the consequences of redundancies in the coal and steel industries.

Article 2 of the Treaty of Paris emphasized that a major responsibility of the ECSC, was to 'ensure the most rational distribution of production . . . while safeguarding continuity of employment and taking care not to provoke fundamental and persistent disturbances in the economies of Member States'. In practice, the ECSC provides conversion loans (at subsidized rates of interest) for investment projects that will create new jobs – in coal, steel or other sectors – for workers made redundant by restructuring. As would be expected, most of the resources have been channelled to the UK, which experienced the largest share of restructuring and of job losses in coal and steel in the 1970s and 80s, followed by France and West Germany (see Table 49 and Figure 35).

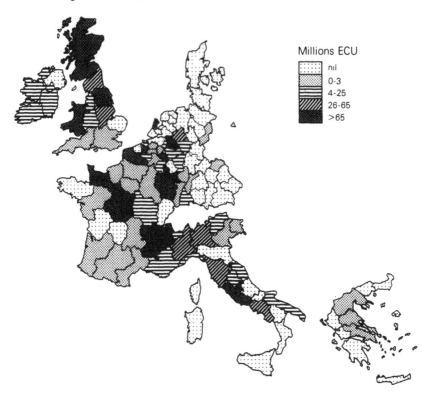

Figure 35   European Coal and Steel Community and European Atomic Energy Agency regional allocations, 1981–4

SOURCE: Eurostat (1986).

Social Fund
The Treaty of Rome provided for the Social Fund as a policy instrument to help ease some of the employment problems which were expected to result from formation of the EC. These were considered to be short-term adjustments, for it was assumed that market operations would eventually produce their own equilibrium between demand for, and supply of, labour. The aims have been to provide resources for the weakest groups in society – mostly defined as young people, women, migrants and the handicapped – and, especially after 1977, for less-favoured regions.

The priority areas for assistance have been broadly similar to those favoured by the ERDF (see Table 49). While the SF can normally cover up to 50 per cent of the eligible costs of a scheme, in the poorest regions this proportion can be up to 70 per cent. In terms of priority social groups, more than half of the total expenditure has been on young people, a recognition of the general plight of unemployed school leavers in recession-affected Europe. In 1982 grants from the SF are estimated to have helped train or retrain 501,910 young people and to have established or maintained jobs for 290,500 (Commission of the European Community 1984a). Although conceived of largely as a social measure, the SF also has an economic function in refining labour markets (through training, etc.) to meet the needs of capital.

Regional expenditures
There are variations in the expenditure patterns of the different branches of 'regional' policy, but there is also a high degree of congruence in the spatial pattern of resource allocation already discussed and for structural agricultural expenditure (see Tables 48 and 49 and Figure 36). Excepting the ECSC fund, Italy is the largest single beneficiary, receiving between about a third and a half of the funds in most years. The UK is usually the second largest beneficiary (excepting agriculture) followed by Ireland and France. The accession of Greece and, latterly, of Spain and Portugal, has inevitably led to some shifts in resource allocation, but without fundamentally altering this pattern. The broad geographical distribution of priority areas within the EC9 in the late 1970s was not substantially different from that identified by the Thomson Report in 1973.

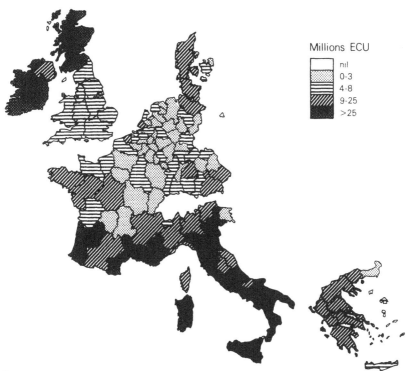

Figure 36 European Agricultural Guidance Fund regional allocations, 1981–4
SOURCE: Eurostat (1986).

Essentially, all the 'regional' funds are seeking to lessen economic and social difficulties arising from the same underlying process of capital accumulation. They may ameliorate some difficulties but hardly offer the prospect of a long-term 'cure' for the problems of these regions, and of the people who live in them. This is not to say that it is in the interests of capital to eliminate all regional inequalities, for capital accumulation may be dependent on the continued ability to exploit labour reserves in less-developed regions.

Further reading

1   Regional inequalities: R. Hudson, D. Rhind and H. Mounsey (1984), *An atlas of EEC affairs*, London: Methuen.
2   Regionalism: C. H. Williams (1986), 'The question of national congruence', in R. J. Johnston and P. J. Taylor (eds.), *A world in crisis*, Oxford: Blackwell.
3   Regional policy: J. R. Lewis (1984), 'Regional policy and planning', in S. Bornstein, D. Held and J. Krieger (eds.), *The State in capitalist Europe*, London: George Allen and Unwin; D. Pinder (1983), *Regional economic development and policy; theory and practice in the European Community*, London: George Allen and Unwin.

8 Selected regional case studies

Introduction

In this chapter, the interrelationships between economic sectors – previously considered separately – are analysed at the regional level. Some of these relationships, such as that between diffuse industrialization and double job-holding in farming, and that between internationalization of manufacturing and services, have already been discussed. Here, further perspectives are provided on the links between different forms of production and labour markets in particular types of regions. While the processes of capital accumulation, of company reorganization and of labour mobility operate at international and national scales, many of the more dramatic features of growth, decline and restructuring are evident most clearly at the regional scale.

The main types of regional economies to be discussed have already been outlined (see pp. 249–50). They are: major metropolitan areas, which exhibit a level of dominance over national and even international economic activities; new centres of rapid accumulation in the post-1945 period, either representing planned developments or private capital adjustments to new conditions of production; existing centres of accumulation undergoing major restructuring; and rural areas which are only partially integrated with advanced urban-industrial capitalism. These represent only a few types among a larger range of regional economies and, for example, regions based on tourism or commercialized agriculture are not discussed here. However, the four types do represent some of the significant features of regional economies in western Europe. In each case, limitations of space preclude discussion of more than one or two examples. These have been selected to represent a diverse set of national as well as regional contexts:

- Metropolitan centres – London, Paris
- New centres of rapid accumulation – Mezzogiorno (planned), northeast and central Italy (diffuse growth)
- Restructuring of existing centres of accumulation – Ruhr, Wallonia
- Partly integrated labour reserve regions – northern Norway, eastern Portugal

Metropolitan centres of accumulation

Metropolitan centres are characterized by their central roles in the division of labour in both the private and public sectors. While their manufacturing bases frequently have been in relative and/or absolute decline, they remain the locations of the headquarters and the research and development facilities of the larger companies. The largest metropolitan centres are also the bases for major MNCs. Their economies have become increasingly tertiarized, but they are also highly specialized within this sector. They are the headquarters of consumer service firms, have significant employment in producer services, and are centres of central government (and, sometimes, international agency) activities. Their expansion has been sustained by cumulative rounds of state and private-sector investment, and by large-scale net interregional and international migration.

In the 1950s and early 1960s the metropolitan centres experienced largely unbroken growth but, since the late 1960s, they have undergone major economic adjustments. Changes in economic and occupational structures have led to differential growth in their outer and inner zones, leading to what has been termed 'an inner city crisis' – that is, inner-urban concentrations of poverty, deprivation and unemployment (see Hall and Hay 1980). At the same time, attempts have been made to replan comprehensively the metropolitan regions so as to provide a more-ordered framework for expansion and further rounds of investment.

Almost every western European country has a major metropolitan centre, but these range in size and importance from mainly national capitals, such as Lisbon and Helsinki, to cities of global significance (see also p. 211). Three western European metropoli are considered by Hall (1984) to be 'world cities'. These are London, Paris and Randstadt (Netherlands), and the first two (and most important of these) are discussed here. Other important examples are Brussels (with the headquarters of the EC and of NATO), Milan and Madrid with significant concentrations of company HQs and important stock exchanges, and Frankfurt which is emerging as the dominant centre in the relatively geographically dispersed economy of West Germany. However, none of these match London and Paris in importance.

London and the South-East

In 1945 the South-East was the leading region in the UK. It had been a focus for expansion of consumer goods industries in the inter-war period, was the headquarters of many national and MNC companies, was still the centre of an Empire, had 50 per cent of national government activities, and the City of London was a world financial centre. The 1950s and 60s in the UK was a period of recovery and of expansion of existing regional economic structures (Hudson and Williams 1986). This meant that the South-East's dominance of the tertiary sector, and of many modern consumer goods industries, was extended (see Figure 37). The South-East also attracted more than 50 per cent of foreign investment in the period to

A

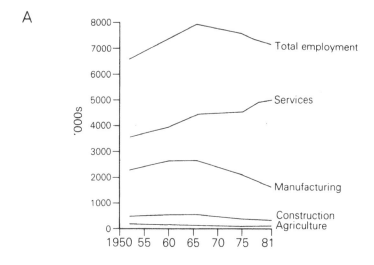

B

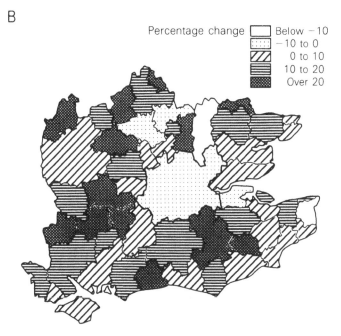

Figure 37 Employment changes in the south-east: *a* sectoral changes 1952–81; *b* total employment change 1971–81

SOURCE: *a* Hudson and Williams (1986); *b* data provided by A.R. Townsend.

1959 (Law 1980). Employment in agriculture had already dwindled to about 125,000 by the end of the decade, so that future transfers of labour to the tertiary sector would have to come from migration or the manufacturing sector. In fact, accumulation was generally sustained by in-migration from which the South-East had a net annual population gain of 43,800 in the 1950s.

After 1962 the UK economy displayed increasing signs of frailty and the South-East was not entirely untouched by this, although it remained the dominant region. Manufacturing employment went into absolute decline after about 1965, although the South-East remained the centre of control and of research and development. Service employment continued to expand, although at a lesser rate after 1965.

Since the economic crisis of the mid-1970s, these patterns have intensified. Agricultural employment has 'bottomed-out' at a level of about 70–80,000, including an element of 'hobby' and/or part-time farming by those in well-paid jobs in London. Manufacturing employment fell in aggregate terms, although new nodes such as the M4 corridor of high-technology growth (see Figure 38) are also emerging. Service employment has expanded but public-sector jobs have declined in the 1980s. The extent of the South-East's dominance is indicated by the location of 55 per cent of all UK office space in this single region (Bateman 1985). Its monopoly of higher-order functions is even greater, with the City of London remaining one of the three leading world financial centres (see p. 211).

Planned decentralization

There has been a geographical reorganization of employment in the South-East as a result of restructuring of production. This has been brought about by changing conditions of production, partly assisted by the process of metropolitan and regional planning. Abbercrombie's 1944 *Greater London Plan* was an early advocate of planned decentralization from what was then considered to be congested London. This was partly facilitated by a programme of New Town development although, in fact, most growth subsequently occurred elsewhere in the outer South-East (Hall *et al.* 1973). Subsequently, the 1964 South-East Study and the 1970 Strategic Plan for the South-East provided for further planned decentralization from London to major growth areas in the periphery of the region (see Figure 39). These areas were conceived as large-scale (populations over one million) counter-attractions to London, although, as Hall (1984, p. 49) emphasizes: 'In its own terms, the Strategic Plan offered a contemporary version of the same policies that animated the Abbercrombie plan of 1974.'

The planning system was backed by a parallel system of controls and incentives to influence the location of economic activities. New manufacturing developments in the South-East were regulated by a system of Industrial Development Certificates and regional development grants, first introduced in 1947 and successively extended until a major cutback in regional policy

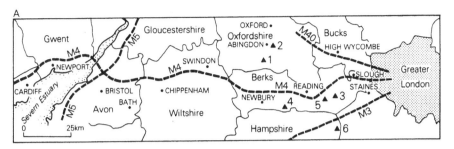

Major research stations

1 Harwell (atomic energy)   4 Aldermaston (weapons)   —— County boundary
2 Culham (nuclear fusion)   5 Wokingham (transport)
3 Bracknell (meteorology)   6 Farnborough (aircraft)

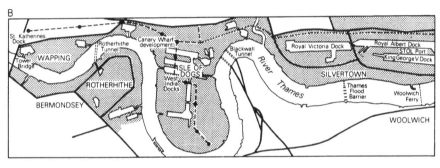

—•— Docklands Light Railways lines and stations under construction

------- Docklands Light Railway Stage 2 proposals

▨ Docklands area

Figure 38 Contrasting sub-regions in the south-east: *a* the M4 corridor;
b London's Docklands

in 1979. Office developments in the South-East were controlled by Office
Development Permits after 1965, while the Location of Offices Bureau
promoted decentralization from London.

There has been a major employment shift within the South-East from
the GLC to the outer parts of the region (see Figure 37). While the South-
East continued to expand in the 1970s, London lost 378,000 jobs in this
decade. This stems from an earlier and deep-rooted decline of manufacturing
in the capital amounting to a 51 per cent job loss between 1960 and 1981
(Fothergill *et al.* 1985). Much of the growth has been in smaller towns and
rural areas within this region (5.8 and 44.9 per cent respectively), reflecting
both the types of new firms (small-scale and high-tech) and their
production needs. Consequently, manufacturing as a proportion of all
employment in the capital has declined from 33 per cent in 1961 to just 18
per cent in 1982. The decrease has been particularly sharp in inner

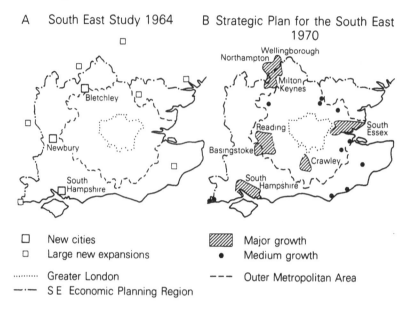

A   South East Study 1964   B Strategic Plan for the South East 1970

☐	New cities	
▱	Large new expansions	
··········	Greater London	
—·—	S E Economic Planning Region	

▨	Major growth	
•	Medium growth	
— — —	Outer Metropolitan Area	

Figure 39   Regional strategies for the south-east.

London, associated with a running down of the older docks and related industries, of smaller firms in traditional sectors such as clothing, and of some modern (inter-war) consumer-goods firms.

Firm closures and redundancies (accounting for about three-quarters of jobs lost) have been more important than firm relocation in industrial decline (see Elias and Keogh 1982). This is fundamentally a consequence of the age of fixed capital and the requirements of new conditions of production. As plant and machinery in inner London became obsolete, many firms closed. At the same time, new firms found it more profitable to invest in the outer parts of the South-East, or in peripheral regions of the UK. These had the advantages of labour availability and, especially in the peripheral regions, of a more flexible, often female, labour force.

The shift in industrial locations was also partly influenced by regional policy which, effectively, made regions more attractive to new firm formation, especially for incoming MNCs in the 1960s. The boom in office development in the late 1960s and early 1970s also militated against manufacturing interests by forcing up the costs of the limited supply of land available in inner areas. That supply, especially for manufacturing, had been severely constrained by land-use planning stemming from the 1947 Town and Country Planning Act and the implementation of Green Belt controls (Hall *et al.* 1973).

There was a long wave of office development in post-war London, culminating initially in a crash in property values in the mid-1970s. This growth reflected expansion of central government activities, and London's role as the headquarters of many manufacturing and commercial services companies. Above all, London was a global centre for international finance, evident in an increase in the number of international banks with offices in the City from 73 in 1960 to 428 in 1982 (Bateman 1985). This forced up office rents (and skilled labour costs), notably in the late 1960s and, in anticipation of financial deregulation, the mid-1980s. In order to diffuse the demand for office space, a system of office controls (ODPs) and the Location of Offices Bureau was set up in the mid-1970s. Between 1963/4 and 1979 the mixture of controls and promotional activity is estimated to have led to some 160,000 jobs being decentralized from central London, of which 60 per cent relocated within 40 miles (Hall 1984). Even so, the tertiary sector continued to expand and, in 1982, Greater London still had 38 per cent of all UK office jobs. This was boosted by London's role as a major international tourist centre.

London has also experienced decentralization of population. Net out-migration per annum from the GLC was 61,000 in the 1950s, 97,000 in the 1960s and 83,000 in the 1970s. In part this was an instance of labour following jobs (although the reverse is also occurring, see Spence *et al.* 1982); but population was also moving out to the suburbs in the search for cheaper land and dwellings, and the establishment of more individual and home-based consumption and life-styles (see Hudson and Williams 1986). Further disaggregated for the 1970s, the data reveal far-heavier losses in inner London (− 17.6 per cent) than in outer London (− 5.0 per cent). At the same time, inner London continued to receive New Commonwealth immigrants who provided a replacement for the native-born population in the labour force in the 1950s and 60s.

The net effect of these changes was to produce a severe imbalance between labour supply and job opportunities in inner London. Jobs declined and the composition of the labour market changed, with non-manual office and female jobs replacing manual, manufacturing and male jobs. As a result, a sizeable part of the local population was excluded from the labour market, and unemployment differentials between inner and outer London sharpened, reaching 14.4 and 8.0 per cent respectively by 1981. Non-white immigrants, the unskilled, males and young school-leavers were particularly adversely affected. Many people in this group effectively became 'trapped' as a pool of long-term unemployed in inner London. This was the core of 'the inner-city problem', which was compounded by poor access to housing and services into acute multiple deprivation in parts of inner London. Civil unrest – notably in Brixton in 1981 and Tottenham in 1985 – made the inner-city problems of London a matter of national political concern.

Reversing decline and decentralization
From the late 1960s central government launched a series of initiatives to

tackle some of the problems of the inner cities, such as Educational Priority Areas and Community Development Projects. The Inner City Partnerships established in 1978 – in Lambeth and Islington in London – provided a mixture of social and economic measures coordinated jointly by central and local government. With a change of government in 1979 and a deepening recession, policy became more concerned with economic objectives. Enterprise Zones were introduced in 1981 – at the Isle of Dogs (in dockland) in the case of London (Figure 38). They were areas of reduced government controls and taxation, which were supposed to stimulate creation of new enterprises. The aim was as much to revitalize the national as the local economy, and these were potent symbols of the Thatcher government's attempts to deregulate the British economy. There was also a general revision of macro regional policies from the mid-1970s. The Labour government, in 1976, set out the aim of halting decentralization from the larger metropolitan areas and diverting resources to the inner cities. Meanwhile, the 1976 Review of the Strategic Plan for the South-East shifted some emphasis from development of new growth areas to reinvigorating inner London.

Given that the supply of land was critical in this, it was inevitable that London's docklands would hold centre stage in any strategy of economic renewal. The debates over the use of the derelict docks (see Figure 38) mirrored much of the political debate over economic policy in the UK after the early 1970s. The Conservative government in 1971 launched a plan to commercialize docklands, 'bringing the West End into the East End', via large-scale private investment in offices and in medium/high-income housing. Local councils strongly objected to this because it offered little to local residents, either in terms of employment or housing. The post-1974 Labour government rejected the plan but was unable to reach agreement with local interests over a suitable alternative. In 1979, one of the first acts of the newly elected Conservative government was to create the Dockland Development Corporation (DDC) with powers to overrule local, demo-cratically elected councils. The DDC adopted a general strategy of commercialising the docks area. Some £1 billion of private investment was attracted by 1986, mainly in offices and warehousing, complemented by state investments in infrastructure, such as the light railway to connect the docks and the City. This connection will be further strengthened if proposals to build 40,000 square feet of office space at Canary Wharf are approved. This has largely been conceived as an extension to the City of London. Docklands, therefore, is becoming reintegrated in the global economy; having lost its pivotal role in world trade (following the relative decline of the UK economy), it is becoming an important centre for international finance. While this is important in sustaining London's role in international finance, it offers only limited local employment opportunities. Some 150,000 jobs have been lost in the docklands since the early 1970s while, by 1986, the DDC had only managed to create 5000 new jobs, many of which have been filled by young professionals newly resident in the area or commuting from outside it.

Paris and the Ile de France

Planning and disorder

Paris is London's main rival as a global city within Europe. The Paris region (Ile de France) containing 23 per cent of the country's active population and about 40 per cent of tertiary employment, has almost the same weight within the French economy as the South-East has in the UK. Like the South-East, its growth was based on the ability to attract migrant labour – both from within France and from abroad – as an essential condition of sustained and rapid capital accumulation after 1945. Net annual migration into the region averaged about 43,000 during 1954–62, giving rise to fears that growth in the Paris region would not only retard but actually damage other regional economies. This was highlighted by J. F. Gravier's provocative book *Paris et le désert français*, raising the spectre of an economic desert outside Paris.

In some ways that vision proved false. After 1962 inmigration fell to only about 11,000 per annum while, by the 1970s, this had become a net loss of 20,000 (Scargill 1983). Nevertheless, the spatial reorganization of the French economy, and the creation of new spatial divisions of labour in the 1970s (see Savey 1983), led to a change in the role of the Paris region. Above all, its importance as a tertiary centre increased and the proportion employed in this sector increased from 54 per cent in 1962 to 62 per cent by 1975. Manufacturing employment in the region declined but the region's dominance of control functions did not lessen.

To some extent, the reorganization of the Paris region was planned. Centralization of political power in France facilitates rapid implementation of state investment programmes, as in development of the Fos growth pole or the office complex at La Defénse; but it also means that there is less capacity at local level to monitor and regulate development in detail. In consequence, while grand schemes may be implemented with great efficiency, the quality of individual schemes and their integration in larger regional structures may be poor. There has certainly been an element of disorder in the rapid suburban expansion of Paris which accompanied economic recovery after 1945. This was evident in the creation of the *grands ensembles*, high-density and high-rise housing suburbs lacking both employment and services locally. Sarcelles is a spectacular example, its population having grown from 8397 in 1954 to 35,430 by 1962.

Largely in response to this disorder, a series of plans was created for the Paris region in the 1960s and 70s (see Figure 40). The broad aim was to encourage the growth of Paris (unlike London there was little emphasis on constraint) but in a more orderly manner than hitherto. Suburban *poles*

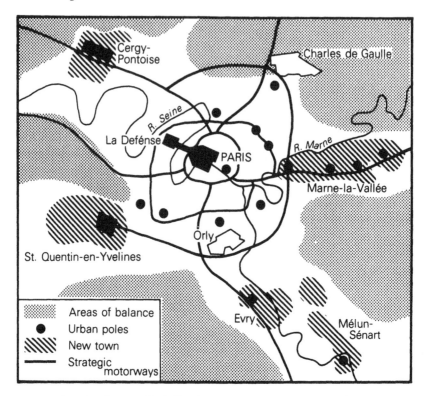

Figure 40 The Greater Paris Region

SOURCE: after Moseley (1980).

restructurateurs were planned as nodes for a more regulated expansion of services and employment. The 1965 *Schéma Directeur* proposed eight New Towns as zones of expansion, although, with downwards revision of demographic predictions, this was reduced to only five New Towns in 1969. As unrestrained growth had contributed to chronic and costly traffic congestion in the region, the *Schemas* also planned a new transport network.

The redistribution of population within Ile de France is shown in Figure 41. The proportion resident in the city of Paris fell from 41 per cent in 1946 to 22 per cent by 1982, while the share in the inner ring – the *Petite Couronne* – expanded until 1968 and only then declined. Population has increased in the outer ring – the *Grande Couronne* – throughout. In absolute terms the population of the city of Paris fell from 2,795,000 in 1962 to 2,172,000 in 1982. Despite these developments, there is still a massive spatial disequilibrium between the locations of housing and of jobs, leading to some 900,000 people commuting to the centre of Paris from the suburbs every day (Thompson 1981). The average journey-to-work for commuters is one hour and twenty minutes (Scargill 1983).

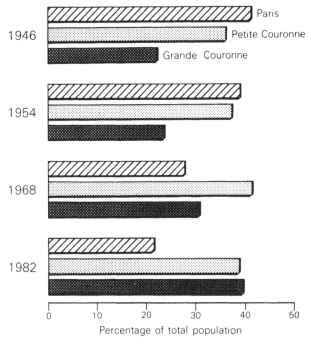

1946

1954

1968

1982

Paris

Petite Couronne

Grande Couronne

Percentage of total population

Figure 41 Percentage distribution of population in Paris and the Couronnes, 1946–82

SOURCE: based on Scargill (1983).

Economic restructuring and inner-city crisis

Agriculture accounts for less than 1 per cent of total employment in the Paris region and so is not considered further here. However, as with London, changes in manufacturing and service employment have been substantial. The Paris region of Ile de France has been losing industrial jobs in absolute terms since the early 1960s (see Figure 42). After that date, many of the industries which had been cornerstones of Paris' industrial recovery in the 1950s – electrical goods, cars, aircraft manufacture, etc. – experienced locational shifts to more peripheral regions. Whereas 60 per cent of the car industry was located in the Paris region in the early 1950s, by 1981 this proportion had fallen to 30 per cent, partly encouraged by regional policies. Tuppen (1983) considers that these policies account for the decentralization of about 200,000 manufacturing jobs from the Paris region. In addition, the requirements of new technology and of new forms of production led to many firms decentralizing parts of the manufacturing process to those regions in the west, south and north-east which had ready supplies of flexible and relatively low-cost labour. It was mainly unskilled production which was decentralized, while senior management remained concentrated in Paris (Lipietz 1980a; 1980b). For example,

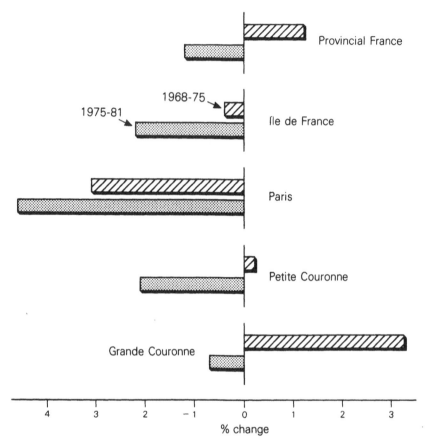

Figure 42   Industrial employment changes in the Ile-de-France, 1968–75–81
SOURCE: based on Tuppen (1983).

highly skilled personnel constituted only 8.3 per cent of the total manufacturing workforce in France but 13.5 per cent in Paris.

Manufacturing decline was particularly severe in the Paris Department, while the *Petite* and *Grande Couronne* experienced gains until the mid-1970s (Figure 42). By the late 1970s manufacturing job losses had become generalized throughout the Paris region, symbolized by some dramatic closures, such as the Renault works at Boulogne–Billancourt. Manufacturing decline was particularly severe in some of the inner areas of Paris, such as Nanterre. Outright closures (46 per cent) were more important than relocations (42 per cent) in this process (Scargill 1983). Decline affected both modern and traditional industries, and such diverse products as aircraft, clothing and printing.

To some extent, manufacturing decline has been compensated for by expansion of tertiary activities, even though since 1958 there had been a system of controls and incentives to relocate offices outside of the Paris region. However, employment growth rates were less in Paris than in the *Petite* and *Grande Couronnes*: between 1968 and 1975 employment increased by only 6.5 per cent in Paris compared with over 20 per cent in most parts of the inner and outer rings (Moseley 1980). These differentials reflect a degree of success for the *Schemas* in the decentralization of services to the suburbs. Nevertheless, office space in Paris expanded by 30 per cent during 1968–75, exceeding the growth rate in all parts of the *Petite* and *Grande Couronnes*, except Hauts-de-Seine where the La Defénse office complex is sited (Bateman 1985). Paris' dominance of higher-order functions is even greater than London's. It is the location of 31 per cent of all office workers but 60 per cent of all research personnel, more than 40 per cent of all financial services, and 50 per cent of all company headquarters. The *Schemas* had proposed decentralization of such functions but this has patently failed, one reason being the decision to create a new office complex at La Defénse. Some 860,000 square feet of office space had been built at La Defénse by 1982, providing 45,000 jobs and attracting the headquarters of several MNCs.

As in London, the spatial reorganization of production in Paris, combined with socially selective population shifts, has produced severe imbalances in the labour market. However, the disequilibrium assumes a different form in the French metropolis. While employment has remained more centralized than in London, Paris itself has also remained more of a middle-class city. Instead, it is low-income groups who have been distilled into cheap housing estates in the suburbs, encouraged by the state programme of subsidized housing, Habitations à Loyer Modére (HLM). This sought to resolve a major crisis, that is, how to house the labour force that was required to sustain the growth of Paris (and thereby the national economy). The 'solution' was construction of HLM subsidized dwellings on low-cost land in the outer zones of the Paris region. While 68 per cent of unsubsidized housing in Ile de France has been in Paris, only 18 per cent of HLM construction has been in the city (Moseley 1980). For low-income households in the suburbs, the costs are manifold: poor access to services and to labour markets, prospects of long journeys to work and, frequently, poorly planned residential areas.

Regions of rapid accumulation

While the metropolitan centres dominated the initial recovery of the European economy, new centres of capital accumulation were emerging by the 1960s. These assumed a variety of forms, examples being Mediterranean tourist zones or agricultural regions in France or Denmark which were brought more fully into the system of capitalist agriculture. There were also new centres of investment and production in the manufacturing sector;

these were both centrally planned and private-sector adjustments to new conditions of production.

Planned decentralization, often to growth centres, involved devalorization of capital by the state in support of private capital accumulation. The state underwrote private investment by undertaking massive investments in housing, roads and serviced land so as to open up new areas for production. Examples can be found in such diverse locations as Fos (France), Taranto (Italy), Sines (Portugal), Tarragona (Spain), Shannon (Ireland) and Washington New Town (UK). The second type of area involves decentralization of production and/or diffuse industrialization as the private sector was reorganized in response to changing conditions for profit realization. Examples include parts of Thessalonika, north-west Portugal, eastern Ireland, Bavaria and south-west France (Wabe 1986). The two types of industrialization are very different in the types of employment generated, labour migration, ownership of capital, opportunities for local entrepreneurship, and linkages with agriculture and other activities within and beyond the region.

Examples of both types of region are here drawn from Italy, a country which industrialized very rapidly after 1945. The traditional core of the Italian economy was the north-west triangle of Milan–Turin–Genoa, and this provided the initial focus for post-war growth. However, as conditions for accumulation were threatened and profit levels fell in the 1960s, two new foci of industrial growth emerged. The Mezzogiorno received a series of state investments to create new industrial complexes, while more spontaneous growth emerged in the north-east/centre regions. Together these three distinctive regional formations gave rise to the notion of the 'three Italy's', most effectively argued in Bagnasco's (1977) *Tre Italie*. An idea of the changing balance of industrial employment changes in the major regions is given in Figure 43. The growth of traditional industries was consistently higher in the north-east/centre of Italy than in other regions. Expansion of modern industries, up to 1963, was pronounced in the north-west which, thereafter, experienced relative decline. The south passed from having the lowest growth rates in modern industries during 1958–63, to having the highest rates in 1970–76, although at the expense of traditional industries. These differences have continued; between 1971 and 1981 employment levels were stagnant in the north-west but increased by about one-quarter in both the Mezzogiorno and the north-east/centre regions (Saraceno 1983).

The Mezzogiorno: planned industrial expansion

Structural weaknesses in the southern economy
Italy has experienced some of the deepest regional inequalities in western Europe (see Williamson 1965), with the south lagging considerably behind

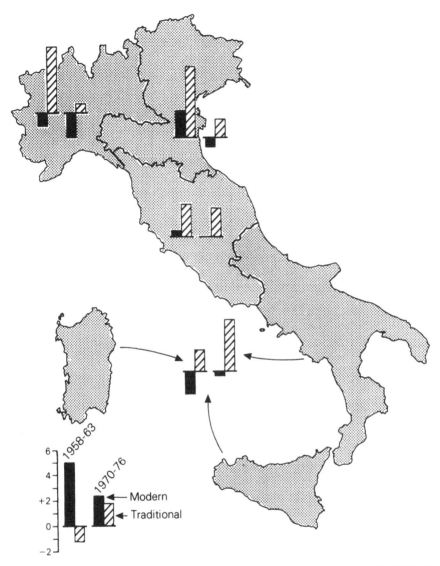

Figure 43 Employment changes in modern and traditional firms in Italy, 1958–63 and 1970–76

SOURCE: based on Arcangeli *et al*. (1980).

the north of Italy. For example, in 1951 per capita incomes in the north were about two and a half times higher than in the south, while the proportions employed in agriculture, respectively, were 35 and 56 per cent. No wonder that King (1981, p. 119) described Italy as 'the most classic model of economic dualism' in western Europe.

The lack of development in the south (the Mezzogiorno) stemmed from both its weak resource base and from its relationship with northern Italy. There was a lack of mineral resources in the south, while agriculture was limited by unfavourable relief and climate. Irrigation could have boosted agriculture but the latifundia land system (see pp. 134–5) militated against this. As a result, the majority of the population in 1945 were still engaged in relatively unproductive agriculture. This constrained incomes and the size of local markets and, hence, the scope for industrial growth. Low returns from farming, some of which were siphoned off for investment in the more productive north, also limited the possibilities for capital accumulation and modernization of the regional economy.

Southern industry has also been weakened by intense competition from northern firms following political (and market) unification in the mid-nineteenth century. Southern firms were small (only 3 per cent had more than ten employees), there was a shortage of many types of industrial skills, local markets were limited and access to other European markets was difficult. In short, there seemed little prospect for any form of private-sector-led industrial growth. The tertiary sector was also poorly developed. Most service employment was concentrated in a few large settlements and there was a lack of small and medium-sized towns (such as existed in the north-east/centre) which could provide an infrastructural base for rural expansion. The tertiary sector also contained a grossly overblown public sector, swollen by the requirements of political patronage, while producer services were poorly developed.

During the first fifteen years after 1945, the Mezzogiorno remained a weak and fragmented regional economy. Schneider *et al.* (1972) considered that extra-regional links via exports of primary materials and imports of manufactured goods were more significant than intra-regional links. Effectively, there was no regional market but a series of relatively isolated local economies. Probably the major role of the Mezzogiorno in this period was as a reserve of labour for the expanding industries of the north and for other European countries, notably West Germany. Between 1950 and the late 1970s some four million workers are estimated to have left the region. In this respect, the Mezzogiorno at this stage was similar to other rural economies (such as the interior of Iberia, to be discussed later) which were poorly integrated with urban-industrial capitalism. The Mezzogiorno differed, however, because it would receive large-scale state investment.

Policies for regional transformation
The depths of poverty in the south were such that no Rome government could ignore the needs of the region for long, especially as conditions seemed ripe for expansion of Communist Party influence. Therefore, an attempt was made to transform the productive base of the region. At first the state favoured agricultural rather than industrial development, and so a land-reform programme was launched in the late 1940s, with six of the eight reform zones being located in the Mezzogiorno. These involved state-

funded schemes to redistribute under-used latifundia lands among landless workers. Typically, the state provided each farmer with irrigation, a farmhouse and basic infrastructure. However, as King (1973) has argued, land reform came too late for it established a peasantry in the south, just as market forces were about to commercialize agriculture and reduce the need for labour. Nevertheless, in breaking the power of the latifundist class, it was an essential prerequisite for modernization.

The Cassa per il Mezzogiorno, established in 1950, was to be the vehicle for implementing state policies. During the 1950s the *Cassa* concentrated on further investment in agriculture as part of what Martinelli (1985) terms 'pre-industrialization policies'; some 77 per cent of its budget was devoted to agriculture at this time. This assisted a steady growth of incomes – by some 4.4 per cent a year – while still permitting labour to be released for the industries of the north-west. However, the relative gap between the Mezzogiorno and the north actually widened during this period, so that pressure grew for a new policy initiative. This came with the launching of an industrialization programme in 1957.

A framework for industrialization was established in the period 1957–64, and it had three major elements. Firstly, incentives were made available to private firms which relocated in the south. The various tax concessions, preferential loans and grants in this financial package could amount to as much as 30 per cent of a firm's relocation costs. However, of greater importance was a second measure, a directive that state holdings – which are significant in the Italian economy (see p. 81) – had to locate up to 60 per cent of new investment in the Mezzogiorno. Thirdly, under the 1957 Industrial Areas Act incentives were to be concentrated on twelve growth areas and thirty growth nuclei. Owing to local political pressures this actually encompassed almost a half of the region's population, a larger proportion than would normally be expected in a programme of spatially concentrated investment.

This basic policy framework remained in operation during the remaining lifetime of the *Cassa*, with minor modifications. Among the more important changes was a redefinition of the growth areas so as to concentrate investment in just six major nodes (see Figure 44). These were to be the loci of major state investments to create large-scale industrial complexes, especially for steel and oil-petrochemicals. This was backed up by a surcharge on building costs to limit industrial development in the Milan area.

These new rounds of investment and sustained emigration resulted in considerable changes in the employment structure of the south. By 1970 agriculture accounted for only about one-third of employment, with important gains in manufacturing and a substantial shift to the tertiary sector. New industries and the increased availability of emigrants' remittances contributed to rising living standards in the south. By the mid-1970s per capita consumption had doubled, infant mortality had been

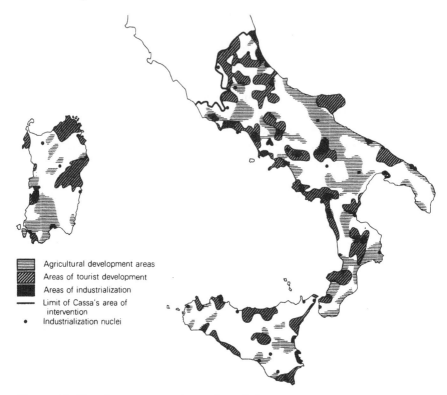

Figure 44   Development areas in southern Italy

SOURCE: various.

halved and per capita incomes had increased by about 4 per cent per annum (King 1981). A number of very large industrial complexes had also been formed in the south. Perhaps the most successful was Bari–Brindisi–Taranto which attracted eight major industrial plants, as well as a number of smaller works. As a result of the productive decentralization strategies of companies such as Fiat (Amin 1985), the south had also acquired a number of large private-sector industrial investments. Consequently, the proportion of national industrial investment in the Mezzogiorno grew from 16 per cent in the 1950s to 24 per cent in 1961–4, at which level it has remained subsequently (Martinelli 1985).

A flawed programme

Southern Italian experiences also reveal a number of potential weaknesses in this type of planned industrialization strategy. It has remained a dependent region, as measured by interregional trade flows; Wade (1979), for example, has estimated that the region's balance-of-trade deficit was equivalent to about one-quarter of its total income and that capitalist development since 1945 had increased this deficit. The composition of the trade balance was also unbalanced, although in a form different from 1945; some two-thirds of exports were composed of agricultural and basic industrial products, while imports were dominated by higher-value-added consumer goods.

The Mezzogiorno has become heavily dependent on external investment, whether by private corporations or the state. State investment has been particularly important, and the Mezzogiorno received some 30–40 per cent of all public investment in most years between 1960 and 1979, compared with north-east/central Italy which had 11–18 per cent (Martinelli 1985). By 1980, 60 per cent of southern manufacturing was owned by non-local interests: 24 per cent being public bodies, 26 per cent northern Italian firms and 10 per cent foreign corporations. Non-southern investment was predominantly in large and modern plant, while southern-owned plants were small and in traditional sectors (see Figure 45). Since the mid-1960s, cutbacks in state investment have meant that southern-owned firms have become relatively more important and, in this period, accounted for about 60 per cent of all new jobs.

Since the early 1970s, the south has attracted more consumer goods production, but this has mainly involved limited productive decentralization by northern companies (Graziani 1978). Decentralization of car production has been particularly important, with Alfa-Sud setting up a large plant at Pomigliano di Arco near Naples and Fiat establishing a number of small and medium-sized works at such places as Cassino, Termoli and Sulmona. The Fiat plants were set up as part of that company's deliberate policy of productive decentralization of component manufacture to rural areas where flexible labour market conditions permitted introduction of more capital-intensive technology and more intensive labour routines (Amin 1985; see also pp. 191–2).

The type of industry attracted to the south, especially in large-scale state-owned industrial complexes, has been criticized. They have been characterized as 'cathedrals in the desert', having weak linkages with local industries. They have also created relatively few jobs considering the scale of investment; between 1951 and 1973 the south's share of national investment increased from 16 to 44 per cent while its share of national industrial employment fell from 20 to 18 per cent. This apparent contradiction is explained by the type of investment for, as Wade (1979, p. 203) emphasizes, 'Southern Italy, plagued by problems of under-employment and unemployment, has a more capital-intensive, labour-saving set of new heavy industries than the North.' Hence, the Mezzogiorno can be characterized as having experienced modernization without either development or employment.

Ultimately, the industrialization of the south appears to be skin-deep. The region is supported by emigrants' remittances and by welfarism rather than by earnings in the industrial sector (Pugliese 1985). The south receives about twice as much in social security payments as it contributes,

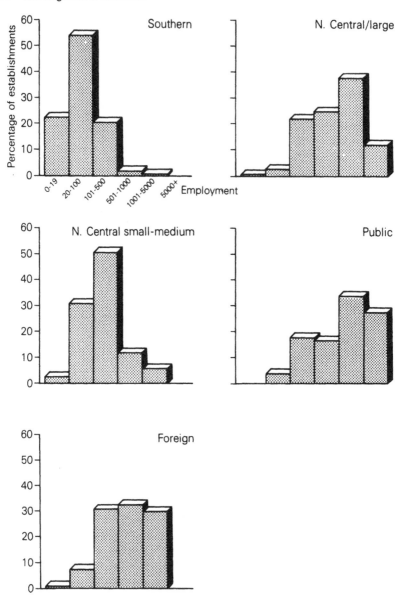

Figure 45 Employment size distribution of firms in southern Italy according to their ownership, 1980

SOURCE: based on Martinelli (1985).

and also benefits from a large number of public-sector service jobs established for political reasons. This is not to say that the south has no role in the Italian economy, but its role is very specific. It acts as a reserve of labour and as a base for cheap production of basic industrial products, while welfare transfers maintain the market which is essential for expansion of northern industry (Pugliese 1985). This is a role which offers little scope for self-sustained industrial growth.

North-east/central Italy: productive decentralization

The NEC model
The first signs of a major locational shift in industrial employment within northern Italy came in the mid-1960s when, following a sharp rise in real wages in the north-west, there was productive decentralization to the north-east/centre (NEC). Following the 'Hot Autumn' of 1969, and another round of sharp wage rises, there was further industrial decentralization to the NEC in the 1970s, accompanied by growth of indigenous small firms. As a result, there was a 58 per cent expansion of small manufacturing firms in the 1970s (Saraceno 1983). The distribution of small firms in Italy is shown in Figure 46, which highlights the importance of the NEC regions.

Industrial growth in the NEC regions has a number of distinctive features, notably diffuse industrialization. Employment is concentrated in small and medium-sized towns to a far larger extent than in the other Italian regions; these accounted for 77 per cent of all industrial employment in the region in 1980 (Saraceno 1983). It is a settlement pattern which offers particular advantages for capital accumulation. Land is available and is relatively cheap, there is a supply of older industrial buildings or farmhouses suitable for conversion to industrial premises, the workforce is already adequately housed, and there is a reasonable level of infrastructure in terms of roads, water supplies, etc.

The firms are implanted in a region of small-scale family-owned farms which provide the basis for a system of double job-holding in agriculture and industry. This offers a number of advantages to employers. Workers can be taken on or laid off as product demand fluctuates, with the family farm providing a safety-net to absorb those temporarily without industrial work. Furthermore, the system tends to weaken workplace conflicts, for there is 'a continuum from the worker who, having family resources, is not entirely proletarian, to the working unit, to the craft industries, to the small and medium sized enterprises, and this cultural continuum blurs class relationships' (Bagnasco 1983). Partial reliance on farm incomes and farmhouse accommodation also means that relatively low wages can be paid to the workforce, especially to women employees. This is a relationship which, of course, also has implications for the agricultural sector, especially in maintaining a system of small farms, in the face of commercial pressures for larger units. Thus, in Marche only one-fifth of the smaller family farms are worked full-time (Zacchia 1983).

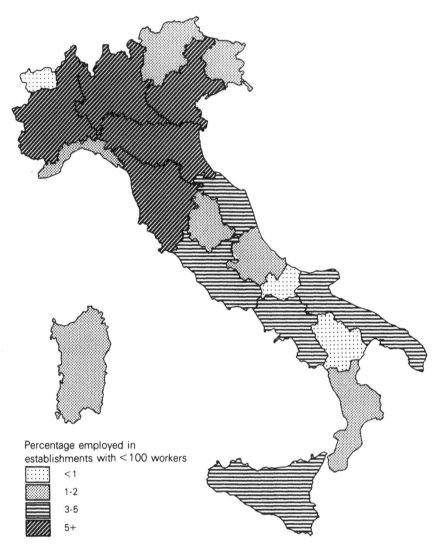

Figure 46 Percentage of manufacturing employees in establishments with less than 100 workers, 1981

SOURCE: Saraceno (1983).

Two interpretations of productive decentralization
Industrial growth in the NEC regions is not a homogeneous process. Instead, there are three distinctive types of firms (Brusco and Sabel 1981): backward, labour-intensive firms serving local markets; firms working under contract to larger companies to produce particular components and hence occupying a distinctive niche in a larger spatial division of labour; and advanced, specialized firms serving directly a large number of markets.

A debate has arisen over the relative importance of these different types of firms and, dependent on this, there are sharply diverging views concerning the potential of the NEC industrial system for self-sustaining growth.

One view is epitomized by Cooke and Rosa Pires (1985) in their review of industrialization in Emilia-Romagna. In the first half of the twentieth century an industrial system evolved whereby highly unionized, medium-sized factories came to depend on non-unionized, female outworkers for low-cost sub-contracting. After the 1960s there was a further wave of industrial expansion in which three types of production were dominant: customized small-batch production in efficient, small, indigenous firms; branch plants of north-western companies; and local firms putting out sub-contracts. Brusco (1983) reports an extreme example of the latter in the Prato wool industry, where 'the leading enterprise often has no employees whatever, consisting only of the entrepreneur himself. It is he who designs the material, has the yarn spun, and commissions the weaving, finishing and mending from different firms.' The overall conclusion of Cooke and Rosa Pires is that in Italy the prime motor of growth is productive decentralization, based on flexible, labour-market conditions.

A different view is held by commentators such as Bagnasco (1982), Fuá (1983) and Garofoli (1981; 1984). For example, Garofoli states (1984, p. 50):

> In the Centre and North-East regions, moveover, in the 70's the transfer of the typical structures of big firms and modern sectors has stopped completely. In these regions a more mature and self-governing process of development, therefore, has taken place though it is mainly based on the so-called traditional sectors.

These authors accept that small firms may be dependent on large national companies or even MNCs, but that, especially in the 1970s, the modernization of artisanal production can be a source of indigenous and independent growth.

An important feature of these areas is the existence of a social structure which is conducive to local entrepreneurship. Garofoli (1984, p. 56) considers that the most important features are:

> . . . homogeneity of cultural behaviour and expectations, a cultural structure aiding social mobility, the presence of many self-employed workers and of artisans, agricultural relations of production mainly based on share cropping and small family farms, and that particular type of business know-how, the need to prove one's own 'social success'.

In other words, there is an occupational structure which provides opportunities for acquiring managerial expertise (whether of small farms or of artisanal production) and, possibly, some capital, and a cultural climate which encourages self-advancement. Over time the source of the entrepreneurs may change as the area industrializes, but the possibilities for enterprise formation remain. For example, in the Pesaro furniture industry, most entrepreneurs in the 1950s had begun their working lives as artisans but, in the 1960s, it was the ex-employees of these firms who emerged as the main source of entrepreneurship (Zacchia 1983).

Another source of strength in the NEC industrial system is the emergence of area systems of production (Garofoli 1981), being local concentrations of specialized industries. While the firms individually enjoy the flexibility and labour cost advantages of smallness, they also share a number of locational externalities. Garofoli has estimated that there are 70 such area systems in Italy, most of which are located in the NEC regions. The externalities include local firms offering specialized marketing (including exports), transportation, finance and accounting services to the manufacturing enterprises. In this way the industrial performance of the NEC regions in the 1970s and 80s has seemed more robust and to have more capacity for self-sustained growth than the Mezzogiorno.

Restructuring of existing centres of accumulation

A number of traditional industrial sectors which were the bases of economic expansion in the nineteenth or early twentieth centuries have been profoundly restructured since 1945. Examples include shipbuilding and textiles, which tended to be highly concentrated in particular regions such as, respectively, the North Sea ports of the UK and West Germany, and Catalonia and the North-West of England. The traditional, and often narrow, bases of these regional economies have experienced sharp decline both in the dominant sector and in linked industries. For example, decline of textile production can have strong negative effects on textile machinery firms, producers of buttons and zips, manufacturers of specialized packaging materials, and related marketing, financial and other service firms. Within Europe, the largest manufacturing employment losses during 1973–81 have been in regions such as Lorraine (− 28 per cent), Lüneburg (− 42 per cent) and the West Midlands (− 24 per cent).

Restructuring creates new reserves of labour which potentially can be utilized either via migration to new centres of accumulation, or through new rounds of investment in the local region. Availability of labour and of other conditions of production – such as existing factory premises, good communications, and training facilities – may be sufficient to attract private investment, whether national or international capital. However, the extent of the decline, the traditional importance of these regions, and their ability to launch effective local or regional protests against redundancies, can be sufficient to bring about major government intervention in the process of restructuring. This may involve providing subsidies to halt or retard decline in traditional industries, or measures to attract new investment. For example, violent community responses to steel plant closures in Nord

(France) helped secure substantial government aid for new employment measures (Hudson and Sadler 1983). More generally, regional policy in countries such as the UK, Belgium and West Germany has often been developed from specific programmes to tackle the particular problems of these types of regions.

The extent and form of restructuring is dependent on a number of considerations, including the particular sector involved, the overall regional configuration, the strength of the national economy, and the form of state intervention. In some regions, such as South Wales or the North-East in the UK, the period since 1958 has seen a series of state interventions, in changing form, responding to recurrent crises in the local economy. Here, the Ruhr and Wallonia, two of Europe's classic areas of nineteenth-century industrialization, will be considered. Coal and steel industries have undergone sharp decline in both regions since 1945, but the nature of the economic adjustments, and the responses by private capital and the state, have been very different.

The Ruhr

The Ruhr is one of western Europe's oldest capitalist industrial regions, and in 1945 it represented one of the world's largest concentrations of coal and steel production (see Figure 47). Since the late 1950s coal production has declined, although less sharply than in most other major

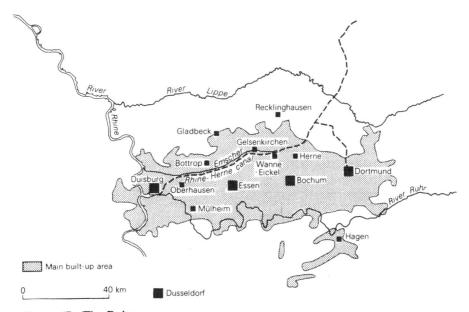

Figure 47 The Ruhr

European coalfields. Nevertheless, the regional economy has undergone considerable restructuring. The fact that this has occurred relatively successfully – although unemployment is above the national average (Blacksell 1987) – is due to effective modernization of the coal industry, the strength of the national economy, and the accessibility of the region. It is near the centre of the major European transport axes, has 30 million people living within a 150 km radius (Hellen 1974) and, according to most indexes of economic potential (for example, Keeble *et al.* 1982), lies close to the geographical centre of the European market. Hence it has a standard of living well above the EC average and an unemployment rate which is low for the EC but high for West Germany as a whole.

Coal and steel: strength in declining industries
After immediate post-war disruption, coal output recovered fairly quickly to a maximum of 125 million tonnes in 1956 but, thereafter, declined to 66 million tonnes by 1984. That the decline was not even sharper was due partly to the good quality of the coal reserves and favourable working conditions. There was also rapid modernization of mining, with mechanization being introduced to 93 per cent of mines by 1972 (Hellen 1974). However, the small scale and fragmented nature of mine ownership had made it difficult to plan the industry coherently. Unlike in the UK and France, where nationalization was the response to such difficulties, the German Government encouraged consolidation of private interests. To this end, Ruhrkole AG was set up in 1969 as a holding company representing 94 per cent of all the mining output of the Ruhr. The company has successfully managed a restructuring programme (involving concentration on the more profitable mines) which has allowed the Ruhr to remain relatively competitive in energy markets. Nevertheless, a reduction in output and the introduction of more capital-intensive methods meant a fall in mining employment from almost one-half million before the 1958 European crisis of overproduction to 128,000 in 1977. This was compounded by the regionally uneven impact of closures, for it was the older mines of the shallower southern part of the coalfield which were least profitable and were the first to close (see Figure 48). Individual towns were badly affected and Bochum, for example, lost about 50,000 mining jobs between 1958 and 1968.

Steel production has been the other staple industry of the Ruhr. The industry was partly dismantled at the close of the Second World War and, at first, highly regulated by the armistice powers. However, it was accepted eventually that recovery of the Ruhr industry was essential for the recovery of the European economy, and the constraints on the industry were relaxed, especially after it became a founder member of the ECSC. Subsequently, output rose from 20 million tonnes in 1951 to about 50 million tonnes in 1971. There were setbacks in 1958 and 1961, which were years of crisis in the European industry; but the Ruhr steel makers managed to overcome these more comfortably than most of their

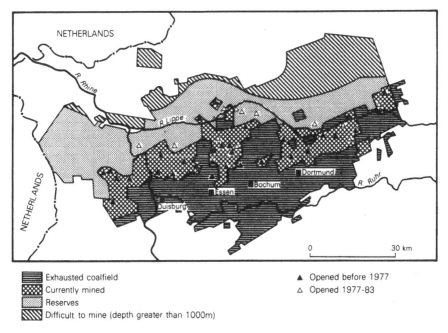

Exhausted coalfield

Currently mined

Reserves

Difficult to mine (depth greater than 1000m)

▲ Opened before 1977

△ Opened 1977–83

Figure 48 The Ruhr coalfield

SOURCE: after Hommel (1984).

competitors. One reason for this is the strong local market which exists in the region, with North-Rhine Westphalia accounting for over a quarter of West Germany's manufacturing capacity in 1984 (Blacksell 1987). Motor vehicle production has been a particularly important market. The strength of this regional industrial complex, containing some of the leading machine-tool, mechanical-engineering and metal-fabrication plants in Europe, has provided a steady expansion of demand, with only limited reliance on (more erratic) export markets. Links with local manufacturing are strengthened by the vertical integration which exists in the region: Thyssen, based at Duisburg, has both steel production and engineering interests, while Krupps has interests in railway-equipment manufacturing at Essen.

Since the mid-1970s the industry has stagnated, but it has survived the global recession in steel production better than rivals such as the British Steel Company and USINOR (France). Indeed, the overall strength of the Ruhr industry is such that it has not been challenged by investments in coastal steelworks as happened, for example, within France and Belgium. Even so, employment in the Ruhr steel industry fell by some 30 per cent in the late 1970s and early 1980s. Companies such as Thyssen and Krupps made large losses in 1983 (DM416 and 287 million respectively) and had only partly recovered by the mid-1980s. Nevertheless, there has been spatial reorganization of the steel industry, with an east–west shift towards locations on the Rhine and the major canals which provide better access to raw materials imported via Rotterdam. Rationalization of existing plant and new investment have also led to capacity reductions in the central Ruhr, especially around Bochum and Essen, while towns on the Rhine, like Duisburg, have increased in relative importance (Mellor 1978).

The economy of the Ruhr is not based only on coal and steel production. Other manufacturing interests include chemicals (especially around Duisburg and Dortmund), textiles (especially at Gelsenkirchen), and electrical engineering and car production. Vehicle production is particularly interesting, with the first major car works being located at Bochum in 1960. This was an Opel car assembly works specifically attracted to the region by the availability of labour released from declining industries; some four-fifths of its labour force were initially coal miners (Hudson 1983a). In addition, considerable immigration in the 1960s and 70s, especially from southern Europe, ensured that the supply of labour was replenished. Taken together, the new and traditional industries have resulted in the Ruhr remaining extraordinarily dependent on manufacturing employment. Even in 1970 some 58 per cent of all jobs were in this sector, although the region had lost about one-quarter of all industrial jobs during 1974–82. One corollary of this has been a relatively poorly developed tertiary sector. The real 'capital' of the Ruhr is probably Dusseldorf, which is located beyond its boundaries in the Rhine valley. This has the headquarters not only of MNCs like the IBM computer company, but also of major Ruhr companies such as Krupp and Thyssen – an example of a clear spatial division of labour. To some extent this has been balanced by the establishment of new universities in the Ruhr in the 1960s and 70s, but the region still lacks a strong tertiary sector.

Small-scale state intervention
In the Ruhr the major problem areas are to be found in the older coal mining zone, especially along the Emscher valley. Towns such as Bochum and Essen had a high level of dependence on steel or coal production and, consequently, have suffered from the reorganization of these industries. In addition, they have an inheritance of extreme environmental pollution and an ageing housing stock; for example, in Bochum in 1968, 29 per cent of dwellings still had no bath or indoor toilet (Thomas and Tuppen 1977). While private investment was renewing the industrial structure of the Ruhr, there were still pockets of economic or social difficulties which required some state intervention.

Early post-war regional policies in West Germany were mainly concerned with the needs of rural areas, and the Ruhr was anyway little affected by unemployment in the first two decades after 1948. However, by the late 1960s coal and steel closures were having a serious effect on the region, so that the central part of the Ruhr, along the Emscher valley, was made a development area in 1967. Central and regional government subsidies were made available to try to attract new investment, while attempts were made to improve infrastructure. The region benefited from this designation during most of the 1970s, and it was only in 1981 that the (more prosperous) western Ruhr lost its status as an assisted area in the course of a general review of, and cut-back of, finance for regional policy. As a result of the late scheduling, and the partial descheduling of the Ruhr in 1981, the region has received far less state aid and less ERDF aid (only 8 per cent of the West Germany total) than most other assisted regions in the country. It therefore remains essentially a regional economy which has been restructured through the dictates of private capital accumulation rather than through state intervention.

Wallonia

Wallonia (or more precisely the Sambre–Meuse valley) was also an early centre of urban industrial capitalism (see Figure 49). There was rapid expansion of coal mining in the mid-nineteenth century and associated

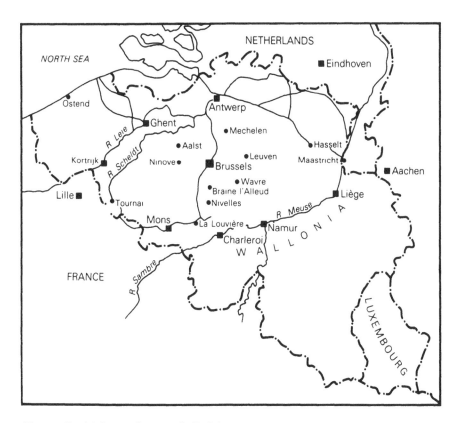

Figure 49 Major settlements in Belgium

growth of steel production. During the early twentieth century, the regional economy was in relative decline; there was little reinvestment and the economic structure became outdated and ill-adapted to the needs of modern production. Despite an immediate post-war revival in the economy, its long-term decline was not halted. Instead, investment in new types of industries within Belgium has shifted to the north (Flanders), giving rise to serious problems of economic adjustment in Wallonia after the late 1950s.

Recovery and decline

These problems were not immediately obvious in 1945. Belgium's economy had emerged relatively unscathed from the Second World War, and Antwerp was the only undamaged major port on the North Sea. Therefore, Belgian industry was in a strong position to benefit from the demand generated by reconstruction in western Europe, while the Congo proved to be a lucrative source of colonial profits (Carney 1980). Given the availability of high profits, existing production facilities were renewed and extended and there was little reinvestment in modern industries. Instead, costs were kept relatively low while the need to invest in more capital-intensive methods was partly averted by reliance on large-scale immigration; there were over 300,000 immigrants in Wallonia in the 1960s.

However, by the early 1950s the high costs of production in the Sambre–Meuse coalfield made it uncompetitive, especially in the face of cheap US coal imports. As a member of the ECSC, subsidies were used to sustain the industry through most of the decade, but the 1958 energy overproduction crisis had a devastating effect on the region – far worse than in the Ruhr or, indeed, any other major EC coalfield. This accelerated closures and between 1955 and 1958 the number of mines in, for example, the Borinage basin fell from 23 to 10 (Carney 1980). Widespread regional demonstrations in 1958 secured greater state aid but, with the availability of cheaper energy supplies in the 1960s (especially of natural gas from Groningen), the coal industry continued to decline. Output was 20.5 million tonnes in 1955, only 6.3 million tonnes in 1968 and had virtually ceased by the mid-1980s. Some 100,000 jobs have disappeared since the peak year of 1955.

There has also been a decline of steel production in Wallonia. In the late nineteenth century this had been one of Europe's premier steel producing areas, based on local coking coal and Lorraine iron-ore. Basic steel production gave rise to a number of related steel-using industries, especially in the Liege and Charleroi areas. Among the more important activities were special steel production for electrical and marine engineering, machine tools and armaments. As with coal, the 1950s was a period of recovery, expansion and prosperity. However, in the 1960s the changing locational requirements of large integrated steel mills favoured sites such as Zelzati in northern Belgium. Industrial decline is epitomized by the case of the Cockerill Sambre steel company which, with its plants in the Liege and Charleroi basins, was one of the totems of the traditional industrial strength of Wallonia. By 1983 the company was on the verge of bankruptcy, having undertaken some disastrous mergers and grossly overestimated its likely market share. It was only saved by a recovery plan which involved selective closures and a cut of one-third in the labour force. Not surprisingly, the Charleroi (− 36 per cent) and Liege (− 33 per cent) basins had some of the largest declines in the EC in manufacturing employment between 1974 and 1982 (Wabe 1986).

The rush to the coast
While the Belgian economy has not been among the more dynamic in western Europe, there has been no lack of new investment. The problem − for Wallonia − has been the shift to the northern coastal region, Flanders, which had two major advantages. One was the availability of deep-water harbour facilities and, not surprisingly, the new hubs of industrialization were Antwerp (especially for petrochemicals) and the Ghent–Terneuzen canal zone. The other advantage was labour-market conditions; Flanders was less unionized and less militant than Wallonia, as had been evident in the uneven regional opposition to government austerity measures, following the 1958 economic crisis (Carney 1980).

The new round of investment in Flanders in the 1960s and 70s was led by private capital, including several MNCs; US companies alone accounted for some 20 per cent of all Belgian investment in the 1960s (Clout 1975b). Foreign investment was highly concentrated in Flanders. As a result, the large numbers of jobs created in modern textiles, electrical engineering, car assembly (especially the General Motors plant at Antwerp) and chemicals production was overwhelmingly located in the north. In the late 1960s some foreign investment was attracted to Wallonia, but it was spatially highly selective, mostly to be found in Liege or at Nivelles, which effectively was becoming a part of the Brussels metropolitan economy. Consequently, it had become possible to talk of a 'two-speed' economy, especially as Flanders accounted for 70 per cent of Belgium's exports.

The relative decline of Wallonia is evident in a number of indicators. Its share of GDP fell from 33 per cent in the late 1950s to 28 per cent in the late 1970s, while its share of employment also declined. Significant variations in unemployment also emerged after the 1958 crisis, and these have persisted until the present. For example, in 1983 unemployment in Wallonia was 17 per cent compared with 15 per cent in Flanders. Per capita incomes in Wallonia have also lagged behind Flanders since 1967 (Gay 1981). Consequently, Wallonia has become less a centre for accumulation, and more a labour reserve region. Out-migration rates of about 25,000 per annum were common in the 1960s (Clout 1975b), mainly to Brussels (expanding rapidly as an international metropolis) and, to a lesser extent, Liege, which was still experiencing some growth.

State policy in a divided society

Restructuring of the Belgian economy has involved some state intervention. In response to the 1958 crisis, the Regional Development Act was passed (in 1959) providing for financial assistance to help establish new firms and to modernize existing plant. When, after the mid-1960s, stagnation in the national economy led to renewed concern about economic development, the more powerful Regional Development Act 1966 was passed, incorporating a variety of grants, loans and tax exemptions for firms locating in designated areas. These represented a mixture of locations in the north (areas with potential) and in the south (areas with severe problems, such as the Bonnage and Campine coalfields). However, this Act reveals a fundamental dilemma of Belgian economic policy, the political need to balance assistance to Wallonia and Flanders. As Gay (1981, p. 189) comments:

> Belgium was not really operating a coherent policy for regional development but was rather implementing a blow-by-blow approach, with any major investment in one socio-cultural region having to be compensated by an equivalent measure in the other.

As a result, despite – or perhaps because of – the problems of the south, Flanders secured 68 per cent of all jobs created by state assistance during 1959–75 (Clout 1975b), although Wallonia has attracted 60 per cent of ERDF funds. Furthermore, investment in Wallonia overwhelmingly concentrated on restructuring old industries rather than establishing new plant, with a few exceptions such as the petrochemical works at Féluy, or the attraction of American MNCs like Liposome and Bioassay.

By the 1970s the difficulties of the national economy further constrained the scope for state intervention. To some extent central government has divested itself of responsibility for these difficult issues. In the early 1980s the regions acquired some financial autonomy but were also to be responsible for financing the traditional industrial sectors such as coal, steel and textiles. This arrangement has caused other problems, however, because the decline of the Kempenland coalfield required expenditure beyond that available to the Flanders government. However, the Wallonia government would not agree to any additional state financial assistance without a reciprocal agreement for the Cockerill Sambre company. The difficulties of regional policy in a divided society are still considerable.

Partly integrated labour-reserve regions

No region in western Europe can be considered not to have been integrated into the capitalist system of production in some way. Even the remotest regions of Finland and Ireland, or the least accessible provinces of Spain's Estremadura or the Greek islands, are linked in some ways with the larger western European economy – whether via trade, some form of decentralized production, or as labour reserves. Such regions are not centres of capital accumulation but – through the provision of labour and of markets – they support this process.

These economies are dominated by primary activities, usually agriculture, but possibly forestry or fishing. Traditional, subsistence agriculture rarely survives intact, but commercialization is limited with only a few farmers supplying regional or national markets. Exceptions do exist for highly specialized products, such as particular types of cheeses or smoked fish which can command a high price in wider markets. But, for most farmers the possibilities of commercialization are limited. Manufacturing is also relatively weak with, again, a few exceptions such as highly valued artisanal products as, for example, hand-knitted woollen clothes or firs glassware. Indeed, local manufacturing has tended to decline in the face of competition from lower-cost regional imports. Tertiary activities are mostly limited to consumer services (especially retailing and personal services) and some collective services such as education, health and public administration.

The lack of local employment opportunities, and the allied features of low income and poor access to services, has meant that many of these regions have become important reserves of labour, sending emigrants to the centres of rapid economic growth. Depopulation may result, leading to further erosion of local services, hence reinforcing a downward spiral of economic and demographic decline, in relative if not in absolute terms. In relation to their national economies, these regions constitute a 'relictual' space' (Kielstra 1985).

Government policies in these regions are faced with hard choices. The regions can be left relatively untouched, so as to maintain their role as reserves of labour. Effectively, this happened in the Mezzogiorno in the early post-war years. However, the legitimation role of the state means that it cannot subscribe to this approach, at least not openly. As a bare minimum, the state has to be seen to try to stabilize population levels and maintain, if not improve, service provision. If the rural population and settlement structure is relatively dispersed, this may lead to an attempt to implement growth-centre or service-centre policies. This may reduce the costs to the state of subsidizing collective consumption, but it is also, effectively, an abandonment of the remotest rural areas outside these centres. Alternatively, government may seek to intervene by subsidizing local primary production, or by attracting new types of economic activities into these regions.

The two examples presented in this section are northern Norway and the interior of Portugal. They have been selected as representing very different levels of national economic development, with Norway and Portugal being at the extremes in any European league table of growth or incomes. This means that somewhat different problems are to be encountered in the two regions and that the resources available for state intervention are very different.

Northern Norway

The three most northerly Norwegian districts (see Figure 50) constitute one of the remotest regions in western Europe. In 1945 the region was

Figure 50   Major regions in Norway

devastated because, for strategic purposes, the occupying German armies had removed much of the population and destroyed many settlements (Wiberg 1984). The Norwegian Government originally planned that repopulation should be concentrated into the major settlements, so as to reduce the costs of service provision. However, repopulation was very rapid, being complete by 1950, and effectively replicated the previous settlement pattern. Consequently, there was a very dispersed population, with some 80 per cent in 1950 resident outside the towns and villages.

The northern economy
Northern Norway constituted one of the most difficult environments for economic development in Europe in the post-war period. Its population lived at exceptionally low densities in a dispersed and inaccessible settlement structure within a remote region. Although it had 12 per cent of Norway's population, it had only 6.2 per cent of GNP. One half of this came from (subsistence) farming and fishing, and only 10 per cent from manufacturing. Enterprises were small-scale, with two-fifths of the labour force being self-employed (Stenstadvold 1981). Manufacturing was particularly weakly developed and, although this sector has since expanded, 72 per cent of all industrial jobs in 1980 were in food processing in Finnmark (Asheim 1986). The reasons for this include the dominance of its manufacturing by foreign and southern Norwegian capital, a traditional social structure wherein merchant capital had little inclination to invest in production, and a failure to develop firms with high linkages and/or advanced technology.

As a result, the north had the classic role of a labour reserve region with high levels of net out-migration: 21,125 in the 1950s, 32,167 in the 1960s and 13,623 in the 1970s (Hansen 1983b). Expansion of the southern regions of Norway meant that most out-migration was interregional, with relatively little emigration. Movement to Oslo was important in the 1950s but, thereafter, it shifted to more diverse destinations (see Figure 51).

Decline in out-migration suggests that some changes have occurred in the region's role, and, indeed, Troms actually experienced net in-migration in the 1970s. In part this reflects indigenous development with gradual capitalist penetration of agriculture. Farmers entered into commercial relationships – selling to local markets – so as to acquire the money they required to purchase farm machinery and motor vehicles (or boats). However, the region also benefited considerably from central state transfers of resources. Maintenance of regional balance is important in political life in Norway so that, not surprisingly, welfare state policies have a strong regional dimension.

Regional policies and the welfare state
The North Norway Plan of 1952 was the first major landmark in regional policy and it sought to achieve greater integration of the region into the

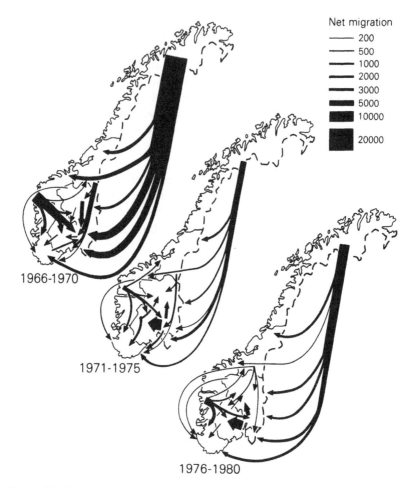

Figure 51   Internal migration in Norway 1966–80

SOURCE: Hansen (1983b).

capitalist economy. This was based on creating a year-round fishing industry and, more importantly, transferring (under-employed) labour from a still largely peasant agriculture to modern manufacturing industries. Investment in fishing aimed to modernize the fleet of vessels and provide a market for year-round catches by building new processing plants. Between 1945 and 1960 some 35 such plants were opened, the largest of which was the Findus factory at Hammerfest with 800 jobs (Wiberg 1984). Alongside this measure, the North Norway Fund provided aid to investments, such as those in aluminium and chemicals, which could provide non-seasonal jobs. This strategy also required spatially selective state investments in infrastructure so as to create minimal operating conditions for the new plants. Permanent jobs were provided in the region but, at the same time, the more rural parts acted as labour reserves for the urban centres. A system of commuting developed which, depending on accessibility conditions, was daily or weekly in character. Nevertheless, the prime function of the region as a whole remained that of a labour reserve for the south (Figure 51).

In the 1960s regional policy intensified in the face of persistent out-migration and unemployment in several regions, especially the north. This policy operated at two levels: a number of measures, such as transport subsidies and infrastructural investments, sought to improve the general conditions for capital formation, while resources were also spatially concentrated in, for example, industrial estates (Stenstadvold 1981). At one stage an even stronger policy of concentration was proposed, based on nine growth centres; but later this was watered-down into a development area policy including more rural areas. The outcome, however, was a reinforcement of the two linked facets of the region's role as a labour reserve. On the one hand, out-migration from the north continued and was even accentuated in the 1960s; per annum out-migration rates were only 2300 in the early 1960s compared with 4100 in the late 1960s. On the other hand, there was also continued concentration, with 414 of the 586 census areas in the north losing population this decade (Stenstadvold 1981). One reaction to this in the 1970s has been a shift in policy to investment in 'anchorage points', often small-population centres which it was hoped would help reduce out-migration (Hansen 1983b).

Regional policy has not been completely ineffective in northern Norway. On the contrary, Bivand (1986), using shift-share analysis, estimates that almost 11,000 additional manufacturing jobs may have been created in the region. However, even with state subsidies and the development of new forms of productive decentralization, the northern regions have only secured a small share of all new employment. There has been relatively little decentralization of private-sector offices or of central administration, although the establishment of Troms University in the 1970s has been locally important in reducing population outflows. The case of the oil industry is probably symbolic. Although the government directed that the headquarters for Statoil should be located outside Oslo, it was Stavanger which was chosen, not one of the northern towns. Consequently, northern Norway, which has 12 per cent of Norway's population, has only acquired 1 per cent of jobs in the oil industry. Oil wealth has been important in other ways, however, especially in underpinning state welfare expenditure in the north during the 1970s, at a time when other governments have had to cut back public expenditure in the face of economic recession. But even this has not been sufficient to halt net out-migration, which was still running at a rate of about 2300 per annum in the late 1970s (Stenstadvold 1981). Regional policy has failed to significantly alter the development potential of northern Norway; instead it has mainly functioned as an instrument of welfare state transfers.

Eastern Portugal

The interior of Portugal is one of Europe's poorest regions. Despite the near-revolutionary changes which have characterized development of the national economy in the post-1945 period, much of the interior has failed to share in this, except in providing a reserve of labour. Regional policy measures have been, at best, half-hearted in a country with one of Europe's lowest levels of GDP per capita. While some of the larger towns have benefited in recent years from productive decentralization and expansion of local government services (Ferrão 1983), much of the interior remains poorly developed. There are differences between the mountainous, minifundist northern region and the classic Alentejo latifundia region of the south (Figure 52A), but the interior as a whole shares a number of features: unmodernized and little-commercialized agriculture, small

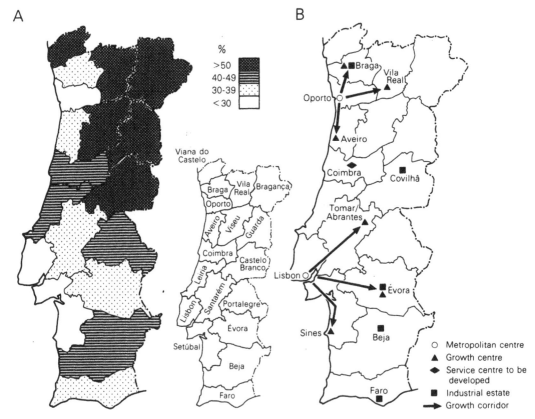

Figure 52 Regional structures in Portugal: *a* percentage of employees in primary sector, 1981; *b* 1974–9, Fourth National Development Plan

SOURCE: Lewis and Williams (1985).

manufacturing firms largely tied to stagnant local markets, and a weak tertiary sector largely devoid of producer services. Both the northern and southern interior have experienced almost unbroken net out-migration throughout the twentieth century, often leading to large population losses. The north has been locked into international emigration while the south has served as a reservoir of labour for rapid expansion of the Lisbon metropolis.

Uneven development of the national economy
In 1945 Portugal was still essentially an agricultural economy; there were small pockets of manufacturing in selected regions and sectors, but the country was still dependent on imports of many essential basic products. The Salazar government was conservative, autarkic and, drawing much of its support from the peasantry of the north and the landowning class in the south, favoured benign neglect of the rural interior. Infrastructure and services were poorly developed and much-needed agrarian reform was studiously avoided. By the late 1950s and 1960s stagnation of the national economy and the costs of escalating colonial wars led to a revised strategy. There were moves to internationalize the economy in a search for greater foreign capital and export-led growth; controls on foreign investment and technology imports were liberalized, membership of EFTA was secured and there was a shift in trade from the African colonies to European markets. Annual growth rates in excess of 6 per cent were recorded in the early 1960s and, following large-scale state investments in infrastructure in the late 1960s, there were even higher growth rates. However, in the mid-1970s growth collapsed following the international oil crisis and disruption of investment after the military coup of 1974. Subsequently, national economic growth has been erratic but generally depressed, especially given the imposition of harsh austerity programmes in the mid-1980s.

The resulting regional inequalities in the national economy are startling (see Lewis and Williams 1981). The coastal regions accounted for more than 80 per cent of total population and had GDP per capita levels well above the national average. They were more urbanized and industrialized, with larger manufacturing firms and higher standards of housing and social provision. The two main foci of industrialization are the north-west (around Oporto) and the Lisbon metropolitan area.

In the interior, however, there was a high dependence on agriculture. In the 1970s, more than 65 per cent of employment in districts such as Beja and Bragança had been in the primary sector and, even in the 1980s, this accounted for between 38 and 62 per cent of jobs in the interior. Family farms dominate in the north but they tend to be small (mean size being between 3 ha and 8 ha in 1979), inefficient in the use of labour, and little commercialized, with some notable exceptions such as port-wine production in the Douro valley. The plains of the south offer greater potential for agriculture but this has been poorly developed. The latifundist system which prevailed until the mid-1970s (the mean farm size was 45 ha in Beja and 48 ha in Évora) was reliant on abundant cheap labour, and consequently there was little investment in the irrigation which is vital to the region, and little specialization in high-value crops. Instead, dry farming methods involving cereals, olives, cork-oaks and sheep prevailed. Popular land seizures and creation of cooperatives after the 1974 coup (Barros 1980) seemed to offer some possibilities for a genuine agrarian reform. However, this was caught up in an intense political struggle which resulted in effective dismemberment of the cooperatives by the mid-1980s. Consequently, the real potential of agriculture is still relatively little utilized.

Manufacturing in the interior is also poorly developed. Some district capitals have benefited from industrial growth in the 1970s; Évora, for example, has attracted Siemens, while ICI has a major plant at Portalegre. However, the mean size of industrial establishments is little more than ten employees in most of the interior districts, and large parts of the interior have low industrialization and low profit rates (Ferrão 1985). Case studies of the textile industry in Gouveia (in Guarda) and of the furniture industry in Carregal do Sal (in Viseu) have revealed how poorly developed industry is in many of these areas (Lewis and Williams 1987). Fixed capital is often outdated, the artisanal sector is rapidly disappearing, there is poor access to export markets and a very low level of producer services. Furthermore, few small firms in the interior have access to bank capital in setting up in business. Indeed, other than basic retail and personal services, the tertiary sector as a whole is weak. Post-1974 revitalization of local government has been accompanied by a sizeable increase in public-sector employment in the cities and towns of the interior, in marked contrast to the excessive centralization of the Salazarist state. However, tourism and most forms of commercial office employment are virtually absent from the region.

The state made no effective attempt to ameliorate regional inequalities. The economic priority was to promote national growth, and social and regional expenditures took a second place to these. Indeed, in the series of national plans produced after 1953, only the Third Development Plan 1968–73 made any real attempt to delineate a set of regional policy objectives and mechanisms (Lewis and Williams 1985). Regional planning commissions were established, but these were subordinate to the ministries in Lisbon and their only effective role was in coordinating what remained essentially sectoral planning. Probably the most coherent programme was the Fourth Development Plan 1974–9. Among its proposals was a series of growth centres linked to Lisbon and Porto by growth corridors (Figure 52B). None of these was ever implemented, except for Sines (see Lewis and Williams 1985), not least because the plan was only published shortly before the 1974 *coup*. In terms of the needs of the interior, the Plan had two fundamental weaknesses. The growth centres were really designed to serve the requirements of decentralization from the metropolitan areas, and only two centres were (just) located in the interior. They were also almost entirely concerned with manufacturing, and ignored agrarian reform and development of the tertiary sector. Between 1974 and the mid-1980s, except for a brief experiment with regionally selective industrial investment grants, there has been no effective regional policy in Portugal. Development of the interior has been largely abandoned to the dictates of the market.

Emigration and return
The major role of the interior has been to provide a reserve of labour. This has been essential for expansion of Portugal's relatively labour-intensive manufacturing industries and, since the 1950s, there has been a migration stream from the interior to the littoral. While migration data are not specific enough to delineate interregional moves in the 1960s, population-change data are suggestive of the broad pattern of movement (see Figure 53A). In addition, the northern interior (and the littoral, too) have been tied into a pattern of emigration (Figure 53B). Traditional destinations were the Americas and the colonies but, in the 1960s and early 1970s, emigration was redirected to the labour-hungry economies of northern Europe. During the period 1960–75, districts such as Bragança and Guarda had annual legal emigration rates in excess of 7 per 1000. As a result, much

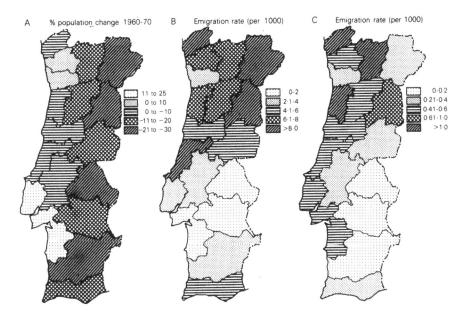

Figure 53 Demographic patterns in Portugal: *a* percentage population change 1960–70; *b* emigration rate per 1000, 1960–75; *c* emigration rate per 1000, 1983

SOURCE: Instituto Nacional de Estatística, and the Secretariado Nacional de Emigração

Table 50 *Returned migrants as employers in central Portugal, 1983*

| | Employers as percentage of returnees in study area | Mean employment created | Sectoral distribution of firms (%) | | | |
			Agriculture	Construction	Industry	Commerce
a Interior						
Remote rural area (Foios)	21	1.8	50	17	0	33
Accessible rural area (Mangualde)	16	1.8	17	0	0	83
b Littoral						
Leiria	40	2.5	17	7	27	53

SOURCE: Lewis and Williams (1986).

of the interior lost more than a fifth of its population in the two decades leading up to the mid-1970s.

Since that date the pattern of emigration and population has been modified. The slow-down in the national and international economies has led to reduced job opportunities outside the region. However, the principal sources of emigration within the country remain the same, as is indicated in Figure 53C. There has also been some redirection of migration to the major settlements within the region. At the same time, the return of emigrants from recession-affected Europe, and from Africa following decolonization, has helped to stabilize population levels in some parts of the interior. The savings and remittances of the returnees have had an important effect in bolstering consumption standards in the interior. Returnees' priorities are usually better housing, followed by purchase of consumer durables. There is little investment in businesses while creation of new manufacturing enterprises is rare (see Table 50). Instead, the preferred business venture is a café or small hotel, offering a leisured life-style rather than a high return on investment (Lewis and Williams 1986). Moreover, returned migrants' investments tend to be far more productive in the coastal region than in the interior. Returnees in the interior are less likely to become employers, create less employment and avoid manufacturing investments. In summary, the role of the region as a labour reserve does result in higher income and savings levels and these are used in various ways, including improving standards of living. However, the essential structure of the regional economy, and in particular its capacity for indigenous growth, has been little changed by this process.

Further reading

1 General: H. D. Clout (ed.) (1981), *Regional development in Western Europe*, Chichester: Wiley.
2 The south-east of Britain: R. Hudson and A. Williams (1986), *The United Kingdom*, London: Harper and Row.
3 Paris and Ile de France: J. N. Tuppen (1983), *The economic geography of France*, London: Croom Helm.
4 Italy: R. King (1987), *Italy*, London: Harper and Row.
5 The Ruhr: J. A. Hellen (1974), *North Rhine–Westphalia*, Oxford: Oxford University Press.
6 Wallonia: H. D. Clout (1975), *The Franco-Belgian border region*, Oxford: Oxford University Press.
7 Northern Norway: J. C. Hansen (1983), 'Regional policy in an oil economy: the case of Norway', *Geoforum*, **14**, pp. 353–61.
8 Portugal's interior: J. R. Lewis and A. M. Williams, 'Regional uneven development on the European periphery: the case of Portugal', *Tijdschrift voor Economische en Sociale Geografie*, **72**, pp. 81–98.

9 Concluding remarks

The western European economy had been severely weakened by the Second World War and in 1945 there was little to suggest the speed at which, eventually, economic recovery would occur. In the 1960s, in the middle of a phase of rapid growth, it would have been equally difficult to foresee the extent of the global recession which was to occur in the 1970s and 1980s. These two phases of growth and stagnation have been the major features of the evolution of the post-1945 economy. However, the process of recovery and adjustment has been highly uneven, and has been far more effective in West Germany and Scandinavia than in, say, the UK or Belgium. There has also been the, at least partial, transformation of the economies of southern Europe, leading to much-publicized claims in 1987 that the Italian economy had finally overtaken that of the UK.

The process of adjustment is, however, much more acutely observed at the regional than at the national level. In Chapter 8 a number of case studies illustrated how different types of regions had fared in the course of the two major phases of post-war economic change. Until at least the mid-1960s, the more advanced economic regions all had a share in economic growth, although there were clear differences in performance between, say, the Ruhr or the Frankfurt regions and Wallonia and South Wales. In the 1970s, however, differences in competitivity were cruelly exposed and some of the older economic regions have been severely affected by industrial closures and redundancies. The roles of the labour reserve regions have also been affected by these two phases of change, with a surge of labour demand in the 1960s being followed by sharp reductions in emigration and migration in the 1970s. A particular problem for these regions has been that the recession has also weakened the ability of national governments to continue to support them on a welfare basis.

There is no simple key to interpreting the performances of individual regions. In part it depends on the economic structure of the regional economies. Those which specialize in manufacturing have been most open to international competition, especially from Japan and the NICs. In contrast, agricultural regions have been highly subject to national and/or EC policies and have been largely isolated from world market trends. Much the same is true of the tertiary sector although there is increasing evidence of its internationalization. Of course, in all these cases there are considerable variations depending on the precise sector or sub-sector involved. However, regional economic performance depends on far more than the structure of the economic base. It also depends on how each region fits into the economic strategies of particular companies, that is, how Club Méditerranée or Fiat evaluate the role of existing or potential establishments in different regions at home or abroad. It also depends on the structure of social relationships, the effectiveness of organized labour and of social protests, and the willingness and ability of central or local government to intervene in the process of regional restructuring. In other words, there is nothing fixed or immutable in the economic resource base of any region which prescribes how it will develop. Instead, this depends on the conjuncture of sectoral, national and international developments.

This previous discussion and, indeed, the structure of this book, should not be taken to imply that there is a uni-causal set of relationships from the international to the national to the regional scale. These relationships do exist but they are also complemented by the influence of regional changes on the national and international context. In one sense, the national and international economies are simply the sum of the regional parts and, therefore, must be influenced by regional adjustments. More specifically, it can be shown that the ability of, say, the UK to compete in world manufacturing trade depends very much on developments in particular regions such as the M4 corridor or Silicon Glen. Similarly, the greater robustness of the German steel industry compared to, say, that of France or the UK, partly depends on the complex inter-industry linkages which exist in the Ruhr. At a broader scale, it can be shown that the international economic framework can also be shaped by regional influences. For example, West Germany's attitude to the CAP and the EC's attitude to protectionism of the clothing sector are both considerably influenced by the requirements of a few regions. This emphasizes the need for a broadly-based analysis of regional and national restructuring and avoidance of undue generalizations.

Nevertheless, the process of regional adjustment has been a major theme of post-war economic change, whether in terms of growth or stagnation. On the one hand, some of the major coal and steel regions experienced sharp recessions as early as the late 1950s, while most of the older manufacturing regions were struggling by the mid-1970s. On the other hand, some agricultural regions which had acted as labour reserves in the 1950s became, in the 1970s, the focus of rapid capital accumulation, whether for tourism or manufacturing. None of these processes – even in the most free-market economies – has occurred in isolation from government intervention. Despite the attempt in some countries 'to roll back the state' in the 1980s, government policies still provide an influential context for economic change. Fiscal and monetary policies influence the operating environments for private businesses, while important sections of forestry, the utilities, transport and manufacturing are under direct state ownership and management.

In the 1970s and 1980s, the requirements of restructuring have increased the pressures for state intervention at precisely the time when resources for this purpose have been constrained. This is most starkly evident in terms of regional policy which has been considerably scaled-down in most western European countries at a time when unemployment levels in many regions have been at record post-war levels. Together with a disillusionment with macro-economic planning policies and a lack of mobile enterprises, this has encouraged many local areas to pursue local economic initiatives which aim to promote indigenous growth potential to achieve self-sustained development. The range of such initiatives is considerable (see Bassand *et al*. 1986) but, at this time, the rhetoric has seemed stronger than implementation. However, some important examples do exist, notably the Mondragon 'cooperative' in the Basque Country, which has over 18,000 members, although it is questionable whether this model can be transferred to other regions. It is also significant to note that, in the late 1980s, even this cooperative was shifting from its earlier focus on electrical 'white goods' manufacture to diversify into financial services and its Eroskil hypermarket chain.

At the time of writing, some of the more painful problems of economic adjustment in the European regions have been eased by a fall in oil prices in 1986 and by appreciation of the Japanese yen in 1986–7. However, a number of difficult problems lie ahead. Agricultural policy is under pressure both because European protectionism threatens to bring about retaliatory actions from other countries and because the CAP is undermining cohesion in the EC. The manufacturing sector also faces continued adjustment both to competition from the NICs and to the consequences of the rapid introduction of micro-electronics. Finally, the tertiary sector probably faces a major upheaval and the changes observed in the City of London are likely to be repeated in other types of services and in other countries. While the events of 1986–7 provided a breathing space, it is clear that further difficult adjustments lie ahead for western Europe, for particular countries, and for its regions.

Bibliography

Abreu, A. T. de (1980), 'Agrarian reform and redevelopment issues in Portugal 1974–8', in J. de Brandt (ed.), *European studies in development*, London: Macmillan.

Aceves, J. B. and Douglass, W. A. (eds.) (1976), *The changing faces of rural Spain*, Cambridge Mass.: Schenkman.

Adelman, M. A. (1983), 'The multinationals in the world oil market: the 1970s and 1980s', in C. P. Kindleberger and D. B. Audretsch (eds.), *The multinational corporation in the 1980s*, Cambridge Mass.: MIT Press.

Aglietta, M. (1979), *A theory of capitalist regulation*, London: New Left Books.

Agostini, D. (1977), 'Part-time farming in the rural–urban fringe in Italy', in R. Gasson (ed.), *Place of part-time farming in rural and regional development*, Kent: Wye College.

Airey, D. (1978), 'Tourism and the balance of payments', *Tourism International Research–Europe*, Third Quarter, 2–16.

Albrow, M. (1970), *Bureaucracy*, London: Macmillan.

Aldcroft, D. H. (1980), *The European economy 1914–1980*, London: Croom Helm.

Allum, P. (1981), 'Thirty years of southern policy in Italy', *The Political Quarterly*, **52**, 314–23.

Altvater, E. (1978), 'Some problems of state interventionism', in J. Holloway and S. Picciotto (eds.), *State and capital: a Marxist debate*, London: Edward Arnold.

Amin, A. (1983a), *Industrial restructuring, state intervention and regional growth: the example of Alfa Sud in southern Italy*, Reading: University of Reading, Geographical Papers.

Amin, A. (1983b), 'The State and uneven regional development in advanced capitalism', in A. Gillespie (ed.), *Technological change and regional development*, London: Pion, London Papers in Regional Science 12.

Amin, A. (1985), 'Restructuring in Fiat and the decentralization of production into southern Italy', in R. Hudson and J. Lewis (eds.), *Uneven development in Southern Europe*, London: Methuen.

Amin, S. (1976), *Unequal development: an essay on the social formations of peripheral capitalism*, Hassocks: Harvester Press.

Anastassopoulos, J.-P., C. (1981), 'The French experience: conflicts with government', in R. Vernon and Y. Aharoni (eds.), *State-owned enterprises in the Western economies*, London: Croom Helm.

Andersen, P. S. and Akerholm, J. (1982), 'Scandinavia', in A. Boltho (ed.), *The European economy: growth and crisis*, Oxford: Oxford University Press.

Andreff, W. (1976), *Profits et structures du capitalisme mondial*, Paris: Calmann–Lévy.

Andreff, W. (1984), 'The international centralization of capital and the reordering of world capitalism', *Capital and Class*, **22**, 58–80.

Arcangeli, F., Borzaga, C. and Goglio, S. (1980), 'Patterns of peripheral development in Italian regions, 1964–1977', *Papers of the Regional Science Association*, **44**, 19–34.

Archer, B. H. (1977), *Tourism multipliers: the state of the art*, Cardiff: Bangor Occasional Papers in Economics 11, University of Wales Press.

Archetti, E. P. and Aass, S. (1978), 'Peasant studies: an overview', in H. Newby (ed.), *International perspectives in rural sociology*, Chichester: Wiley.

Ardagh, J. (1982), *France in the 1980s*, Hardmondsworth: Penguin.

Arnell, L. and Nygren, B. (1980), *The developing countries and the world economic order*, London: Methuen.

Asheim, B. J. (1986), Peripheral industrialization and capitalist development: a macrogeographical comparison of Norway and Italy. Paper presented to the IGU Commission on Industrial Change, Chinchon, Spain, 24–30 August.

Auty, R. M. (1983), 'Multinational resource corporations, the product life-cycle and product strategy: the oil majors' response to heightened risk', *Geoforum*, **14**, 1–13.

Bagnasco, A. (1977), *Tre Italie*, Bologna: Il Mulino.

Bagnasco, A. (1982), 'Economia e società della piccola impresa', in S. Goglio (ed.), *Italia: Centri e Periferie*, Milan: Franco Angeli.

Bagnasco, A. (1983), 'Il contesto sociale', in G. Fuá and C. Zacchia (eds.), *Industrializzazione e senza fratture*, Bologna: Il Mulino.

Bairoch, P. (1976), 'Europe's gross national product: 1800–1975', *The Journal of European Economic History*, **5**, 273–340.

Balassa, B. (1981), *The newly industrializing countries in the world economy*, New York: Pergamon.

Ballance, R. H., Ansari, J. A. and Singer, H. W. (1982), *The international economy and industrial development: the impact of trade and investment on the Third World*, London: Wheatsheaf Books.

Ballance, R. and Sinclair, S. (1983), *Collapse and survival: industrial strategies in a changing world*, London: George Allen and Unwin.

Banaji, J. (1976), 'Summary of selected parts of Kautsky's "The Agrarian Question"', *Economy and Society*, **5**, 2–49.

Bannon, M. J. (1979), 'Office concentration in Dublin and its consequences for regional development in Ireland', in P. W. Daniels (ed.), *Spatial patterns of office growth and location*, Chichester: Wiley.

Bannon, M. J. (1985), 'Service activities in national and regional development: trends and prospects for Ireland', in M. J. Bannon and S. Ward (eds.), *Services and the new economy*, Dublin: Regional Studies Association Irish Branch.

Barbaza, Y. (1970), 'Trois types d'intervention du tourisme dans l'organisation de l'espace littoral', *Annales de Geographie*, **434**, 446–69.

Barker, M. L. (1982), 'Traditional landscapes and mass tourism in the Alps', *Geographical Review*, **72**, 395–415.

Barros, A. de (1980), 'Portuguese agrarian reform and economic and social development', *Sociologia Ruralis*, **20**, 82–96.

Bassand, M., Brugger, E. A., Bryden, J. M., Friedmann, J. and Stuckey, B. (1986), *Self-reliant development in Europe: theory, problems, actions*, Aldershot: Gower.

Bassett, K. and Short, J. R. (1980), *Housing and residential structure: alternative approaches*, London: Routledge and Kegan Paul.

Bateman, M. (1985), *Office development: a geographical analysis*, London: Croom Helm.

Bell, C. and Newby, H. (1974), 'Capitalist farmers in the British class structure',

Sociologia Ruralis, **14**, 86–107.

Bell, D. (1974), *The coming of post-industrial society*, London: Heinemann.

Best, R. H. (1979), 'Land-use structure and change in the EEC', *Town Planning Review*, **50**, 395–411.

Bethemont, J. and Pelletier, J. (1983), *Italy: a geographical introduction*, London: Longman.

Bivand, R. S. (1986), 'The evaluation of Norwegian regional policy: parameter variation in regional shift models', *Government and Policy*, **4**, 71–90.

Blackaby, F. (ed.), (1979), *De-industrialisation*, London: Heinemann.

Blackbourn, A. (1978), 'Multinational enterprises and regional development: a comment', *Regional Studies*, **12**, 125–7.

Blackbourn, A. (1982), 'The impact of multinational corporations on the spatial organisation of developed nations: a review', in M. Taylor and N. Thrift (eds.), *The geography of multinationals*, London: Croom Helm.

Blacksell, M. (1981), *Post-war Europe: a political geography*, London: Hutchinson.

Blacksell, M. (1984), 'The European Community and the Mediterranean region: two steps forward, one step back', in A. Williams (ed.), *Southern Europe transformed*, London: Harper and Row.

Blacksell, M. (1987), 'West Germany', in H. Clout (ed.), *Regional development in Western Europe*, Chichester: Wiley. Third edition.

Bleitrach, D. and Chenu, A. (1982), 'Regional planning: regulation or deepening of social contradictions? The example of Fos-sur-Mer and the Marseilles metropolitan area', in R. Hudson and J. R. Lewis (eds.), *Regional planning in Europe*, London: Pion.

Bloomfield, G. T. (1981), 'The changing spatial organisation of multinational corporations in the world automative industry', in F. E. I. Hamilton and G. J. R. Linge (eds.), *International industrial systems*, Chichester: Wiley.

Böhning, W. R. (1979), 'Les migrations internationales en Europe occidentale: réflexions sur les cinq dernières années', *Revue Internationale du Travail*, **118**, 425–37.

Boissevain, J. (1979), 'Tourism and the European periphery: the Mediterranean case', in D. Seers, B. Schaffer and M. L. Kiljunen (eds.), *Underdeveloped Europe: case studies in core-periphery relations*, Hassocks: Harvester Press.

Boltho, A. (1982), 'Growth', in A. Boltho (ed.), *The European economy: growth and crisis*, Oxford: Oxford University Press.

Bolton Report (1971), *Committee of inquiry on small firms: small firms*, London: HMSO.

Bornstein, S., Held, D. and Krieger, J. (eds.) (1984), *The state in capitalist Europe*, London: George Allen and Unwin.

Borzaga, C. (1982), 'Lo sviluppo regionale Italiano durante gli Anni '70', in S. Goglio, (ed.), *Italia: Centri e Periferie*, Milan: Franco Angeli.

Bowler, I. (1976), 'The CAP and the space-economy of agriculture in the EEC', in R. Lee and P. Ogden (eds.), *Economy and society in the EEC: spatial perspectives*, Farnborough: Saxon House.

Bowler, I. (1979), *Government and agriculture: a spatial perspective*, London: Longman.

Bowler, I. (1982), 'Direct marketing in agriculture: a British example', *Tijdschrift voor Economische en Sociale Geografie*, **73**, 22–31.

Bowler, I. (1983), 'Structural change in agriculture', in M. Pacione (ed.), *Progress in rural geography*, London: Croom Helm.

Boyer, R. and Petit, P. (1981), 'Employment and productivity in the EEC', *Cambridge Journal of Economics*, **5**, 47–58.

Bradford, C. I. (1982), 'The rise of the NICs as exporters on a global scale', in L. Turner and N. McMullen (eds.), *The newly industrializing countries: trade and adjustment*, London: George Allen and Unwin.

Breathnach, P. (1982), 'The demise of growth-centre policy: the case of the Republic of Ireland', in R. Hudson and J. Lewis (eds.), *Regional planning in Europe*, London: Pion, London Papers in Regional Science 11.

Bresso, M. (1980), 'Les paradoxes de la politique agricole commune', *L'Espace Géographique*, **3**, 173–82.

Britton, D. K. (1974), 'The structure of agriculture', in A. Edwards and A. Rogers (eds.), *Agricultural resources: an introduction to the farming industry of the United Kingdom*, London: Faber and Faber.

Brusco, S. (1983), 'Flessibilità e solidità del sistema: l'esperienza emiliana', in G. Fuá and C. Zacchia (eds.), *Industrializzazione senza fratture*, Bologna: Il Mulino.

Brusco, S. and Sabel, C. (1981), 'Artesan production and economic growth', in F. Wilkinson, (ed.),*The dynamics of labour market segmentation*, London: Academic Press.

Bryden, J. and Houston, G. (1976), *Agrarian change in the Scottish Highlands: the role of the HIDB in the agricultural economy of the crofting counties*, London: Martin Robertson.

Burkart, A. J. and Medlik, S. (1981), *Tourism: past, present and future*, London: Heinemann.

Burtenshaw, D. (1974), *Economic geography of West Germany*, London: Macmillan.

Buttel, F. H. (1980), 'Agricultural structure and rural ecology: towards a political economy of rural development', *Sociologia Ruralis*, **20**, 44–62.

Cabouret, M. (1982), 'Traits permanents et tendences récentes de l'agriculture finlandaise', *Annales de Géographie*, **91**, 87–118.

Calmès, R. (1981), 'L'évolution des structures d'exploitation dans les pays de las C.E.E.', *Annales de Géographie*, **90**, 401–27.

Camagni, R. and Cappellin, R. (1981), 'European regional growth and policy issues for the 1980s', *Built Environment*, **7**, 162–71.

Carney, J. (1980), 'Regions in crisis: accumulation, regional problems and crisis formation', in J. Carney, R. Hudson and J. Lewis (eds.), *Regions in crisis: new perspectives in European regional theory*, London: Croom Helm.

Carney, J., Hudson, R. and Lewis, J. (1980), 'New perspectives in European regional theory: some introductory remarks', in J. Carney, R. Hudson and J. Lewis (eds.), *Regions in crisis: new perspectives in European regional theory*, London: Croom Helm.

Carson, R. (1969), *Silent spring*, Harmondsworth: Penguin.

Castles, S., Booth, H. and Wallace, T. (1984), *Here for good: Western Europe's new ethnic minorities*, London: Pluto Press.

Castles, S. and Cosack, G. (1973), *Immigrant workers and class structure in Western Europe*, London: Oxford University Press.

Cavaco, C. (1980), *Turismo e demografia no Algarve*, Lisbon: Editorial Progresso Social e Democracia.

Cavazzani, A. (1977), 'Part-time farming and the Common Agricultural Policy', in

R. Gasson (ed.), *Place of part-time farming in rural and regional development*, Kent: Wye College.

Cesarini, G. (1979), *Rural production cooperatives in Southern Italy*, Langholm: Arkleton Trust.

Clairmonte, F. and Cavanagh, J. (1984), 'Transnational corporations and services: the final frontier', *Trade and Development*, **5**, 215–73.

Clark, G. and Dear, M. (1981), 'The State in capitalism and the capitalist State', in M. Dear and A. J. Scott (eds.), *Urbanization and urban planning in capitalist society*, London: Methuen.

Clarke, I. M. (1982), 'The changing international division of labour within ICI', in M. Taylor and N. Thrift (eds.), *The geography of multinationals*, London: Croom Helm.

Clout, H. D. (1971), *Agriculture*, London: Macmillan.

Clout, H. D. (1972), *Rural geography: an introductory survey*, New York: Pergamon Press.

Clout, H. D. (1975a), 'Structural change in French farming: the case of the Puy-de-Dôme', *Tijdschrift voor Economische en Sociale Geografie*, **66**, 234–45.

Clout, H. D. (1975b), *The Franco-Belgian border region*, Oxford: Oxford University Press.

Clout, H. D. (1981a), 'Rural space', in H. D. Clout (ed.), *Regional development in western Europe*, Chichester: Wiley. Second edition.

Clout, H. D. (1981b), 'Energy and regional problems', in H. D. Clout (ed.), *Regional development in western Europe*, Chichester: Wiley. Second edition.

Clout, H. D. (1981c), 'Regional development in western Europe', in H. D. Clout (ed.), *Regional development in western Europe*, Chichester: Wiley. Second edition.

Clout, H., Blacksell, M., King, R. and Pinder, D. (1985), *Western Europe: geographical perspectives*, London: Longman.

Coakley, J. (1984), 'The internationalization of bank capital', *Capital and Class*, **23**, 107–20.

Cohen, R. B. (1981), 'The new international division of labour, multinational corporations and urban hierarchy', in M. Dear and A. J. Scott (eds.), *Urbanization and urban planning in capitalist society*, London: Methuen.

Commins, P. (1980), 'Imbalances in agricultural modernisation – with illustrations from Ireland', *Sociologia Ruralis*, **20**, 63–81.

Commission of the European Community (1982a), *Social Europe No. 1*, Brussels: Commission of the European Community.

Commission of the European Community (1982b), *The economy of the European Community*, Luxembourg: European Community, European Documentation 1/2.

Commission of the European Community (1982c), *The European Community's industrial strategy*, Luxembourg: European Commission, European Documentation 5.

Commission of the European Community (1982d) *The future of the car industry*, Brussels: Commission of the European Community, European File 1/82.

Commission of the European Community (1982e), *A community policy for tourism*, Brussels: Bulletin of the European Community, Supplement 4/82.

Commission of the European Community (1982f), *Changes in the structure of the retail trade in Europe*, Brussels: Studies Collection, Commerce and Distribution Series No. 8.

Commission of the European Community (1983a), *European Economy No. 18*, Brussels: Commission of the European Community.

Commission of the European Community (1983b), *The European Community and the energy problem*, Luxembourg: European Community, European Documentation 1.

Commission of the European Community (1983c), *The common fisheries policy*, Brussels: Commission of the European Community, European File 11/83.

Commission of the European Community (1984a), *The European Social Fund: a weapon against unemployment*, Brussels: Commission of the European Community, European File 2/84.

Commission of the European Community (1984b), *An industrial strategy for Europe*, Brussels: Commission of the European Community, European File 11/84.

Commission of the European Community (1984c), *A European energy strategy*, Brussels: Commission of the European Community, European File 12/84.

Commission of the European Community (1984d), *Social Europe No. 1*, Brussels: Commission of the European Community.

Commission of the European Community (1984e), *The European Community in the world*, Brussels: Commission of the European Community, European File 14/84.

Commission of the European Community (1984f), *Women in the European Community*, Luxembourg: European Community, European Documentation 4.

Commission of the European Community (1984g), *The European Community and new technologies*, Brussels: Commission of the European Community, European File 8/84.

Commission of the European Community (1985a), *The European Community's research policy*, Luxembourg: European Commission, European Documentation 2.

Commission of the European Community (1985b), *European competition policy*, Brussels: Commission of the European Community, European File 6/85.

Commission of the European Community (1985c), *The European Community's textile trade*, Brussels: Europe Information External Relations Series, 76/85.

Commission of the European Community (1985d), *The European steel policy*, Brussels: Commission of the European Community, European File 2/85.

Commission of the European Community (1985e), *Tourism and the European Community*, Brussels: Commission of the European Community, European File 11/85.

Commission of the European Community (1985f), *European regional policy*, Brussels: Commission of the European Community, European File 7/85.

Commission of the European Community (1985g), *Grants and loans from the European Community*, Luxembourg: European Commission, European Documentation.

Commission of the European Community (1985h), *The European Community's fisheries policy*, Luxembourg: European Commission, European Documentation 1.

Commission of the European Community (1986), *Eleventh annual report (1985) to the Council by the Commission: European Regional Development Fund*, Brussels: Commission of the European Community.

Cooke, P. (1983), *Theories of planning and spatial development*, London: Hutchinson.

Cooke, P. and Rosa Pires, A. da (1985), 'Productive decentralization in three European regions', *Environment and Planning A*, **17**, 527–54.

Council of Europe (1975), *The development of central, regional and local finance since 1950*, Strasbourg: Council of Europe, Study Series Local and Regional Authorities in Europe, Study no. 13, vol. 1.

Counter Information Services (1977), *The Ford Motor Company*, London: Counter Information Services.

Cox, A. (ed.), (1982), *Politics, policy and the European recession*, London: Macmillan.

Cuddy, M. (1981), 'European agricultural policy: the regional dimension', *Built Environment*, 7, 200–10.

Cunningham, C. and Young, J. A. (1983), 'The EEC Common Fisheries Policy: retrospect and prospect'. *National Westminster Bank Quarterly Review*, May, 2–14.

Damesick, P. J. (1986), 'Service industries, employment and regional development in Britain: a review of recent trends and issues', *Transactions, Institute of British Geographers*, **11**, 212–26.

Damette, F. (1980), 'The regional framework of monopoly exploitation: new problems and trends', in J. Carney, R. Hudson and J. R. Lewis (eds.), *Regions in crisis*, London: Croom Helm.

Damette, F. and Poncet, E. (1980), 'Global crises and regional crises', in J. Carney, R. Hudson and J. Lewis (eds.), *Regions in crisis*, London: Croom Helm.

Daniels, P. W. (1979), 'Perspectives on office location research', in P. W. Daniels (ed.), *Spatial patterns of office growth and location*, Chichester: Wiley.

Daniels, P. W. (1982), *Service industries: growth and location*, Cambridge: Cambridge University Press.

Daniels, P. W. (1983), 'Service industries: supporting role or centre stage?' *Area*, **15**, 301–9.

Daniels, P. W. (1984), 'Business service offices in provincial cities: sources of input and destinations of output', *Tijdschrift voor Economische en Sociale Geografie*, **75**, 123–40.

Davenport, M. (1982), 'The economic impact of the EEC', in A. Boltho (ed.), *The European economy*, Oxford: Oxford University Press.

Davies, R. and Bennison, D. J. (1979), *British town centre shopping schemes*, Reading: Unit for Retail Planning Information.

Dawson, J. A. (1976), 'Public policy and distribution in the EEC', in R. Lee and P. Ogden (eds.), *Economy and society in the EEC: spatial perspectives*, Farnborough: Saxon House.

Dawson, J. A. (1982), *Commercial distribution in Europe*, London: Croom Helm.

Dawson, J. A. and Kirby, D. A. (1979), *Small unit retailing in the UK*, Farnborough: Saxon House.

Del Rio, A. (1979), 'Filipina domestic workers in Italy', *Migration Today*, **24**, 20–2.

Della Seta, P. (1978), 'Notes on urban struggles in Italy', *International Journal of Urban and Regional Research*, **2**, 303–29.

Denison, E. F. (1967), *Why growth rates differ*, Washington: Brookings Institution.

Dicken, P. (1980), 'Foreign direct investment in European manufacturing industry: the changing position of the United Kingdom as a host country', *Geoforum*, **11**, 289–313.

Dicken, P. (1982), 'Recent trends in international direct investment, with particular reference to the United States and the United Kingdom', in B. T. Robson and J. Rees (eds.), *Geographical agenda for a changing world*, London: SSRC.

Dicken, P. (1983), 'Japanese manufacturing investment in the United Kingdom: a flood or a mere trickle', *Area*, **15**, 273–84.

Dicken, P. (1986), *Global shift: industrial change in a turbulent world*, London: Harper and Row.

Dicken, P. and Lloyd, P. E. (1980), 'Patterns and processes of change in the spatial distribution of foreign-controlled manufacturing employment in the United Kingdom, 1963 to 1975', *Environment and Planning A*, **12**, 1405–26.

Dineen, D. A. (1985), 'The western regions and the development of services' in M. J. Bannon and S. Ward (eds.), *Services and the new economy*, Dublin: Regional Studies Association Irish Branch.

Dunford, M. (1979), 'Capital accumulation and regional development in France', *Geoforum*, **10**, 81–108.

Dunford, M., Geddes, M. and Perrons, D. (1981), 'Regional policy and the crisis in the UK: a long-run perspective', *International Journal of Urban and Regional Research*, **5**, 337–410.

Dunford, M. and Perrons, D. (1983), *The arena of capital*, London: Macmillan.

Dunning, J. H. and Pearce, R. D. (1985), *The world's largest industrial enterprises, 1962–1985*, Farnborough: Gower.

Dyas, G. P. and Thanheiser, H. T. (1976), *The emerging European enterprise*, London: Macmillan.

Elder, N. (1979), 'The functions of the modern state', in J. Hayward and R. N. Berki (eds.), *State and society in contemporary Europe*, Oxford: Martin Robertson.

Elias, P. and Keogh, G. (1982), 'Industrial decline and unemployment in the inner city areas of Great Britain: a review of the evidence', *Urban Studies*, **19**, 1–15.

Emminger, O. (1981), 'West Germany: Europe's driving force?', in R. Dahrendorf (ed.), *Europe's economy in crisis*, London: Weidenfeld and Nicolson.

Eurostat (1984a), *Basic statistics of the community*, Luxembourg: European Community.

Eurostat (1984b), *Employment in 1983, employment and unemployment No. 3*, Luxembourg: European Community.

Eurostat (1985a), *Coal: Eurostat 9*, Luxembourg: European Community.

Eurostat (1985b), *Basic statistics of the community*, Luxembourg: European Community.

Eurostat (1986), *Regions*, Luxembourg: European Community.

Fahrenkrog, G. (1984), 'Questions of method in regional analysis', in P. O'Keefe (ed.), *Regional restructuring under advanced capitalism*, London: Croom Helm.

Farnell, J. and Elles, J. (1984), *In search of a Common Fisheries Policy*, Aldershot: Gower Press.

Fel, A. (1984), 'L'agriculture française en mouvement', *Annales de Géographie*, **517**, 303–25.

Ferrâo, J. M. M. (1983), Alguns aspectos regionais da evoluçâo recente da industria transformada em Portugal. Paper presented to the III Coloquio Iberico de Geografia, Barcelona.

Ferrâo, J. M. M. (1985), 'Regional variations in the rate of profit in Portuguese industry', in R. Hudson and J. Lewis (eds.), *Uneven development in southern Europe*, London: Methuen.

Fielding, A. J. (1975), 'Internal migration in western Europe', in L. A. Kosinski and R. M. Prothero (eds.), *People on the move: Studies on internal migration*, London: Methuen.

Fishwick, F. (1982), *Multinational companies and economic concentration in Europe*, Aldershot: Gower.

Fitzpatrick, J. (1985), 'Technology and economic development: the role of private services', in M. J. Bannon and S. Ward (eds.), *Services and the new economy*, Dublin: Regional Studies Association Irish Branch.

Fleming, D. K. (1967), 'Coastal steelworks in the Common Market countries', *Geographical Review*, **43**, 48–72.

Fothergill, S. and Gudgin, G. (1982), *Unequal growth: urban and regional employment change in the UK*, London: Heinemann.

Fothergill, S., Kitson, M. and Monk, S. (1985), *Urban industrial change: the causes of the urban-rural contrast in manufacturing employment trends*, London: Department of the Environment, Inner Cities Research Programme, Report 11.

Frank, A. G. (1980), *Crisis in the world economy*, London: Heinemann.

Frank, W. (1983), 'Part-time farming, underemployment and double activity of farmers in the EEC', *Sociologia Ruralis*, **23**, 20–6.

Franklin, S. H. (1969), *The European peasantry: the final phase*, London: Methuen.

Franklin, S. H. (1971), *Rural societies*, London: Macmillan.

Franks, L. G. (1976), *The European multinationals*, London: Harper and Row.

Freeman, C. (1984), *The role of technical change in national economic development*, Brighton: University of Sussex, Science Policy Research Unit.

Fridenson, P. (1986), 'The growth of multinational activities in the French motor industry, 1890–1979', in P. Hertner and G. Jones (eds.), *Multinationals: theory and history*, Aldershot: Gower Press.

Friedland, W. H. (1980), 'Technology in agriculture: labour and the rate of accumulation', in F. H. Buttel and H. Newby (eds.), *The rural sociology of the advanced societies: critical perspectives*, Montclair: Allanheld.

Friedmann, H. (1980), 'Household production and the national economy: concepts for the analysis of agrarian formations', *Journal of Peasant Studies*, **7**, 158–84.

Fröbel, F. (1982), 'The current development of the world economy', *Review*, **5**, 507–55.

Fröbel, F., Heinrichs, J. and Kreye, O. (1980), *The new international division of labour*, Cambridge: Cambridge U.P.

Fuá, G. (1983), 'Rural industrialization in later developed countries: the case of north-east and central Italy', *Banco Nazionale del Lavoro*, **147**, 351–77.

Fuguitt, G. V. (1959), 'Part-time farming and the push–pull hypothesis', *American Journal of Sociology*, **65**, 375–9.

Galt, A. H. (1979), 'Exploring the cultural ecology of field fragmentation and scattering in the island of Pantelleria', *Journal of Anthropological Research*, **35**, 95–108.

Garcia-Ramon, D. (1985), 'Old and new in Spanish farming', *Geographical Magazine*, 128–33.

Garofoli, G. (1981), 'Lo sviluppo delle "aree periferiche" nell'economia italiana degli anni settanta', *L'Industria, – Rivista di Economia e Politica Industriale*, **3**, 391–404.

Garofoli, G. (1984), 'Diffuse industrialization and small firms: the Italian pattern in the 70s', in R. Hudson (ed.), *Small firms and regional development*, Copenhagen: Copenhagen School of Economics and Business Administration.

Gaspar, J. (1977), 'A localizaçâo das principais sociedades em Portugal', *Finisterra*, **12**, 160–8.

Gaspar, J. (1984), 'Urbanization: growth, problems and policies', in A. Williams (ed.), *Southern Europe Transformed*, London: Harper and Row.

Gasson, R. (1966), *The influence of urbanization on farm ownership and practice; some aspects of the effects of London on farms and farm people in Kent and Sussex*, Kent: Wye College, Studies in Rural Land Use, Report 7.

Gasson, R. (1977a) (ed.), *Place of part-time farming in rural and regional development*, Kent: Wye College.

Gasson R. (1977b), 'Farmers' approach to co-operative marketing', *Journal of Agricultural Economics*, **28**, 27–37.

Gasson, R. (1980), 'Roles of farm women in England', *Sociologia Ruralis*, **20**, 165–80.

Gay, F. J. (1981), 'Benelux', in H. D. Clout (ed.), *Regional development in western Europe*, Chichester: Wiley. Second edition.

Gershuny, J. I. (1978), *After industrial society: the emerging self-service economy*, London: Macmillan.

Gershuny, J. I. and Miles, I. D. (1983), *The new service economy: the transformation of employment in industrial societies*, London: Frances Pinter.

Gibbs, D. C. (1983), 'The effect of international and national developments on the clothing industry of the Manchester conurbation', in F. E. I. Hamilton and G. J. R. Linge (eds.), *Regional economies and industrial systems*, Chichester: Wiley.

Ginatempo, N. (1979), 'The structural contradictions of the building industry in Italy, and the significance of the new housing legislation', *International Journal of Urban and Regional Research*, **3**, 465–91.

Giner, S. and Sevilla-Guzman, E. (1980), 'The demise of the peasant: some reflections on ideological inroads into social theory', *Sociologia Ruralis*, **20**, 13–27.

Giugni, G. (1971), 'Recent trends in collective bargaining in Italy', *International Labour Review*, **104**, 307–50.

Goddard, J. B. (1973), *Office linkages and location*, Oxford: Pergamon.

Goddard, J. B. (1975), *Office location in urban and regional development*, London: Oxford University Press.

Goddard, J. B. (1979), 'Office development and urban and regional development in Britain', in P. Daniels (ed.), *Spatial patterns of office growth and location*, Chichester: Wiley.

Goddard, J. B. (1983), 'Industrial innovation and regional economic development in Great Britain', in F. E. I. Hamilton and G. J. R. Linge (eds.), *Regional economies and industrial systems*, Chichester: Wiley.

Goddard, J. B. and Morris, D. (1976) 'The communications factor in office decentralization', *Progress in Planning*, **6**, 1–80.

Goddard, J. B. and Pye, R. (1977), 'Telecommunications and office location', *Regional Studies*, **11**, 19–30.

Goddard, J. B. and Smith, I. J. (1978), 'Changes in corporate control in the British urban system, 1972–1977', *Environment and Planning A*, **10**, 1073–84.

Gonen, A. (1981), 'Tourism and coastal settlement processes in the Mediterranean region', *Ekistics*, **48**, 378–81.

Gould, A. and Keeble, D. (1984), 'New firms and rural industrialisation in East Anglia', *Regional Studies*, **18**, 189–202.

Grahl, J. (1983), 'Restructuring in West European industry', *Capital and Class*, **19**, 118–41.

Granados, V. (1984), 'Small firms and rural industrialization in Spain: some results from an O.C.D.E. project', in R. Hudson (ed.), *Small firms and regional development*, Copenhagen: Copenhagen School of Economics and Business Administration.

Graziani, A. (1978), 'The Mezzogiorno in the Italian economy', *Cambridge Journal of Economics*, **1978**, 355–72.

Greenaway, D. (1983), *International trade policy: from tariffs to the new protectionism*, London: Macmillan.

Grigg, D. (1984), *An introduction to agricultural geography*, London: Hutchinson.

Grou, P. (1983), *La structure financière du capitalisme multinational*, Paris: Presses de la Fondation Nationale des Sciences Politiques.

Guedes, M. (1981), 'Recent agricultural land policy in Spain', *Oxford Agrarian Studies*, **10**, 26–43.

Hackman, P. (1976), 'Attitudes of the farmer's union to multiple jobholders in Finland', in A. M. Fuller and J. A. Mage (eds.), *Part-time farming: problem or resource in rural development*, Guelph: University of Guelph.

Hadjimichalis, C. (1983), 'Regional crisis: the State and regional social movements in southern Europe', in D. Seers and K. Östrom (eds.), *The crises of the European regions*, London: Macmillan.

Hadjimichalis, C. and Vaiou, D. (1986), Changing patterns of uneven regional development and forms of social reproduction. Athens: unpublished paper.

Hall, P. (1983), 'Patterns of economic policy: an organizational approach', in S. Bernstein, D. Held and J. Krieger (eds.), *The state in capitalist Europe*, London: George Allen and Unwin.

Hall, P. (1984), *The world cities*, London: Weidenfeld and Nicolson.

Hall, P. and Hay, D. (1980), *Growth centres in the European urban system*, London: Heinemann.

Hall, P., Thomas, R., Gracey, H. and Drewett, R. (1973), *The containment of urban England*, London: Allen and Unwin.

Hamilton, F. E. I. (1976), 'Multinational enterprise and the European Economic Community', *Tijdschrift voor Economische en Sociale Geografie*, **67**, 258–78.

Hamilton, F. E. I. (1978), 'Multinational enterprise and the European Economic Community', in F. E. I. Hamilton (ed.), *Industrial change: international experience and public policy*, London: Longman.

Hamnett, C. (1985), *Inner city decline*, Milton Keynes: Open University Press, Unit 18 D205 Human Geography.

Hansen, J. C. (1983a), 'Oil and the changing geography of Norway', *Geography*, **68**, 162–5.

Hansen, J. C. (1983b), 'Regional policy in an oil economy: the case of Norway', *Geoforum*, **14**, 353–61.

Harrison, J. (1985), *The Spanish economy in the twentieth century*, London: Croom Helm.

Haselen, H. van and Molle, W. (1981), 'Regional patterns of natural population growth and migration in the European community of twelve', *Foundations of Empirical Economic Research No. 12*, Rotterdam: Netherlands Economic Institute.

Haug, P., Hood, H. and Young, S. (1983), 'R & D intensity in the affiliates of US owned electronics manufacturing in Scotland', *Regional Studies*, **17**, 383–92.

Hayter, R. (1982), 'Truncation, the international firm and regional policy', *Area*, **14**, 277–82.

Heal, D. (1974), 'Ownership, control and location decisions: the case of the British steel industry since 1945', in F. E. I. Hamilton (ed.), *Spatial perspectives on industrial organization and decision making*, Chichester: Wiley.

Heap, S. H. (1980), 'World profitability crises in the 1970s: some empirical evidence', *Capital and Class*, **12**, 66–84.

Hellen, J. A. (1974), *North Rhine–Westphalia*, Oxford: Oxford University Press.

Herz, J. H. (ed.), (1982), *From dictatorship to democracy*, Westport: Greenwood Press.

Herzfeld, M. (1980), 'Social tension and inheritance by lot in three Greek villages', *Anthropological Quarterly*, **53**, 91–100.

Hill, B. E. (1984), *The Common Agricultural Policy: past, present and future*, London: Methuen.

Hill, B. E. and Ingersent, K. A. (1977), *An economic analysis of agriculture*, London: Heinemann.

Hirsch, G. P. and Maunder, A. H. (1978), *Farm amalgamation in western Europe*, Farnborough: Saxon House.

Hirsch, J. (1978), 'The state apparatus and social reproduction: elements of a theory of the bourgeois state', in J. Holloway and S. Picciotto (eds.), *State and capital: a Marxist debate*, London: Edward Arnold.

Hirsch, J. (1980), 'Developments in the political system of West Germany since 1945', in R. Scase (ed.), *The state in western Europe*, London: Croom Helm.

Hodge, I. D. and Whitby, M. C. (1981), *Rural employment: trends, options, choices*, London: Methuen.

Holland, S. (1976), *Capital vs. the regions*, London: Macmillan.

Holloway, J. and Picciotto, S. (eds.), (1978), *State and capital: a Marxist debate*, London: Edward Arnold.

Holmes, G. M. and Fawcett, P. D. (1983), *The contemporary French economy*, London: Macmillan.

Hommel, M. (1984), Raumnutzungskonflikte an Nordrand des Ruhrgebietes, *Erdkunde*, **38**, 114–24.

Hörnell, E. and Vahlne, J.-E. (1986), *Multinationals: the Swedish case*, London: Croom Helm.

Hough, J. R. (1979), 'Government intervention in the economy of France', in P. Maunder (ed.), *Government intervention in the developed economy*, London: Croom Helm.

Howells, J. R. L. (1984), 'The location of research and development: some observations and evidence from Britain', *Regional Studies*, **18**, 13–30.

Hudson, R. (1983a), 'Regional labour reserves and industrialization in the EEC', *Area*, **15**, 223–30.

Hudson, R. (1983b), 'Capital accumulation and chemical production in western Europe in the postwar period', *Environment and Planning A*, **15**, 105–22.

Hudson, R. (1983c), 'Capital accumulation and regional problems: a study of North East England, 1945 to 1980', in F. E. I. Hamilton and G. J. R. Linge (eds.), *Regional economies and industrial systems*, Chichester: Wiley.

Hudson, R. and Lewis, J. (1984), 'Capital accumulation: the industrialization of southern Europe?', in A. Williams (ed.), *Southern Europe transformed*, London: Harper and Row.

Hudson, R., Rhind, D. and Mounsey, H. (1984), *An atlas of EEC affairs*, London: Methuen.

Hudson, R. and Sadler, D. (1983), 'Region, class and the politics of steel closures in the European Community', *Environment and Planning D: Society and Space*, **1**, 405–28.

Hudson, R. and Williams, A. (1986), *The United Kingdom*, London: Harper and Row.

Hughes, H. and Waelbroeck, J. (1983), 'Foreign trade and structural adjustment – is there a threat of new protectionism?', in H. C. Brown, H. Laumer, W. Leibfritz and H. C. Sherman (eds.), *The European Economy in the 1980s*, Aldershot: Gower Press.

Hull, C. (1983), 'Federal Republic of Germany', in D. J. Storey (ed.), *The small firm: an international survey*, London: Croom Helm.

Ilbery, B. W. (1981), *Western Europe: a systematic human geography*, Oxford: Oxford University Press.

Illeris, S. (1985), 'How to analyse the role of services in regional development', in M. J. Bannon and S. Ward (eds.), *Services and the new economy*, Dublin: Regional Studies Association Irish Branch.

ILO (1984), *World Labour Yearbook*, Geneva: International Labour Organization.

International Energy Agency (1983), *Coal information report*, Paris: OECD, International Energy Agency.

IWC Motors Group (1978), *A workers' enquiry into the motor industry*, London: CSE Books.

Jacquemin, A. (1981), 'Concentrations et fusions d'enterprises dans la C. E. E.', *Revue d' Economie Politique*, **9**, 251–64.

Jenkins, R. (1984), 'Divisions over the international division of labour', *Capital and Class*, **22**, 28–57.

Jensen, W. G. (1967), *Energy in Europe, 1945–80*, London: Foulis.

Jessop, B. (1982), *The capitalist state*, Oxford: Martin Robertson.

John, B. (1984), *Scandinavia: a new geography*, London: Longman.

Johnson, P. S. and Cathcart, D. G. (1979), 'New manufacturing firms and regional development: some evidence from the northern region', *Regional Studies*, **13**, 269–80.

Johnston, D. M. (1965), *The international law of fisheries: a framework for policy-oriented enquiries*, New Haven: Yale University Press.

Johnston, R. J. (1982), *Geography and the state: an essay in political geography*, London: Macmillan.

Jones, A. R. (1984), 'Agriculture: organization, reform and the EEC', in A. Williams (ed.), *Southern Europe transformed*, London: Harper and Row.

de Kadt, E. (1979), *Tourism: passport to development?* Oxford: Oxford University Press.

Kafkalas, G. (1984), 'Small firms and the development of a peripheral region: the case of Thraki, Greece', in R. Hudson (ed.), *Small firms and regional development*, Copenhagen: Copenhagen School of Economics and Business Administration.

Kaldor, N. (1966), *Causes of the slow rate of economic growth of the United Kingdom*, Cambridge: Cambridge University Press.

Kane, T. T. (1978), 'Social problems and ethnic change: Europe's guest workers', *Intercom*, **6**, 7–9.

Keeble, D. (1976), *Industrial location and planning in the United Kingdom*, London: Methuen.

Keeble, D. and Kelly, T. (1985), *New firms and high technology industry in the United Kingdom: the case of computer electronics*. Unpublished paper presented to the conference on New Firms and Area Development in the European Community, University of Utrecht, May 1985.

Keeble, D., Owens, P. L. and Thompson, C. (1982), 'Regional accessibility and economic potential in the European Community', *Regional Studies*, **16**, 419–32.

Keeble, D., Owens, P. L. and Thompson, C. (1983), 'The urban–rural manufacturing shift in the European Community', *Urban Studies*, **20**, 405–18.

Keesing, D. B. (1979), *World trade and output of manufacturing*, Washington: World Bank Staff Working Paper 316.

Kemper, N. J. and Smidt, M. de (1980), 'Foreign manufacturing establishments in the Netherlands', *Tijdschrift voor Economische en Sociale Geografie*, **71**, 21–40.

Kielstra, N. (1985), 'The rural Languedoc: periphery to "relictual space"', in R. Hudson and J. R. Lewis (eds.), *Uneven development in southern Europe*, London: Methuen.

Kindleberger, C. P. (1967), *Europe's post-war growth: the role of labour supply*, Cambridge, Mass: Harvard University Press.

King, R. (1971), 'Italian land reform: critique, effects, evaluation', *Tijdschrift voor Economische en Sociale Geografie*, **62**, 368–82.

King, R. (1973), *Land reform: the Italian experience*, London: Butterworths.

King, R. (1976), 'The evaluation of international labour migration movements concerning the EEC', *Tijdschrift voor Economische en Sociale Geografie*, **67**, 66–82.

King, R. (1977), *Land reform: a world survey*, London: Bell.

King, R. (1981), 'Italy', in H. Clout (ed.), *Regional development in western Europe*, Chichester: Wiley. Second edition.

King, R. (1984) 'Population mobility: emigration, return migration and internal migration', in A. Williams (ed.), *Southern Europe transformed*, London: Harper and Row.

King, R. (1985), *The industrial geography of Italy*, London: Croom Helm.

King, R. (1987), *Italy*, London: Harper and Row.

King, R. and Burton, S. (1981), *An introduction to the geography of land fragmentation and consolidation*, Leicester: University of Leicester, Department of Geography, Occasional Paper 8.

King, R. and Burton, S. (1982), 'Land fragmentation: notes on a fundamental rural spatial problem', *Progress in Human Geography*, **6**, 475–94.

King, R., Mortimer, J., Strachan, A. and Trono, A. (1985), 'Return migration and rural economic change: a south Italian case study', in R. Hudson and J. R. Lewis (eds.), *Uneven development in southern Europe*, London: Methuen.

King, R. and Took, L. (1983), 'Land tenure and rural social change: the Italian case', *Erdkunde*, **37**, 186–98.

Kirk, M. (1981), *Demographic and social change in Europe: 1975–2000*, Liverpool: Liverpool University Press.

Klatzmann, J. (1978), *L'agriculture française*, Paris: Seuil.

Knapp, H. (1981), 'Incomes policy:Austria's secret weapon', in R. Dahrendorf, (ed.), *Europe's economy in crisis*, London: Weidenfeld and Nicolson.

Knox, P. (1984), *The geography of western Europe*, London: Croom Helm.

Kofman, E. (1985), 'Dependent development in Corsica', in R. Hudson and J. R. Lewis (eds.), *Uneven development in southern Europe*, London: Methuen.

Kok, J. A. A. M. (1986), Regional-economic development and product–market relationships. Paper presented to the IGU Commission on Industrial Change at Chinchon, Spain, 24–30 August.

Kondratieff, N. D. (1935), 'The long waves in economic life', *Review of Economic Statistics*, **17**, 105–15.

Krejci, J. (1978), 'Ethnic problems in Europe', in S. Giner and M. S. Archer (eds.), *Contemporary Europe: social structures and cultural patterns*, London: Routledge and Kegan Paul.

Laganà, G., Pianta, M. and Segre, A. (1982), 'Urban social movements and urban restructuring in Turin, 1969–76', *International Journal of Urban and Regional Research*, **6**, 223–45.

Lambert, A. M. (1961), 'Farm consolidation and improvement in the Netherlands, with an example from the Land van Mass en Waal', *Economic Geography*, **32**, 115–23.

Lanfant, M.-F. (1980), 'Introduction: tourism in the process of internationalization', *International Social Science Journal*, **23**, 14–43.

Läpple, D. and Van Hoogstraten, P. (1980), 'Remarks on the spatial structure of capitalist development: the case of the Netherlands', in J. G. Carney, R. Hudson and J. R. Lewis (eds.), *Regions in crisis*, London: Croom Helm.

Law, C. M. (1980), 'The foreign company's location investment decision and its role in British regional development', *Tijdschrift voor Economische en Social Geografie*, **71**, 15–20.

Law, C. M. (1985a), 'The geography of industrial rationalisation: the British motor car assembly industry, 1972–1982', *Geography*, **70**, 1–12.

Law, C. M. (1985b), 'The British conference and exhibition business', Manchester: University of Salford, Department of Geography, Urban Tourism Project Working Paper 2.

Lawless, R. I. and Findlay, A. M. (1984), 'Algerian emigration to France and the Franco-Algerian accord of 1980', in M. Lacher and W. Ruf (eds.), *Transnational mobility of labour and regional developments in the Mediterranean*, Kassel: Gesamthoschule Kassel.

Lebon, A. (1981), 'Les jeunes migrants dans la vie active en Europe occidentale', *Etudes Migrations*, **61**, 29–39.

Lewis, J. R. (1984), 'Regional policy and planning', in S. Bornstein, D. Held and J. Krieger (eds.), *The state in capitalist Europe*, London: George Allen and Unwin.

Lewis, J. R. (1986), 'International labour migration and uneven regional development in labour exporting countries', *Tijdschrift voor Economische en Sociale Geografie*, **77**, 22–41.

Lewis, J. R. and Williams, A. M. (1981), 'Regional uneven development on the European periphery: the case of Portugal', *Tijdschrift voor Economische en Sociale Geografie*, **72**, 81–98.

Lewis, J. R. and Williams, A. M. (1982), 'Portugal' in M. Wynn (ed.), *Housing in Europe*, London: Croom Helm.

Lewis, J. R. and Williams, A. M. (1984), 'Where are the innovators and job creators? An empirical study of returned migrants in Central Portugal', in M. Lacher and W. Ruf (eds.), *Transnational mobility of labour and regional developments in the Mediterranean*, Kassel: Gesamthoschule Kassel.

Lewis, J. R. and Williams, A. M. (1985), 'The Sines project: Portugal's growth centre or white elephant', *Town Planning Review*, **56**, 339–66.

Lewis, J. R. and Williams, A. M. (1986), 'Economic development and return migration: regressados, retornados, and non-migrants in the Região Centro, Portugal', in R. King (ed.), *Return migration and regional economic problems*, London: Croom Helm.

Lewis, J. R. and Williams, A. M. (1987), 'The role and formation of small firms in Portugal', *Regional Studies*, **21**.

Lindmark, L. (1983), 'Sweden', in D. J. Storey (ed.), *The small firm: an international survey*, London: Croom Helm.

Linge, G. J. R. and Hamilton, F. E. I. (1981), 'International industrial systems', in F. E. I. Hamilton and G. J. R. Linge (eds.), *Spatial analysis, industry and the industrial environment: international industrial systems*, Chichester: Wiley.

Lipietz, A. (1980a), 'The structuration of space, the problem of land, and spatial policy', in J. Carney, R. Hudson and J. Lewis (eds.), *Regions in crisis: new perspectives in European regional theory*, London: Croom Helm.

Lipietz, A. (1980b), 'Inter-regional polarisation of society', *Papers of the Regional Science Association*, **40**, 3–18.

Lipietz, A. (1984), 'Imperialism or the beast of the apocalypse', *Capital and Class*, **22**, 81–109.

Lloyd, P. E. and Mason, C. M. (1984), 'Spatial variations in new firm formation in the United Kingdom: comparative evidence from Merseyside, Greater Manchester and South Hampshire', *Regional Studies*, **18**, 207–220.

Loeve, A., de Vries, J. and de Smidt, M. (1985), 'Japanese firms and the gateway to Europe: the Netherlands as a location for Japanese subsidiaries', *Tijdschrift voor Economische en Sociale Geografie*, **76**, 2–8.

Lowe, P. and Goyder, J. (eds.), (1983), *Environmental groups in politics*, London: Allen and Unwin.

Luciani, G. (1984), 'The Mediterranean and the energy picture', in G. Luciani (ed.), *The Mediterranean region: economic interdependence and the future of society*, London: Croom Helm.

Lundberg, E. (1981), 'The rise and fall of the Swedish economic model', in R. Dahrendorf (ed.), *Europe's economy in crisis*, London: Weidenfeld and Nicolson.

McDermott, P. J. (1977), 'Overseas investment and the industrial geography of the United Kingdom', *Area*, **9**, 200–7.

MacKay, D. I. and MacKay, G. A. (1975), *The political economy of North Sea Oil*, London: Martin Robertson.

Maddison, A. (1964), *Economic growth in the West*, London: Allen and Unwin.

Malerba, F. (1985), *The semiconductor business: the economics of rapid growth and decline*, London: Frances Pinter.

Mandel, E. (1975), *Late capitalism*, London: New Left Books.

Manitea-Tsapatsaris, V. (1986), 'Crisis in Greek agriculture: diagnosis and an alternative strategy', *Capital and Class*, **27**, 107–30.

Mann, S. A. and Dickinson, J. A. (1978), 'Obstacles to the development of capitalist agriculture', *Journal of Peasant Studies*, **5**, 466–81.

Mann, S. A. and Dickinson, J. A. (1980), 'State and agriculture in two eras of American capitalism', in F. H. Buttel and H. Newby (eds.), *The rural sociology of the advanced societies: critical perspectives*, Montclair: Allanheld.

Manners, G. (1984), 'North Sea oil: benefits, costs and uncertainties', *Geoforum*, **15**, 57–64.

Marcelloni, M. (1979), 'Urban movements and political struggles in Italy', *International Journal of Urban and Regional Studies*, **3**, 251–67.

Marsden, T. (1984), 'Land ownership and farm organisation in capitalist agriculture', in T. Bradley and P. Lowe (eds.), *Locality and rurality: economy and society in rural regions*, Norwich: Geo Books.

Marsh, J. S. and Swanney, P. J. (1980), *Agriculture and the European Community*, London: George Allen and Unwin.

Martinelli, F. (1985), 'Public policy and industrial development in southern Italy: anatomy of a dependent industry', *International Journal of Urban and Regional Research*, **9**, 47–81.

Mason, C. (1982), 'Foreign-owned manufacturing firms in the United Kingdom: some evidence from south Hampshire', *Area*, **14**, 7–17.

Massey, D. B. (1978), 'Regionalism: some current issues', *Capital and Class*, **6**, 106–25.

Massey, D. B. (1979), 'In what sense a regional problem?', *Regional Studies*, **13**, 233–44.

Massey, D. B. (1984), *Spatial divisions of labour: social structures and the geography of production*, London: Macmillan.

Massey, D. B. and Meegan, R. A. (1979), 'The geography of industrial reorganisation: the spatial effects of the restructuring of the electrical engineering sector under the Industrial Reorganisation Corporation', *Progress in Planning*, **10**, 155–237.

Mathias, P. (1969), *The first industrial nation: an economic history of Britain 1700–1914*, London: Methuen.

Mathieson, A. and Wall, G. (1982), *Tourism: economic, physical and social impacts*, London: Longman.

Maunder, P. (1979), 'Government intervention in the economy of the United Kingdom', in P. Maunder (ed.), *Government intervention in the developed economy*, London: Croom Helm.

Mayhew, A. (1970), 'Structural reform and the future of West German agriculture', *Geographical Review*, **60**, 54–68.

Mazier, J. (1982), 'Growth and crisis – a Marxist interpretation', in A. Boltho (ed.), *The European economy: growth and crisis*, Oxford: Oxford University Press.

Mead, W. R. (1951), 'The cold farm in Finland: resettlement of Finland's displaced farmers', *Geographical Review*, **41**, 529–43.

Mellor, R. E. H. (1978), *The two Germanies: a modern geography*, London: Harper and Row.

Mellor, R. E. H. and Smith, E. A. (1979), *Europe: a geographical survey of the Continent*, London: Macmillan.

Mérigo, E. (1982), 'Spain', in A. Boltho (ed.), *The European economy: growth and crisis*, Oxford: Oxford University Press.

Metton, A. (1979), 'Centres commerciaux périphériques: regards sur l'expérience Parisienne', *Revue Géographique des Pyrenées et du Sud-Ouest*, **50**, 107–13.

Miles, R. (1986), 'Labour migration, racism and capital accumulation in western Europe since 1945: an overview', *Capital and Class*, **28**, 49–86.

Miliband, R. (1973), *The state in capitalist society*, New York, Basic Books.

Milward, A. S. (1984), *The reconstruction of western Europe 1945–51*, London: Methuen.

Mingione, E. (1983), 'Informalization, restructuring and the survival strategies of the working class', *International Journal of Urban and Regional Research*, **7**, 311–39.

Minshull, G. N. (1980), *The New Europe: an economic geography of the EEC*, London: Hodder and Stoughton.

Mintz, S. W. (1973) 'A note on the definition of peasantries', *Journal of Peasant Studies*, **1**, 91–106.

Mitter, S. (1986), 'Industrial restructuring and manufacturing homework: immigrant women in the UK clothing industry' *Capital and Class*, **27**, 37–80.

Molle, W., Holst, B. van and Smit, H. (1980), *Regional disparity and economic development in the European Community*, Farnborough: Saxon House.

Monsted, M. (1984), 'Small enterprises in Denmark – What are the conditions, and how do they adapt to them?', in R. Hudson (ed.), *Small firms and regional development*, Copenhagen: Copenhagen School of Economics and Business Administration.

Morgan, K. (1983), 'Restructuring steel: the crises of labour and locality in Britain', *International Journal of Urban and Regional Research*, **7**, 175–201.

Morgan, K. and Sayer, A. (1983), *The international electronics industry and regional development in Britain*, Brighton: University of Sussex, Urban and Regional Studies, Working Paper 34.

Moseley, M. J. (1974), *Growth Centres in Spatial Planning*, Oxford: Pergamon Press.

Moseley, M. J. (1980), 'Strategic planning and the Paris agglomeration in the 1960s and 1970s: the quest for balance and structure', *Geoforum*, **11**, 179–223.

Moss, J. E. (1980), *Part-time farming in Northern Ireland: a study of small scale beef and sheep farms*, Belfast: Department of Agriculture for Northern Ireland, Studies in Agricultural Economics.

Mottura, G. and Pugliese, E. (1980), 'Capitalism in agriculture and capitalistic agriculture: the Italian case', in F. Buttel and H. Newby (eds.), *The rural sociology of advanced societies*, London: Croom Helm.

Munton, R. C. (1977), 'Financial institutions: their ownership of agricultural land', *Area*, **9**, 29–37.

Murphy, R. E. (1985), *Tourism: a community approach*, London: Methuen.

Murray, F. (1983), 'The decentralization of production – the decline of the mass-collective worker', *Capital and Class*, **19**, 74–99.

Murray, R. (1978), 'Value and theory of rent: part two', *Capital and Class*, **4**, 11–33.

Nakase, T. (1981), 'Some characteristics of Japanese-type multinationals today', *Capital and Class*, **13**, 61–98.

Nalson, J. S. (1968), *Mobility of farm families: a study of occupational and residential mobility in an upland area of England*, Manchester: University Press.

Naylon, J. (1959), 'Land consolidation in Spain', *Annals of the Association of American Geographers*, **49**, 361–73.

Naylon, J. (1961), 'Progress of land consolidation in Spain', *Annals of the Association of American Geographers*, **51**, 335–8.

Naylon, J. (1966), 'The Badajoz Plan', *Erdkunde*, **20**, 44–60.

Naylon, J. (1967), 'Tourism – Spain's most important industry', *Geography*, **52**, 23–40.

Naylon, J. (1973), 'An appraisement of Spanish irrigation and land settlement policy since 1939', *Iberian Studies*, **2**, 12–18.

Netting, R. M. (1972), 'Of men and meadows: strategies of Alpine land use', *Anthropological Quarterly*, **45**, 145–57.

Newby, H. (1977), *The deferential worker*, London: Allen Lane.

Newby, H. (1978), 'The rural sociology of advanced capitalist societies', in H. Newby (ed.), *International perspectives in rural sociology*, Chichester: Wiley.

Newby, H. (1979), *Green and pleasant land?* London: Hutchinson.

Nicol, W. R. and Yuill, D. (1980), *Regional problems and policies in Europe: the post-war experience*, Glasgow: University of Strathclyde, Studies in Public Policy no. 53.

Nore, P. (1978), 'The international oil industry and national economic development: the case of Norway', in J. Faundez and S. Picciotto (eds.), *The nationalisation of multinationals in peripheral economies*, London: Macmillan.

Oakey, R. P., Thwaites, A. T. and Nash, P. A. (1980), 'The regional distribution of innovative manufacturing establishments in Britain', *Regional Studies*, **14**, 235–53.

O'Connor, J. (1973), *The fiscal crises of the State*, New York: St Martin's Press.

Odell, P. R. (1976), 'The EEC energy market: structure and integration', in R. Lee and P. Ogden (eds.), *Economy and society in the EEC*, Farnborough: Saxon House.

Odell, P. R. (1977), 'The potential for natural gas from the North Sea in relation to western Europe's market for gas by the mid-1980s', *Geoforum*, **8**, 155–68.

Odell, P. R. (1978), 'North Sea oil and gas resources: their implications for the location of industry in Western Europe', in F. E. I. Hamilton (ed.), *Industrial change: international experience and public policy*, London: Longman.

Odell, P. R. (1981), 'The energy economy of western Europe: a return to the use of indigenous resources', *Geography*, **66**, 1–14.

Odell, P. R. (1986), 'Energy and regional development: a European perspective', *Built Environment*, **11**, 9–21.

Odell, P. R. and Rosing, K. E. (1983), *The future of oil: world oil resources and use*, London: Kogan Page.

OECD (1964), *Low incomes in agriculture*, Paris: OECD.

OECD (1969), *Agricultural development in southern Europe*, Paris: OECD.

OECD (1972a), *The traditional farmer: maximization and mechanisation*, Paris: OECD.

OECD (1972b), *Structural reform measures in agriculture*, Paris: OECD.

OECD (1980), *Review of agricultural policies in OECD member countries, 1979*, Paris: OECD.

OECD (1983a), *Les enfants des migrants et l'emploi dans les pays d'Europe*, Paris: OECD.

OECD (1983b), *Review of fisheries in OECD member countries*, Paris: OECD.

OECD (1983c), *Industry in transition: experience of the 70s and prospects for the 80s*, Paris: OECD.

OECD (1985a), *Tourism policy and international tourism in OECD countries*, Paris: OECD.

OECD (1985b), *The semi-conductor industry*, Paris: OECD.

OEEC (1959), *The small family farm: a European problem*, Paris: OEEC.

O'Farrell, P. N. (1980), 'Multinational enterprises and regional development: Irish evidence', *Regional Studies*, **14**, 141–50.

O'Farrell, P. N. and Crouchley, R. (1984), 'An industrial and spatial analysis of new firm formation in Ireland', *Regional Studies*, **18**, 221–36.

Offe, C. (1975), 'The theory of the capitalist state and the problem of policy formation', in L. Lindberg *et al.* (ed.), *Stress and contradictions in modern capitalism*, Lexington: D. C. Heath.

O'Flanagan, T. P. (1980), 'Agrarian structures in north-western Iberia: reponses and their implications for development', *Geoforum*, **11**, 158–69.

O'Flanagan, T. P. (1982), 'Land reform and rural modernisation in Spain: a Galician perspective', *Erdkunde*, **36**, 48–53.

Owen-Smith, E. (1979), 'Government intervention in the economy of the Federal Republic of Germany', in P. Maunder (ed.), *Government intervention in the developed economy*, London: Croom Helm.

Owen-Smith, E. (1983) *The West German economy*, London: Croom Helm.

Owens, P. R. (1980), 'Direct foreign investment – some spatial implications for the source economy', *Tijdschrift voor Economische en Sociale Geografie*, **71**, 50–62.

Paine, S. (1977), 'The changing role of migrant labour in the advanced capitalist economies of western Europe', in R. T. Griffiths (ed.), *Government, business and labour in European capitalism*, London: Europotentials Press.

Paine, S. (1979), 'Replacement of the western European migrant labour system by investment in the European periphery', in D. Seers, B. Schaffer and M. L. Kiljunen (eds.), *Underdeveloped Europe*, Hassocks: Harvester Press.

Paine, S. (1982), 'International investment, migration and finance: issues and policies', in J. A. Girâo (ed.), *Southern Europe and the enlargement of the EEC*, Lisbon: Economia.

Palloix, C. (1975), *L'economie mondiale capitaliste et les firmes multinationals*, Paris: Maspero.

Parker, G. (1981), *The logic of unity: a geography of the European Economic Community*, London: Longman.

Peach, C. (1968), *West Indian migration to Britain*, London: Oxford University Press.

Pearce, D. G. (1981), *Tourist development*, London: Longman.

Perrons, D. C. (1981), 'The role of Ireland in the new international division of labour: a proposed farmework for regional analysis', *Regional Studies*, **15**, 81–100.

Perry, P. J. (1969), 'The structural reform of French agriculture: the role of the

Sociétés d'Aménagement Foncier et d'Etablissement Rural', *Revue Géographique de Montreal*, **23**, 137–51.

Philipponneau, M. (1975), 'Breton farmyard politics', *Geographical Magazine*, **47**, 289–94.

Phillips, D.R. and Williams, A. M. (1984), *Rural Britain: a social geography*, Oxford: Blackwell.

Pickvance, C. G. (1981), 'Policies as chameleons: an interpretation of regional policy and office policy in Britain', in M. Dear and A. J. Scott (eds.), *Urbanization and urban planning in capitalist society*, London: Methuen.

Pinder, D. (1976), *The Netherlands*, Folkestone: Dawson.

Pinder, D. (1983), *Regional economic development and policy: theory and practice in the European Community*, London: George Allen and Unwin.

Podbielski, G. (1981), 'The Common Agricultural Policy and the Mezzogiorno', *Journal of Common Market Studies*, **19**, 333–50.

Pollard, S. (1981), *Peaceful conquest: the industrialization of Europe 1760–1970*, Oxford: Oxford University Press.

Poulantzas, N. (1969), 'The problem of the capitalist state', *New Left Review*, **58**, 67–78.

Poulantzas, N. (1978), *State, power, socialism*, London: New Left Books.

Préau, P. (1970), 'Principe d'analyse des sites en montagne', *Urbanisme*, **116**, 21–5.

Pugliese, E. (1985), 'Farm workers in Italy: agricultural working class, landless peasants, or clients of the welfare state?', in R. Hudson and J. Lewis (eds.), *Uneven development in southern Europe*, London: Methuen.

Pye, R. (1977), 'Office location and the cost of maintaining contact', *Environment and Planning A*, **9**, 149–68.

Rainnie, A. (1985), 'Small firms, big problems: the political economy of small businesses', *Capital and Class*, **25**, 140–68.

Ray, G. F. (1983), 'The European energy outlook', in H.-G. Braun, H. Laumer, W. Leibfritz and H. C. Sherman (eds.), *The European economy in the 1980s*, Aldershot: Gower.

Reed, H. C. (1983), 'World city formation', in D. Audretsch (ed.), *The multinational corporation in the 1980s*, Cambridge, Mass: MIT Press.

Rees, D. J. (1978), 'Price uncertainty – the case for government intervention: a comment', *Journal of Agricultural Economics*, **29**, 85–8.

Rey, G. M. (1982), 'Italy', in A. Boltho (ed.), *The European economy: growth and crisis*, Oxford: Oxford University Press.

Rhodes, J. and Kan, A. (1971) *Office dispersal and regional policy*, London: Cambridge University Press.

Robertson, A. H. (1973), *European institutions*, London: Stevens.

Rokkan, S. (1980), 'Territories, centres and peripheries', in J. Gottmann (ed.), *Centre and periphery: spatial variations in politics*, Sage: Beverly Hills.

Ross, R. J. S. (1983), 'Facing Leviathan: public policy and global capitalism', *Economic Geography*, **59**, 144–60.

Ross, R. and Trachte, K. (1983), 'Global cities, global classes: the peripheralization of labour in New York City', *Review*, **6**, 393–431.

Roth, A. (1973), 'The business backgrounds of MPs' in J. Urry and J. Wakeford (eds.), *Power in Britain*, London: Heinemann.

Rothwell, R. (1982), 'The role of technology in industrial change: implications for regional policy', *Regional Studies*, **16**, 361–9.

Russell, J. (1983), *Geopolitics of natural gas*, Cambridge, Mass: Ballinger.

Rybczynski, T. M. (1978), 'Structural changes in the world economy', *Three Banks Review*, **120**, 21–35.

Sadler, D. (1984), 'Works closure at British Steel and the nature of the state', *Political Geography Quarterly*, **3**, 297–311.

Safvestad, V. (1976), 'International panel discussion one: part-time farming in Sweden', in A. M. Fuller and J. A. Mage (eds.), *Part-time farming: problem or resource in rural development*, Guelph: University of Guelph.

Salt, J. and Clout, H. D. (eds.), (1976), *Migration in post-war Europe: geographical essays*, Oxford: Clarendon Press.

Sampson, A. (1977), *The seven sisters*, London: Corgi.

Saraceno, E. (1983), Rural industrialization in the North East and Centre, Paper presented to the OECD Intergovernmental Meeting on Rural Entrepreneurial Capacities, Senigallia, 7–10 June.

Sautter, C. (1982), 'France', in A. Boltho (ed.), *The European economy: growth and crisis*, Oxford: Oxford University Press.

Sauvray, J. (1984), *French multinationals*, London: Frances Pinter.

Savey, S. (1981), 'Pechiney Ugine Kuhlmann: a French multinational corporation', in F. E. I. Hamilton and G. J. R. Linge (eds.), *International industrial systems*, Chichester: Wiley.

Savey, S. (1983), 'Organization of production and the new spatial division of labour in France', in F. E. I. Hamilton and G. J. R. Linge (eds.), *Regional economics and industrial systems*, Chichester: Wiley.

Sawyer, M. (1982) 'Income distribution and the welfare state', in A. Boltho (ed.), *The European economy: growth and crisis*, Oxford: Oxford University Press.

Sayer, A. (1984), *Method in social science: a realist approach*, London: Hutchinson.

Scammell, W. M. (1980), *The international economy since 1945*, London: Macmillan.

Scargill, D. I. (1983), *Urban France*, London: Croom Helm.

Schneider, P., Schneider, J. and Hansen, E. (1972), 'Modernization and development: the role of regional elites and non-corporate groups in the European Mediterranean', *Comparative Studies in Society and History*, **14**, 328–50.

Schumpeter, J. A. (1939), *Business cycles*, New York: McGraw-Hill.

Seers, D., Schaffer, B. and Kiljunen, M. L. (eds.) (1979), *Underdeveloped Europe: case studies in core periphery relations*, Hassocks: Harvester Press.

Seers, D. and Vaitsos, C. (eds.) (1982), *The second enlargement of the EEC*, London: Macmillan.

Selwyn, P. (1979), 'Some thoughts on cores and peripheries', in D. Seers, B. Schaffer and M. L. Kiljunen (eds.), *Underdeveloped Europe: case studies in core–periphery relations*, Hassocks: Harvester Press.

Servan-Schreiber, J.-J. (1968), *The American challenge*, Harmondsworth: Pelican.

Shanin, T. (1979), 'Introduction', in T. Shanin (ed.), *Peasants and peasant societies*, Harmondsworth: Penguin.

Shepherd, D., Silberston, A., and Strange, R. (1985), *British manufacturing investment overseas*, London: Methuen.

Short, J. (1981), *Public expenditure and taxation in the UK*, Farnborough: Gower.

Sinclair, R. and Walker, D. F. (1982), 'Industrial development via the multinational corporation', *Regional Studies*, **16**, 433–42.

Slater, M. (1984), 'Italy: surviving into the 1980s', in A. Williams (ed.), *Southern Europe transformed*, London: Harper and Row.

Slattery, M. (1966), 'The relative income of farmers', *Quarterly Review of Agricultural Economics*, **19**, 115–27.

Smidt, M.de (1983), 'Regional locational cycles and the stages of locating foreign manufacturing plants – the case of the Netherlands', *Tijdschrift voor Economische en Sociale Geografie*, **74**, 2–11.

Smidt, M. de (1984), 'Office location and the urban functional mosaic: a comparative study of five cities in the Netherlands', *Tijdschrift voor Economische en Sociale Geografie*, **75**, 110–22.

Smidt, M.de (1985a), New foreign establishments in the Netherlands. Unpublished paper presented to the conference on New Firms and Area Development in the European Community, University of Utrecht, May 1985.

Smidt, M.de (1985b), 'Window on the Netherlands: relocation of government services in the Netherlands, *Tijdschrift voor Economische en Sociale Geografie*, **76**, 232–36.

Smith, I. J. (1979), 'The effect of external takeovers on manufacturing employment change in the Northern Region between 1963 and 1973', *Regional Studies*, **13**, 421–37.

Solinas, G. (1982), 'Labour market segmentation and workers' careers', *Cambridge Journal of Economics*, **6**, 331–52.

Spence, N., Gillespie, A., Goddard, J., Kennett, S., Pinch, S. and Williams, A. M. (1982), *British cities: an analysis of urban change*, Oxford: Pergamon.

Steed, G. P. F. (1978), 'Product differentiation, locational protection and economic integration: western Europe's clothing industries', *Geoforum*, **9**, 307–18.

Steed, G. P. F. (1981), 'International location and comparative advantage: the clothing industries and developing countries', in F. E. I. Hamilton and G. J. R. Linge (eds.), *International industrial systems*, Chichester: Wiley.

Stenstadvold, K. (1981), 'Northern Europe', in H. D. Clout (ed.), *Regional development in western Europe*, Chichester: Wiley.

Storey, D. J. (1982), *Entrepreneurship and the new firm*, London: Croom Helm.

Streit, M. E. (1977), 'Government and business: the case of West Germany', in R. T. Griffiths (ed.), *Government, business and labour in European capitalism*, London: Europotentials Press.

Sundin, E. (1984), 'Regional development and the role of small firms in Sweden', in R. Hudson (ed.), *Small firms and regional development*, Copenhagen: Copenhagen School of Economics and Business Administration.

Susman, P. H. (1984), 'Capital restructuring and the changing regional environment', in P. O'Keefe (ed.), *Regional restructuring under advanced capitalism*, London: Croom Helm.

Tapinos, G. P. (1982), 'European migration patterns: economic linkages and policy experiences', *Etudes Migrations*, **62**, 339–57.

Teulings, A. W. M. (1984), 'The internationalization squeeze: double capital movement and job transfer within Philips worldwide', *Environment and Planning A*, **16**, 597–614.

Thomas, S. and Tuppen, J. (1977), 'Readjustment in the Ruhr – the case of Bochum', *Geography*, **62**, 168–75.

Thompson, I. B. (1981), *The Paris Basin*, Oxford: Oxford University Press.

Thrift, N. (1985), 'The internationalization of producer services and the integration of the Pacific Basin property market', in M. J. Taylor and N. J. Thrift (eds.), *Multinationals and the restructuring of the world economy*, London: Croom Helm.

Thrift, N. (1986), 'The geography of international economic disorder', in R. J. Johnston and P. J. Taylor (eds.), *A world in crisis: geographical perspectives*, Oxford: Basil Blackwell.

Toepfer, H. (1985), 'The economic impact of returned emigrants in Trabzon, Turkey', in R. Hudson and J. Lewis (eds.), *Uneven development in southern Europe*, London: Methuen.

Toft Jensen, H. (1982), 'The role of the state in regional development, planning, and implementation: the case of Denmark', in R. Hudson and J. Lewis (eds.), *Regional planning in Europe*, London: Pion, London Papers in Regional Science 11.

Toyne, B., Arpan, J. S., Barnett, A. H., Richs, D. A. and Shrimp, T. A. (1984), *The global textile industry*, London: Allen and Unwin.

Tracy, M. (1982), *Agriculture in western Europe: challenge and response 1880–1980*, St Albans: Granada.

Trilling, J. (1981), 'French environmental politics', *International Journal of Urban and Regional Research*, **5**, 67–82.

Tsoukalis, L. (1981), *The European Community and its Mediterranean enlargement*, London: Allen and Unwin.

Tumlir, J. (1983), 'The world economy today: crisis or a new beginning?', *National Westminster Bank Quarterly*, 26–44.

Tuppen, J. N. (1977), 'Redevelopment of the city centre: the case of Lyon – Le Part Dieu', *Scottish Geographical Magazine*, **93**, 151–8.

Tuppen, J. N. (1983), *The economic geography of France*, London: Croom Helm.

Tuppen, J. N. (1985), Urban tourism in France: a preliminary assessment. Working Paper 3, Urban Tourism Project, Department of Geography, University of Salford.

Turner, L. (1982), 'Western Europe and the NICs', in L. Turner and N. McMullen (eds.), *The newly industrializing countries: trade and adjustment*, London: George Allen and Unwin.

Turner, L. and McMullen N. (1982), *The newly industrializing countries: trade and adjustment*, London: George, Allen and Unwin.

Turnock, D. (1979), *The new Scotland*, Newton Abbott: David and Charles.

UNCTC (1982), *Transnational corporations in international tourism*, New York: United Nations Centre on Transnational Corporations.

Vaitsos, C. (1982), 'Transnational corporate behaviour and the enlargement', in D. Seers and C. Vaitsos (eds.), *The second enlargement of the EEC: the integration of unequal partners*, London: Macmillan.

Valenzuela, M. (1985), 'Everything under the sun', *Geographical Magazine*, **57**, 274–78.

Van Rijckeghem, W. (1982), 'Benelux', in A. Boltho (ed.), *The European economy: growth and crisis*, Oxford: Oxford University Press.

Vilaça, J. L. da C. (1984), 'Movimentos de trabalhadores na CEE', in M. C. L.

Porto (ed.), *Emigração e retorno na regiâo centro*, Coimbra: Commissâo de Coordenaçâo da Regiâo Centro.

Ville, Ph.de and Leroy, R. (1973), 'Processus et facteurs de l'evolution régionale de l'emploi', *Recherche Economiques*, **2**, 219–36.

Vincent, J. (1980), 'The political economy of Alpine development: tourism or agriculture in St Maurice', *Sociologia Ruralis*, **20**, 250–71.

Wabe, J. S. (1986), 'The regional impact of de-industrialization in the European Community', *Regional Studies*, **20**, 27–36.

Wade, R. (1979), 'Fast growth and slow development in Southern Italy', in D. Seers, B. Schaffer and M. L. Kiljunen (eds.), *Underdeveloped Europe*, Hassocks: Harvester Press.

Wallerstein, I. (1979), *The capitalist world economy*, Cambridge: Cambridge University Press.

Wallerstein, I. (1984), *The politics of the world economy: the states, the movements, and the civilizations*, Cambridge: Cambridge University Press.

Ward, M. F. (1982), 'Political economy, industrial location and the European motor car industry in the post-war period', *Regional Studies*, **16**, 443–54.

Wärneryd, D. (1968), *Interdependence in urban systems*, Gothenburg: Regionkansult Aktiebolog.

Warren, K. (1978), 'Regional policy and industrial complexes: reflections on British experience', in F. E. I. Hamilton (ed.), *Industrial change: international experience and public policy*, London: Longman.

Warren, K. (1985), 'World steel: change and crisis', *Geography*, **70**, 106–17.

Watts, H. D. (1979), 'Large firms, multinationals and regional development: some new evidence from the United Kingdom', *Environment and Planning A*, **11**, 71–81.

Watts, H. D. (1980), 'The location of European direct investment in the United Kingdom', *Tijdschrift voor Economische en Sociale Geografie*, **71**, 3–14.

Watts, H. D. (1981), *The branch plant economy*, London: Longman.

Watts, H. D. (1982), 'The inter-regional distribution of West German multinationals in the United Kingdom', in M. Taylor and N. Thrift (eds.), *The geography of multinationals*, London: Croom Helm.

Westaway, E. J. (1974), 'The spatial hierarchy of business organizations and its implications for the British urban system', *Regional Studies*, **8**, 145–55.

Whitby, M. C. (1968), 'Lessons from Swedish farm structure policy', *Journal of Agricultural Economics*, **19**, 279–99.

White, P. E. (1976), 'Tourism and economic development in the rural environment', in R. Lee and P. Ogden (eds.), *Economy and society in the EEC: spatial perspectives*, Farnborough: Saxon House.

Wiberg, P. (1984), *Regional policy and spatial planning in Norway 1952–1980*, Edinburgh: Edinburgh College of Art/Heriot-Watt University, Department of Town and Country Planning, Research Paper 5.

Williams, A. M. (1984), 'Introduction', in A. M. Williams (ed.), *Southern Europe transformed*, London: Harper and Row.

Williams, A. M. (1986), 'Economic landscapes of the Mediterranean', *Landscape Research*, **11**, 8–10.

Williams, C. H. (1980), 'Ethnic separatism in western Europe' *Tijdschrift voor Economische en Sociale Geografie*, **71**, 142–58.

Williams, C. H. (1986), 'The question of national congruence', in R. J. Johnston and P. J. Taylor (eds.), *A world in crisis*, Oxford: Blackwell.

Williams, R. (1979), 'The multinational enterprise: a 1977 perspective', in J. Hayward and R. N. Berki (eds.), *State and society in contemporary Europe*, Oxford: Martin Robertson.

Williamson, J. G. (1965), 'Regional inequality and the process of national development; a description of the patterns', *Economic Development and Cultural Change*, **13**, 3–82.

Wilson, G. (1978), 'Farmers' organizations in advanced societies', in H. Newby (ed.), *International perspectives in rural sociology*, Chichester: Wiley.

Winter, M. (1984), 'Agrarian class structure and family farming', in T. Bradley and P. Lowe (eds.), *Locality and rurality: economy and society in rural regions*, Norwich: Geo Books.

Wise, M. (1984), *The Common Fisheries Policy of the European Community*, London: Methuen.

World Bank (1981), *International migrant workers' remittances: issues and prospects*, Washington: World Bank, Staff Working Paper 481.

World Bank (1984), *World development report 1984*, New York: Oxford University Press.

World Bank (1985), *World development report 1985*, New York: Oxford University Press.

World Bank (1986), *World development report 1986*, New York: Oxford University Press.

Yannopoulos, G. N. and Dunning, J. H. (1976), 'Multinational enterprises and regional development: an explanatory paper', *Regional Studies*, **10**, 389–99.

Young, S. and Hood, N. (1976), 'The geographical expansion of US firms in western Europe: some survey evidence', *Journal of Common Market Studies*, **14**, 223–9.

Yuill, D. and Allen, K. (1986), *European regional incentives*, Glasgow: University of Strathclyde, Centre for the Study of Public Policy.

Zacchia, C. (1983), Rural industrialization in the North East and the Centre: summary of case studies. Paper presented to the OECD Intergovernmental Meeting on Rural Entrepreneurial Capacities, Senegallia, 7–10 June.

Ziebura, G. (1983), 'Internationalization of capital, international division of labour and the role of the European Community', in L. Tsoukalis (ed.), *The European Community: past, present and future*, Oxford: Basil Blackwell.

Place index

Subject index